40- $52.00

40- $52.00

D1154583

NONLINEAR ANALYSIS IN
CHEMICAL ENGINEERING

McGraw-Hill Chemical Engineering Series

Editorial Advisory Board

James J. Carberry, *Professor of Chemical Engineering, University of Notre Dame*
James R. Fair, *Professor of Chemical Engineering, University of Texas, Austin*
Max S. Peters, *Professor of Chemical Engineering, University of Colorado*
William R. Schowalter, *Professor of Chemical Engineering, Princeton University*
James Wei, *Professor of Chemical Engineering, Massachusetts Institute of Technology*

BUILDING THE LITERATURE OF A PROFESSION

Fifteen prominent chemical engineers first met in New York more than 50 years ago to plan a continuing literature for their rapidly growing profession. From industry came such pioneer practitioners as Leo H. Baekeland, Arthur D. Little, Charles L. Reese, John V. N. Dorr, M. C. Whitaker, and R. S. McBride. From the universities came such eminent educators as William H. Walker, Alfred H. White, D. D. Jackson, J. H. James, Warren K. Lewis, and Harry A. Curtis. H. C. Parmelee, then editor of *Chemical and Metallurgical Engineering*, served as chairman and was joined subsequently by S. D. Kirkpatrick as consulting editor.

After several meetings, this committee submitted its report to the McGraw-Hill Book Company in September 1925. In the report were detailed specifications for a correlated series of more than a dozen texts and reference books which have since become the McGraw-Hill Series in Chemical Engineering and which became the cornerstone of the chemical engineering curriculum.

From this beginning there has evolved a series of texts surpassing by far the scope and longevity envisioned by the founding Editorial Board. The McGraw-Hill Series in Chemical Engineering stands as a unique historical record of the development of chemical engineering education and practice. In the series one finds the milestones of the subject's evolution: industrial chemistry, stoichiometry, unit operations and processes, thermodynamics, kinetics, and transfer operations.

Chemical engineering is a dynamic profession, and its literature continues to evolve. McGraw-Hill and its consulting editors remain committed to a publishing policy that will serve, and indeed lead, the needs of the chemical engineering profession during the years to come.

THE SERIES

Bailey and Oils: *Biochemical Engineering Fundamentals*
Bennett and Myers: *Momentum, Heat, and Mass Transfer*
Beveridge and Schechter: *Optimization; Theory and Practice*
Carberry: *Chemical and Catalytic Reaction Engineering*
Churchill: *The Interpretation and Use of Rate Data—The Rate Concept*
Clarke and Davidson: *Manual for Process Engineering Calculations*
Coughanowr and Koppel: *Process Systems Analysis and Control*
Danckwerts: *Gas Liquid Reactions*
Finlayson: *Nonlinear Analysis in Chemical Engineering*
Gates, Katzer, and Schuit: *Chemistry of Catalytic Processes*
Harriott: *Process Control*
Johnson: *Automatic Process Control*
Johnstone and Thring: *Pilot Plants, Models, and Scale-up Methods in Chemical Engineering*
Katz, Cornell, Kobayashi, Poettmann, Vary, Elenbaas, and Weinaug: *Handbook of Natural Gas Engineering*
King: *Separation Processes*
Knudsen and Katz: *Fluid Dynamics and Heat Transfer*
Lapidus: *Digital Computation for Chemical Engineers*
Luyben: *Process Modeling, Simulation, and Control for Chemical Engineers*
McCabe and Smith, J. C.: *Unit Operations of Chemical Engineering*
Mickley, Sherwood, and Reed: *Applied Mathematics in Chemical Engineering*
Nelson: *Petroleum Refinery Engineering*
Perry and Chilton (Editors): *Chemical Engineers' Handbook*
Peters: *Elementary Chemical Engineering*
Peters and Timmerhaus: *Plant Design and Economics for Chemical Engineers*
Ray: *Advanced Process Control*
Reed and Gubbins: *Applied Statistical Mechanics*
Reid, Prausnitz, and Sherwood: *The Properties of Gases and Liquids*
Resnick: *Process Analysis and Design for Chemical Engineers*
Satterfield: *Heterogeneous Catalysis in Practice*
Sherwood, Pigford, and Wilke: *Mass Transfer*
Slattery: *Momentum, Energy, and Mass Transfer in Continua*
Smith, B. D.: *Design of Equilibrium Stage Processes*
Smith, J. M.: *Chemical Engineering Kinetics*
Smith, J. M., and Van Ness: *Introduction to Chemical Engineering Thermodynamics*
Thompson and Ceckler: *Introduction to Chemical Engineering*
Treybal: *Mass Transfer Operations*
Van Winkle: *Distillation*
Volk: *Applied Statistics for Engineers*
Walas: *Reaction Kinetics for Chemical Engineers*
Wei, Russell, and Swartzlander: *The Structure of the Chemical Processing Industries*
Whitwell and Toner: *Conservation of Mass and Energy*

**McGRAW-HILL INTERNATIONAL
BOOK COMPANY**

New York
St. Louis
San Francisco
Auckland
Bogotá
Hamburg
Johannesburg
London
Madrid
Mexico
Montreal
New Delhi
Panama
Paris
São Paulo
Singapore
Sydney
Tokyo
Toronto

Bruce A. Finlayson

Professor of Chemical Engineering and Applied Mathematics
University of Washington,
Seattle

NONLINEAR ANALYSIS IN CHEMICAL ENGINEERING

This book was set in Times Roman 327.

British Library Cataloging in Publication Data

Finlayson, Bruce Alan
  Nonlinear analysis in chemical engineering.
  1. Chemical engineering—Mathematics
  2. Differential equations, Nonlinear
  I. Title
  515'.35   TP149   79-41698
  ISBN 0-07-020915-4

**NONLINEAR ANALYSIS IN
CHEMICAL ENGINEERING**

Copyright © 1980 McGraw-Hill Inc. All rights reserved.
Printed in the United States of America. No part of this publication
may be reproduced, stored in a retrieval system, or transmitted, in any form or
by any means, electronic, mechanical, photocopying or otherwise, without
the prior written permission of the publisher.

5 6 7 8 9 0   **MPMP** 7 9 8 7 6 5

Printed and bound in the United States of America.

CONTENTS

PREFACE

This book provides an introduction to many methods of analysis that arise in engineering for the solution of ordinary and partial differential equations. Many books, and often many courses, are oriented towards linear problems, yet it is nonlinear problems that frequently arise in engineering. Here many methods— finite difference, finite element, orthogonal collocation, perturbation—are applied to nonlinear problems to illustrate the range of applicability of the method and the useful results that can be derived from each method. The same problems are solved with different methods so that the reader can assess these methods in practical and similar cases. The examples are from the author's own experience: fluid flow (including polymers), heat transfer, and chemical reactor modeling.

The level of the book is introductory, and the treatment is oriented toward the nonspecialist. Even so the reader is introduced to the latest, most powerful techniques. The course is based on a successful graduate course at the University of Washington, and most chemical engineers taking the course are experimentalists. The reader desiring to delve deeper into a particular technique or application can follow the leads given in the bibliography of each chapter.

The author especially thanks the class of 1979, who tested the first written version of the book, and especially Dan David and Mike Chang, who were diligent about providing corrections. The draft was expertly typed by Karen Fincher and Sylvia Swimm.

The author is also thankful to his family for supporting him during the project—both fiscally and psychologically. Writing a book really involves the whole family. Special thanks go to the author's children Mark, Cady, and Christine, who gave up some of their father–child time to make this book possible, and to the author's wife, Pat, for her continued support and encouragement.

Seattle 1980   Bruce A. Finlayson

x

ONE
INTRODUCTION

The goal of this book is to bring the reader into contact with the efficient computation tools that are available today to solve differential equations modeling physical phenomena, such as diffusion, reaction, heat transfer, and fluid flow. After mastering the material in this book you should be able to apply a variety of methods—finite difference, finite element, collocation, perturbation, etc.—although you will not be an expert in any of them. When faced with a problem to solve you will know which methods are suitable and what information can be easily determined by which method. The emphasis is on numerical methods, using a computer, although some of the approaches can also yield powerful results analytically. The author's philosophy is to use preprogrammed computer packages when available because they allow the reader to sample, peruse, and solve difficult problems with less effort. The reader is, however, introduced to the theory and techniques used in these computer programs.

1-1 CLASSIFICATION OF EQUATIONS

Equations modeling physical phenomena have different characteristics depending on how they model evolution in time and the influence of boundary conditions. When confronted with a model, expressed in the form of a differential equation, the analyst must decide what type of equation is to be solved. That characterization determines the methods that are suitable.

Consider a closed system (i.e. no interchange of mass with the surroundings) containing three chemical components whose concentrations are given by c_1, c_2, and c_3. The three components can react (say when the system is illuminated with

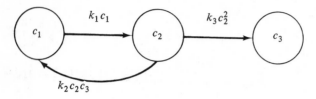

Figure 1-1 Reaction system.

light of a specified frequency), and the goal is to predict the concentration of each species as a function of time. The rates of reaction are known as a function of the concentrations. The reaction system is shown in Fig. 1-1, and the differential equations governing this system are

$$\frac{dc_1}{dt} = -k_1c_1 + k_2c_2c_3$$

$$\frac{dc_2}{dt} = k_1c_1 - k_2c_2c_3 - k_3c_2^2 \tag{1-1}$$

$$\frac{dc_3}{dt} = k_3c_2^2$$

Initially the concentrations of components two and three are zero, and the initial concentration of component one is given as c_0. We thus wish to solve Eqs. (1-1) subject to the initial conditions

$$c_1(0) = c_0 \qquad c_2(0) = c_3(0) = 0 \tag{1-2}$$

Note that the conditions apply only at time zero, not to later times t. The reaction proceeds in time; if we know where to start we can integrate the equations indefinitely. This evolution property yields equations that are called initial-value problems. In this case Eqs. (1-1) are ordinary differential equations, since there is only one independent variable, time t. Thus Eqs. (1-1) and (1-2) are governed by a system of ordinary differential equations that are initial-value problems. In this text this is abbreviated to ODE–IVP.

Consider next diffusion and reaction in a porous medium. We have a heterogeneous system (solid material with pores through which the reactants and products diffuse), but here we model the system as simple diffusion using an effective diffusion coefficient. A mass balance on a volume of the porous medium gives

$$\frac{\partial c}{\partial t} = -\left(\frac{\partial J_x}{\partial x} + \frac{\partial J_y}{\partial y} + \frac{\partial J_z}{\partial z}\right) + R(c) \tag{1-3}$$

where R is the rate of reaction per unit volume (solid plus void volume), $\mathbf{J}$ is the flux of material (in units of concentration per time per unit area—including both solid and void area), and J_x is the x component of the vector $\mathbf{J}$. By using an effective diffusion coefficient we express the flux $\mathbf{J}$ in a form similar to Fick's law

$$\mathbf{J} = -D_e \nabla c$$

or

$$J_x = -D_e \frac{\partial c}{\partial x} \qquad J_y = -D_e \frac{\partial c}{\partial y} \qquad J_z = -D_e \frac{\partial c}{\partial z} \tag{1-4}$$

This equation assumes equimolar diffusion (one mole of reactant diffuses in and one mole of product diffuses out), and all the microscopic details of the porous medium are lumped into the diffusion coefficient. Obviously to model a specific physical situation the diffusion coefficient must either be measured or deduced from similar systems. With this approximation the equation becomes

$$\frac{\partial c}{\partial t} = \frac{\partial}{\partial x}\left(D_e \frac{\partial c}{\partial x}\right) + \frac{\partial}{\partial y}\left(D_e \frac{\partial c}{\partial y}\right) + \frac{\partial}{\partial z}\left(D_e \frac{\partial c}{\partial z}\right) + R(c) \tag{1-5}$$

or

$$\frac{\partial c}{\partial t} = \nabla \cdot D_e \nabla c + R(c)$$

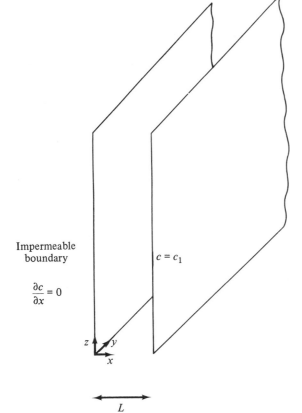

Impermeable
boundary

$$\frac{\partial c}{\partial x} = 0$$

$c = c_1$

Figure 1-2. Diffusion in a slab.

L

Let us next assume that the diffusion occurs in a porous slab that is infinite in extent in two directions, giving a large plane sheet with diffusion through the thickness of the sheet (see Fig. 1-2). We simplify Eq. (1-5) to one dimension by assuming negligible variation of the concentrations in the y and z directions. Also we assume steady-state reaction and diffusion so that the time derivative is zero

$$\frac{d}{dx}\left(D_e \frac{dc}{dx}\right) + R(c) = 0 \qquad (1\text{-}6)$$

As we go from Eq. (1-5) to (1-6) we go from a partial differential equation (the concentration depends on at least two independent variables) to an ordinary differential equation (the concentration depends on only one independent variable). Equation (1-6) is second-order and the theory of linear second-order ordinary differential equations says that we must specify two constants in the general solution. We do that by stating two boundary conditions, one at each side of the slab. Here we consider one side of the slab as impermeable (no flux) and the concentration is held fixed at the other side. These boundary conditions are

$$x = 0 \qquad -D_e \frac{dc}{dx} = 0 \qquad (1\text{-}7)$$

$$x = L \qquad c = c_1 \qquad (1\text{-}8)$$

The problem in Eqs. (1-6) to (1-8) is an ordinary differential equation and a boundary-value problem ODE–BVP. It is also called a two-point boundary-value problem because the two conditions are expressed at different positions x. If they had both been specified at the same point, say $x = 0$, then the problem would have been an initial-value problem. This nature of boundary-value problems—having conditions at each end of the domain—complicates the solution techniques but is characteristic of diffusion, heat transfer, and fluid-flow problems.

Retracing our steps back to Eq. (1-5) describing diffusion and reaction in a three-dimensional space, this time let us simplify the equation for one space dimension, as before, but include transient phenomena, such that

$$\frac{\partial c}{\partial t} = \frac{\partial}{\partial x}\left(D_e \frac{\partial c}{\partial x}\right) + R(c) \qquad (1\text{-}9)$$

This is a partial differential equation, because the solution c depends on two independent variables x and t. The character of the dependence on x and on t is different, however. Only a single derivative in t occurs, and the dependence on t is an evolution phenomenon. We require an initial value of the concentration at each position

$$c(x, 0) = c_0(x) \qquad (1\text{-}10)$$

The dependence on x is like a boundary-value problem, and two conditions are necessary. Conditions like Eqs. (1-7) and (1-8) are feasible, but the concentration c_1 could now be a function of time, corresponding to variations in the bulk-stream concentration. We call this system in Eqs. (1-7) to (1-10) a parabolic partial differential equation in one space dimension (1-D PDE). The term parabolic refers to the fact that one variable is evolutionary in character.

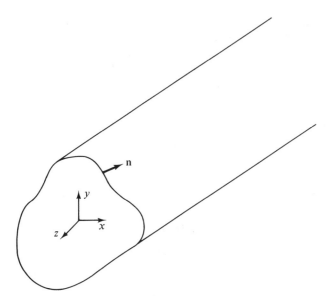

Figure 1-3 Diffusion in a long catalyst pellet.

If we solve Eq. (1-5) in two or three space dimensions we also have a parabolic partial differential equation, with the t variable being evolutionary and the x, y, and z variables being of boundary-value type. In two dimensions we have

$$\frac{\partial c}{\partial t} = \frac{\partial}{\partial x}\left(D_e\frac{\partial c}{\partial x}\right) + \frac{\partial}{\partial y}\left(D_e\frac{\partial c}{\partial y}\right) + R(c) \tag{1-11}$$

If we include two space dimensions but allow only steady-state situations then the equation reduces to

$$\frac{\partial}{\partial x}\left(D_e\frac{\partial c}{\partial x}\right) + \frac{\partial}{\partial y}\left(D_e\frac{\partial c}{\partial y}\right) + R(c) = 0 \tag{1-12}$$

This equation would model both diffusion and reaction in a catalyst particle that is very long in the z direction, so that z variations are negligible (see Fig. 1-3). The type of boundary conditions allowed are Dirichlet-type or boundary conditions of the first kind

$$c = c_B \tag{1-13}$$

Neumann-type or boundary conditions of the second kind

$$-D_e\frac{\partial c}{\partial n} = -D_e\mathbf{n}\cdot\nabla c = f_B \tag{1-14}$$

and Robin-type or boundary conditions of the third kind or mixed conditions

$$-D_e\frac{\partial c}{\partial n} = k_m(c-c_B) \tag{1-15}$$

where **n** is the outward pointing normal, f_B is the specified mass flux, and c_B is the concentration external to the porous medium. The mass transfer coefficient is k_m. Similar boundary conditions apply to heat transfer, in which case D_e is replaced by the thermal conductivity k, f_B is the specified heat flux, k_m is replaced by the heat transfer coefficient h, and the value c_B becomes the external temperature. These boundary conditions are

$$T = T_B$$

$$-k\frac{\partial T}{\partial n} = f_B \tag{1-16}$$

$$-k\frac{\partial T}{\partial n} = h(T - T_B)$$

Equation (1-12) is an elliptic partial differential equation, and the independent variables are of the boundary-value type.

Generally diffusion problems are elliptic in nature; if the problem is unsteady-state or evolutionary the added accumulation term makes them parabolic. This classification is deduced for the following general linear second-order equation

$$A\frac{\partial^2 c}{\partial x^2} + B\frac{\partial^2 c}{\partial x \partial y} + C\frac{\partial^2 c}{\partial y^2} = 0 \tag{1-17}$$

The type of equation is deduced from the discriminant

$$D = B^2 - 4AC$$

$$\begin{aligned} D &< 0 && \text{elliptic} \\ D &= 0 && \text{parabolic} \\ D &> 0 && \text{hyperbolic} \end{aligned} \tag{1-18}$$

For example, the heat transfer equation

$$\frac{\partial T}{\partial t} = \frac{k}{\rho C_p}\frac{\partial^2 T}{\partial x^2} \tag{1-19}$$

would have $A > 0$, $C = 0$, $B = 0$,† and $D = 0$ and is therefore parabolic. The steady-state equation

$$\frac{\partial^2 T}{\partial x^2} + \frac{\partial^2 T}{\partial y^2} = 0 \tag{1-20}$$

has $A = C = 1$, $B = 0$, and D is negative. The equation is therefore elliptic whereas

$$\frac{\partial T}{\partial t} = \frac{k}{\rho C_p}\left(\frac{\partial^2 T}{\partial x^2} + \frac{\partial^2 T}{\partial y^2}\right) \tag{1-21}$$

must be tested for each variable. The time variable is parabolic whereas the spatial dimensions x and y are elliptic in character.

† Identify y with t.

Other differential equations are introduced below in the context of specific applications. Mathematical problems are most easily solved in nondimensional form, and we illustrate here the procedure for turning a model equation into nondimensional form. Take Eqs. (1-6) to (1-8) for the case $R = -kc^2$ and D_e is a constant. We define $c' = c/c_1$ and $x' = x/L$ and introduce these new variables into the differential equation, noting that c_1 and L are constants that can be brought outside the differential

$$\frac{D_e c_1}{L^2} \frac{d^2 c'}{dx'^2} - kc_1^2 (c')^2 = 0 \tag{1-22}$$

$$\frac{-D_e c_1}{L} \frac{dc'}{dx'} = 0 \qquad \text{at } x'L = 0 \tag{1-23}$$

$$c_1 c' = c_1 \qquad \text{at } x'L = L \tag{1-24}$$

We multiply Eqs. (1-22) to (1-24) by $L^2/(D_e c_1)$ and simplify the equations to

$$\frac{d^2 c'}{dx'^2} - \frac{kc_1 L^2}{D_e} (c')^2 = 0 \tag{1-25}$$

$$\frac{dc'}{dx'} = 0 \qquad \text{at } x' = 0 \tag{1-26}$$

$$c' = 1 \qquad \text{at } x' = 1 \tag{1-27}$$

Equations (1-25) to (1-27) are then the nondimensional form of the problem, which is solved mathematically. In this case the parameter $\phi^2 = kc_1 L^2/D_e$ has a specific meaning (ratio of reaction to diffusion phenomena) and name (Thiele modulus squared). Insights obtained from the nondimensional form of the equation are left to the appropriate section treating that problem. If no characteristic parameter suggests itself (as do c_1 and L above) then we just assign a standard c_s and proceed. This situation is actually more suggestive than it seems. The implications are explored in Sec. 5-1.

The remainder of the book is organized according to the type of problem: ODE–IVP, ODE–BVP, 1-D PDE, 2-D PDE, elliptic and parabolic. When solving problems in each category, however, systems of nonlinear algebraic equations must be considered. The next chapter reviews methods for doing this.

TWO

ALGEBRAIC EQUATIONS

Systems of nonlinear algebraic equations must be solved. Two useful techniques—successive substitution and Newton–Raphson—are reviewed here. The first method is considered because it is simple and sometimes very useful, and the second because it is an excellent method, although not fool-proof. In other sections of the book more specialized techniques are considered—see Sec. 4-8 for lower-upper decomposition of matrices, which is important for large sets of equations.

2-1 SUCCESSIVE SUBSTITUTION

Consider the set of nonlinear algebraic equations

$$F_1(x_1, x_2, \ldots, x_n) = 0$$
$$F_2(x_1, x_2, \ldots, x_n) = 0$$
$$\vdots$$
$$F_n(x_1, x_2, \ldots, x_n) = 0$$

(2-1)

which we write in compact form as

$$F_i(x_j) = 0 \qquad i, j = 1, \ldots, n$$

or

$$F_i(\mathbf{x}) = 0 \qquad i = 1, \ldots, n$$

(2-2)

or

$$\mathbf{F}(\mathbf{x}) = 0$$

We wish to find the set of x_i satisfying Eq. (2-1). The notation $\mathbf{x}$ means the set of x_i, $\{x_i | i = 1, n\}$. Reformulating the equations by adding x_i to the ith equation gives

8

$$x_i + F_i(\mathbf{x}) = x_i \tag{2-3}$$

The iterative scheme is defined from Eq. (2-3) as

$$x_i^{k+1} = x_i^k + F_i(\mathbf{x}^k) \tag{2-4}$$

where the superscript k denotes the iterate number. We merely guess x^0 and apply Eq. (2-4) to find x^1, repeat to find x^2, and so forth. The scheme is simple to apply and quick to use. A programmable calculator can be used for small problems. Let us apply the method to a simple example.

The goal is to find the x satisfying the equation

$$F(x) = x^2 - 2 = 0 \tag{2-5}$$

We apply successive substitution in the form

$$x = x^2 - 2 + x$$

or

$$x^{k+1} = x^k(x^k + 1) - 2 \tag{2-6}$$

Starting at $x^0 = 1$ we get successive values of x^k of $0, -2, 0, -2, 0, \ldots$. The method does not converge. If we try $x^0 = 1.4$ we get successive iterate values of 1.36, 1.21, 0.67, $-0.88, \ldots$, and again the method does not converge. Even if we insert the exact answer in a hand calculator the method diverges. Obviously for this example convergence of the successive iterates is a problem.

Next we apply the successive substitution method when the equation is written in the form

$$F(x) = \frac{1}{x} - \frac{x}{2} = 0 \tag{2-7}$$

Now the successive iterates are calculated by

$$x^{k+1} = \tfrac{1}{2}x^k + \frac{1}{x^k} \tag{2-8}$$

Starting with $x^0 = 1.6$ we get values of $x^k = 1.425,\ 1.41425,\ 1.414213563,$ 1.414213562, $\ldots$, or the first 10 digits of the exact answer with only 4 iterations. Starting from $x^0 = 1.2$ gives similar results. Obviously Eq. (2-8) is a better iteration scheme than Eq. (2-4), and we would like to know this in advance. The needed information is given by the following convergence theorem, which we prove.

Theorem 2-1 Let $\boldsymbol{\alpha}$ be the solution to $\alpha_i = f_i(\boldsymbol{\alpha})$. Assume that given an $h > 0$ there exists a number $0 < \mu < 1$ such that

$$\sum_{j=1}^n \left| \frac{\partial f_i}{\partial x_j} \right| \leqslant \mu \qquad \text{for } |x_i - \alpha_i| < h \qquad i = 1, \ldots, n \tag{2-9}$$

$$x_i^k = f_i(\mathbf{x}^{k-1})$$

Then x_i^k converges to α_i as k increases.

PROOF We apply a Taylor series and the mean-value theorem to the equation

$$x_i^k - \alpha_i = f_i(\mathbf{x}^{k-1}) - f_i(\boldsymbol{\alpha})$$

$$= f_i(\boldsymbol{\alpha}) + \sum_{j=1}^{n} \left. \frac{\partial f_i}{\partial x_j} \right|_{x_i = \alpha_i + \zeta_i(x_i^{k-1} - \alpha_i)} (x_j^{k-1} - \alpha_j) - f_i(\boldsymbol{\alpha}) \qquad (2\text{-}10)$$

which holds exactly for some $0 < \zeta_i < 1$. If each term in the summation is made positive the result will be larger than if some of the terms are negative and offset the positive ones, thus

$$|x_i^k - \alpha_i| \leqslant \sum_{j=1}^{n} \left| \frac{\partial f_i}{\partial x_j} \right| |x_j^{k-1} - \alpha_j| \qquad (2\text{-}11)$$

The maximum norm is defined as

$$\|\mathbf{x}\|_\infty \equiv \max_{1 \leqslant i \leqslant n} |x_i| \qquad (2\text{-}12)$$

Then

$$\|\mathbf{x}^k - \boldsymbol{\alpha}\|_\infty \leqslant \left\| \sum_{j=1}^{n} \left| \frac{\partial f_i}{\partial x_j} \right| |x_j^{k-1} - \alpha_j| \right\|_\infty \qquad (2\text{-}13)$$

If we replace $|x_j^{k-1} - \alpha_j|$ by $\|\mathbf{x}^{k-1} - \boldsymbol{\alpha}\|_\infty$ on the right-hand side we get

$$\|\mathbf{x}^k - \boldsymbol{\alpha}\|_\infty \leqslant \left\| \sum_{j=1}^{n} \left| \frac{\partial f_i}{\partial x_j} \right| \right\|_\infty \|\mathbf{x}^{k-1} - \boldsymbol{\alpha}\|_\infty \qquad (2\text{-}14)$$

$$\leqslant \mu \|\mathbf{x}^{k-1} - \boldsymbol{\alpha}\|_\infty \qquad (2\text{-}15)$$

We apply this for $k = 1, 2, \ldots$

$$\|\mathbf{x}^1 - \boldsymbol{\alpha}\|_\infty \leqslant \mu \|\mathbf{x}^0 - \boldsymbol{\alpha}\|_\infty < \mu h \qquad (2\text{-}16)$$

$$\|\mathbf{x}^2 - \boldsymbol{\alpha}\|_\infty \leqslant \mu \|\mathbf{x}^1 - \boldsymbol{\alpha}\|_\infty < \mu^2 h \qquad (2\text{-}17)$$

Combining the results gives

$$\|x_i^k - \alpha_i\|_\infty \leqslant \mu^k h \qquad i = 1, \ldots, n \qquad (2\text{-}18)$$

and if $\mu < 1$, as assumed, the right-hand side goes to zero as k increases, proving the theorem.

We note two things about this theorem. First it gives conditions under which the iteration will converge, but says nothing about what happens if the conditions of the theorem are not met. In that case the iteration may converge or diverge, and the theorem is not applicable. It may converge because the conditions of the theorem are too restrictive and were only needed to prove the theorem, rather than being needed to ensure convergence. The second point is that to apply the theorem we must ensure that Eq. (2-9) is satisfied. This may restrict the allowable choices of x^0, and finding the limits on x^0 may not be a trivial task. However, we can learn some interesting things from the theorem. Suppose the problem we wish to solve is

$$x = \beta f(x) \qquad (2\text{-}19)$$

where β is a parameter, and we apply successive substitution.

$$x^{k+1} = \beta f(x^k) \tag{2-20}$$

We need to look at $\beta df/dx$. Clearly for large β the conditions of the theorem will not be met because $\beta df/dx > 1$ and convergence is not assured, whereas if β is small, $\beta df/dx < 1$ and the iteration scheme converges. Knowing this ahead of time, and knowing the range of β for which we desire solutions, can influence the iteration strategy.

We now apply the theorem to the example tried in Eq. (2-4). Here

$$\frac{df}{dx} = 2x + 1 \tag{2-21}$$

and for

$$|x - 1.414\ldots| < \varepsilon \tag{2-22}$$

we need

$$\left|\frac{df}{dx}\right| \leqslant 1 + 2(1.414 \pm \varepsilon) \tag{2-23}$$

Clearly we cannot find a $\mu < 1$ and the theorem does not apply. Also we found by example that the method diverges. When we change to Eq. (2-8) we need to look at $f(x) = x/2 + 1/x$ and get

$$\frac{df}{dx} = \tfrac{1}{2} - \frac{1}{x^2} \tag{2-24}$$

For $1.2 < x < 1.6$, $|f'| < 0.20$. Thus for $1.2 < x < 1.6$ the theorem says the iteration converges to the solution, as it does.

Now the theorem on successive substitution can be used to turn a divergent scheme into a convergent one. In place of Eq. (2-4) let us use

$$x^{k+1} = x^k + \beta F(x^k) = f(x^k) \tag{2-25}$$

and make β sufficiently small that

$$\frac{df}{dx} = 1 + \beta 2x < 1 \tag{2-26}$$

We choose $\beta = -0.25$ and apply the iteration scheme

$$x^{k+1} = x^k + \beta[(x^k)^2 - 2] \tag{2-27}$$

Starting with $x^0 = 0$ gives us successive values of 0.5, 0.9375, 1.22, 1.35,..., 1.41416 after 10 iterations. The iteration scheme converges, although it takes many iterations.

2-2 NEWTON–RAPHSON

To apply the Newton–Raphson method we expand Eq. (2-1) in a Taylor series about the x^k iterate. We do this first for a single equation

$$F(x^{k+1}) = F(x^k) + \frac{dF}{dx}\bigg|_{x=x^k}(x^{k+1} - x^k) + \frac{d^2F}{dx^2}\bigg|_{x=x^k}\frac{(x^{k+1} - x^k)^2}{2!} + \ldots \tag{2-28}$$

We neglect derivatives of second and higher orders, and we set $F(x^{k+1}) = 0$, since we wish to choose x^{k+1} so that this is true. The result is rearranged to give

$$x^{k+1} = x^k - \frac{F(x^k)}{dF/dx(x^k)} \tag{2-29}$$

Again we choose x^0 and apply Eq. (2-29) successively. This is the Newton method.

If we have several equations, as in Eq. (2-1), we do the same thing

$$F_i(\mathbf{x}^{k+1}) = F_i(\mathbf{x}^k) + \sum_{j=1}^{n} \left.\frac{\partial F_i}{\partial x_j}\right|_{\mathbf{x}^k} (x_j^{k+1} - x_j^k) + \dots \tag{2-30}$$

define the jacobian matrix

$$A_{ij}^k = \left.\frac{\partial F_i}{\partial x_j}\right|_{\mathbf{x}^k} \tag{2-31}$$

and set $F_i(\mathbf{x}^{k+1}) = 0$. We can write the Newton–Raphson method in alternate forms

$$\sum_{j=1}^{n} A_{ij}^k (x_j^{k+1} - x_j^k) = -F_i(\mathbf{x}^k) \tag{2-32}$$

$$x_j^{k+1} = x_j^k - \sum_{i=1}^{n} (A^k)_{ji}^{-1} F_i(\mathbf{x}^k) \tag{2-33}$$

$$\sum_{j=1}^{n} A_{ij}^k x_j^{k+1} = \sum_{j=1}^{n} A_{ij}^k x_j^k - F_i(\mathbf{x}^k) \tag{2-34}$$

To use this method for a system of equations we must solve the system of equations over and over, either by inverting the matrix A_{ij}^k or by decomposition. Since all computer centers have matrix inversion routines readily available, it is assumed here that the reader can do that. Problem 2-4 is a useful review, and the subroutine INVERT can be used.

The convergence of the Newton–Raphson method can be proved under certain conditions (see Isaacson and Keller, p. 115).

Theorem 2-2 Assume $\mathbf{x}^0$ is such that

$$\|\mathbf{A}^{-1}(\mathbf{x}^0)\| \leqslant a \tag{2-35}$$

and

$$\|\mathbf{x}^1 - \mathbf{x}^0\| = \|\mathbf{A}^{-1}(\mathbf{x}^0)F(\mathbf{x}^0)\| \leqslant b \tag{2-36}$$

and

$$\sum_{k=1}^{n} \left|\frac{\partial^2 f_i}{\partial x_j \partial x_k}\right| \leqslant \frac{c}{n} \quad \text{for } \|\mathbf{x} - \mathbf{x}^0\| \leqslant 2b \quad i,j = 1,\dots,n \tag{2-37}$$

Then the Newton iterates lie in the $2b$ sphere

$$\|\mathbf{x}^k - \mathbf{x}^0\| \leqslant 2b \tag{2-38}$$

and

$$\lim_{k \to \infty} \mathbf{x}^k = \boldsymbol{\alpha} \tag{2-39}$$

where

$$F_i(\boldsymbol{\alpha}) = 0$$

$$\|\mathbf{x}^k - \boldsymbol{\alpha}\| \leqslant \frac{2b}{2^k} \qquad (2\text{-}40)$$

and

$$\|\mathbf{x}\| = \max_{1 \leqslant i \leqslant n} |x_i| \qquad (2\text{-}41)$$

$$\|\mathbf{A}\| = \max_i \left(\sum_{j=1}^{n} |a_{ij}| \right) \qquad (2\text{-}42)$$

For example, for $F(x) = x^2 - 2$ we get

$$F'(x) = 2x \qquad (2\text{-}43)$$

$$x^{k+1} = x^k - \frac{(x^k)^2 - 2}{2x^k} = \frac{x^k}{2} + \frac{1}{x^k} \qquad (2\text{-}44)$$

Thus the second iteration scheme, Eq. (2-8), is actually a Newton–Raphson method. Indeed it was prior knowledge of this fact that permitted the selection of the form of Eq. (2-8) which would lead to a convergent iteration scheme.

The Newton–Raphson method, contained in one of the three versions given in Eqs. (2-32), (2-33), or (2-34), requires calculating the jacobian Eq. (2-31). At first glance this means the function must be differentiated analytically. Frequently, however, numerical derivatives are suitable, and they do not affect the answer, only the speed of convergence to get there. Obviously if the numerical approximation is very poor then the Newton–Raphson method would not converge as predicted. We would then use in place of Eq. (2-31) the approximation

$$A_{ij}^k = \frac{F_i(x_l^k(1 + \varepsilon \delta_{lj})) - F_i(x_l^k)}{x_j^k + \varepsilon x_j^k - x_j^k} \qquad (2\text{-}45)$$

where ε is a small number. (Using $\varepsilon = 10^{-6}$ has proved feasible for a CDC computer with a machine accuracy of about 10^{-15}.)

2-3 COMPARISON

The successive substitution method has the advantage of simplicity in that no derivatives need be calculated and no matrices need be inverted. It may not work, however. In the Newton–Raphson method the derivatives must be calculated, and the matrix inversion may take considerable computation time for large problems. However, the chances of success are considerably better. Another feature of the methods is how many iterations are necessary to reach a specified accuracy. Theory says that the successive substitution method converges linearly. Close to the answer, if it takes three iterations to reduce the error from 10^{-2} to 10^{-3} it will take a total of 18 iterations to reduce the error from 10^{-2} to 10^{-8}. By contrast, the Newton–Raphson method converges quadratically. To go from an error of

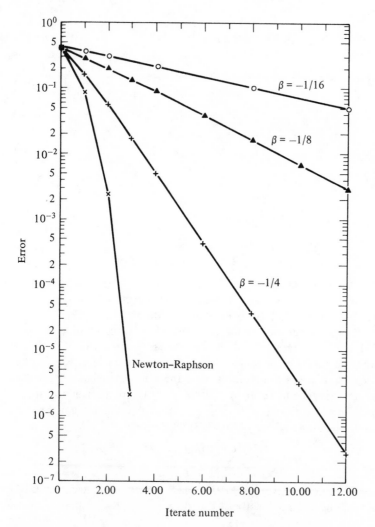

Figure 2-1 Iterate error as function of number of iterations.

10^{-2} to an error of 10^{-8} takes only 2 iterations. The iterate error is roughly the square of the iterate error in the previous iteration. Alternatively the number of significant figures that are correct is doubled at each iteration. Of course each iteration may take more work, since derivatives must be calculated, and perhaps a matrix must be inverted. The final trade-off involves the number of iterations and the work per iteration.

For a sample problem the error is plotted in Fig. 2-1 versus the iterate number and the rapid convergence of the Newton–Raphson method is shown. The speed of convergence of the successive substitution method depends on the value of β; results for several β are shown. A smaller β ensures convergence, but the rate of convergence is slower. For this simple problem the two methods take equivalent work (same number of multiplications) and Newton–Raphson is preferred. For systems of equations the Newton–Raphson method takes a great deal more work since it takes about $n^3/3$ multiplications to solve the linear system of size $n \times n$. Under these circumstances successive substitution may then be preferred, if it works.

STUDY QUESTIONS

Successive substitution and Newton–Raphson methods
1. How to write the iteration scheme
2. Which one has the wider range of convergence
3. Which one converges faster
4. The amount of work necessary to solve each of them
5. What does the convergence depend upon
6. What happens when the problem is linear

PROBLEMS

2-1 Consider the problem

$$-15c_1 + 15c_2 = \phi^2 f(c_1)$$
$$c_2 = 1.0$$

Set up a calculation scheme that is useful for small ϕ^2 (successive substitution) and for large ϕ^2 (Newton–Raphson). Apply your scheme to the following cases (solve within 10^{-6}):

    (a) $\phi^2 = 1, f(c) = c$
    (b) $\phi^2 = 40, f(c) = c$
    (c) $\phi^2 = 1, f(c) = c^2$
    (d) $\phi^2 = 1000, f(c) = c/(1+\alpha c)^2, \alpha = 20$

2-2 Discuss the following points after working problem 2-1. Apply the convergence theorem for successive substitution to problems 2-1a to 2-1d. Does the method converge when the conditions of the theorem are satisfied? When they are not? How many iterations are required to achieve the required accuracy for the two methods and the four cases? What happens in the Newton–Raphson method for linear problems? Comment on the ease of applying the two methods in the four cases.

2-3 Solve $-10.5c + 10.5 = \phi^2 R(c)$

$$R(c) = c \exp\left[\gamma\beta \frac{1-c}{1+\beta(1-c)}\right]$$

for $\gamma = 30, \beta = 0.4, \phi = 0.4$. The solution is in $[0, 1]$.

2-4 Solve using the Newton–Raphson method

$$-13.59530877c_1 + 20.42831009c_2 - 6.833001321c_3 = \phi^2 f(c_1)$$
$$14.57168991c_1 - 91.40469119c_2 + 76.83300129c_3 = \phi^2 f(c_2)$$
$$0.9482702526c_1 - 14.948270256c_2 + 14c_3 = \text{Bi}_m(1 - c_3)$$

for

    (a) $f(c) = c, \phi^2 = 1, \text{Bi}_m = 100$
    (b) $f(c) = c^2, \phi^2 = 1, \text{Bi}_m = 100$
    (c) $f(c) = c/(1+\alpha c)^2, \alpha = 20, \phi = 32, \text{Bi}_m = 100$

2-5 Apply the theory on convergence of Newton–Raphson method to Eq. (2-5). Choose an x_0 that permits satisfaction of the inequalities, Eqs. (2-35) to (2-37). What values of a, b, and c are appropriate to your choice of x_0? If such constants exist then the theorem says that the Newton–Raphson method converges. Try it for your x_0 and verify convergence.

BIBLIOGRAPHY

Complete details and convergence proofs for the iterative schemes discussed here (as well as others, such as the chord method) are in

Isaacson, E., and H. B. Keller: *Analysis of Numerical Methods*, John Wiley & Sons, Inc., New York, 1966.

THREE

ORDINARY DIFFERENTIAL EQUATIONS—
INITIAL-VALUE PROBLEMS

Evolution problems lead to initial-value problems in time. Here we outline some successful and popular methods of solving those problems. After introducing the terminology, interpolation and quadrature schemes are presented, since they lead to many of the methods for solving ordinary differential equations. Special techniques—extrapolation and step-size control—are explained and the important matter of stability is treated in depth. Standard integration packages, such as Gear's and the Runge–Kutta method, are summarized before comparing the methods on some easy and some difficult problems.

3-1 TERMINOLOGY

In this chapter we consider how to solve systems of initial-value problems of the type

$$\frac{dy_i}{dt} = f_i(y_1, y_2, \ldots, y_n) \qquad i = 1, \ldots, n \tag{3-1}$$

$$y_i(0) = g_i = \text{given} \tag{3-2}$$

We note that all the boundary conditions are prescribed at time zero, which is necessary if the problems are initial-value. If the conditions must be applied at two or more times the problem is a boundary-value type. These problems are treated in Chapter 4.

What if we wish to solve higher-order systems? The equation

$$y^{(n)} + F(y^{(n-1)}, y^{(n-2)}, \ldots, y'', y', y) = 0 \tag{3-3}$$

17

can be reduced to the form of Eq. (3-1) by making the substitution

$$y_i = y^{(i-1)} = \frac{d^{(i-1)}y}{dt^{(i-1)}}$$

(The reader is encouraged to write this out in detail.) The initial conditions for the high-order equation may be of the form

$$G_i(y^{(n-1)}(0), y^{(n-2)}(0), \ldots, y'(0), y(0)) = 0 \qquad (3\text{-}4)$$

and we can reduce this system of equations to the form of Eq. (3-2) by solving the system of Eq. (3-4) for $y_i(0) = y^{(i-1)}(0)$ using the techniques of Chapter 2.

Another simplification we have made in Eq. (3-1) is to have the right-hand side depend only on $\{y\}$ and not on t. This is not limiting, because if we wish to solve a problem for which the function $\mathbf{f}$ depends on t we need only append the differential equation

$$\frac{dy_{n+1}}{dt} = 1 \qquad y_{n+1}(0) = 0 \qquad (3\text{-}5)$$

to the system. Of course $y_{n+1} = t$, so the system of equations can be written in the form of Eq. (3-1). Sometimes the notation of Eq. (3-1) is simplified and written in the form of a vector equation, with $\mathbf{y} = \{y_i\}$,

$$\frac{d\mathbf{y}}{dt} = \mathbf{f}(\mathbf{y}) \qquad (3\text{-}6)$$

$$\mathbf{y}(0) = \mathbf{g} \qquad (3\text{-}7)$$

We call a method explicit or implicit depending on whether the function $\mathbf{f}$ is evaluated at known conditions $y_i(t_n)$, or at unknown conditions $y_i(t_{n+1})$. Explicit methods of integration, such as the Euler method, evaluate the function $\mathbf{f}$ with known information

$$\left.\frac{d\mathbf{y}}{dt}\right|_{t=t_n} = \mathbf{f}(\mathbf{y}_n) \qquad \mathbf{y}_n = \mathbf{y}(t_n) \qquad (3\text{-}8)$$

Implicit methods of integration, on the other hand, evaluate the function $\mathbf{f}$ at the unknown solution $\mathbf{y}_{n+1}$. An example is the trapezoid rule

$$\left.\frac{d\mathbf{y}}{dt}\right|_{t=t_n} = \tfrac{1}{2}[\mathbf{f}(\mathbf{y}_n) + \mathbf{f}(\mathbf{y}_{n+1})] \qquad (3\text{-}9)$$

An important characteristic of a system of ordinary differential equations is whether or not they are stiff. The idea of stiffness is easily illustrated. Suppose we wish to solve the problem

$$\frac{du_1}{dt} = -u_1 \qquad u_1(0) = 1.5 \qquad (3\text{-}10)$$

The solution is

$$u_1 = 1.5e^{-t} \qquad (3\text{-}11)$$

Numerical integration may be desired from $t = 0$ to $t = 10$, say. The stable step size we can use with explicit methods is limited by

$$\Delta t \leqslant p \qquad (3\text{-}12)$$

(We see the reason for this restriction in Sec. 3-7.) For one method $p = 2$, thus approximately $10/2 = 5$ time steps are necessary to integrate to $t = 10$. Now we numerically solve the equation

$$\frac{du_2}{dt} = -1000u_2$$

$$u_2(0) = 0.5 \qquad u_2 = 0.5e^{-1000t} \qquad (3\text{-}13)$$

This time the largest step size we can use is

$$\Delta t \leqslant \frac{p}{1000} \qquad (3\text{-}14)$$

and with $p = 2$ $\Delta t \leqslant 0.002$. We generally only want to integrate until $t = 0.01$, and this requires $0.01/0.002 = 5$ steps. If we integrate to $t = 10$ we would need 5,000 integration steps.

Next suppose we are not able to separate out the functions u_1 and u_2, and we must solve for $y_1 = u_1 + u_2$ and $y_2 = u_1 - u_2$. The differential equations governing y are then

$$\frac{d\mathbf{y}}{dt} = \mathbf{A}\mathbf{y} \qquad \mathbf{y}(0) = [2, 1]^T \qquad (3\text{-}15)$$

$$\mathbf{A} = \begin{pmatrix} -500.5 & 499.5 \\ 499.5 & -500.5 \end{pmatrix} \qquad (3\text{-}16)$$

and the solution is

$$y_1 = 1.5e^{-t} + 0.5e^{-1000t}$$
$$y_2 = 1.5e^{-t} - 0.5e^{-1000t} \qquad (3\text{-}17)$$

Now we must integrate to $t = 10$ to see the full evolution of y_1 and y_2. However, the largest step size is limited by

$$\Delta t \leqslant \frac{p}{|\lambda|_{max}} \qquad (3\text{-}18)$$

where $|\lambda|_{max}$ is the largest of the absolute magnitudes of the eigen values. The eigen values of the matrix $\mathbf{A}$ are $\lambda_1 = -1,000$, $\lambda_2 = -1$. This means that the largest step size is limited by

$$\Delta t \leqslant \frac{p}{1000} \qquad (3\text{-}19)$$

For $p = 2$ and integration to $t = 10$, this requires $10/0.002 = 5,000$ integration steps.

We have the unfortunate situation with systems of equations that the largest step size is governed by the largest eigen value and the final time is usually governed by the smallest eigen value. Thus we must use a very small time step (because of the large eigen value) for a very long time (because of the small eigen value). For a single equation we do not have this dichotomy; the eigen value and desired integration time go hand-in-hand. This characteristic of systems of equations is called stiffness. We define the stiffness ratio SR (see Lambert, p. 232) as

$$SR = \frac{\max_i |\mathrm{Re}\,\lambda_i|}{\min_i |\mathrm{Re}\,\lambda_i|} \tag{3-20}$$

Typically $SR = 20$ is not stiff, $SR = 10^3$ is stiff, and $SR = 10^6$ is very stiff.

If the system of equations is nonlinear, Eq. (3-1) instead of Eq. (3-15), we linearize the equation about the solution at that time

$$\frac{dy_i}{dt} = f_i(\mathbf{y}(t_n)) + \sum_{j=1}^{n} \frac{\partial f_i}{\partial y_j} (y_j - y_j(t_n)) \tag{3-21}$$

$$A_{ij} = \frac{\partial f_i}{\partial y_j} \tag{3-22}$$

We calculate the eigen values of the matrix $\mathbf{A}$, the jacobian matrix, and define stiffness, etc. based on the jacobian matrix. The stiffness then applies only to that particular time, and, as the evolution proceeds, the stiffness of the system of equations may change. This, of course, makes the problem both interesting and difficult. We need to be able to classify our problems as stiff or not, however, because some methods of integration work well for stiff problems. Some methods do not work at all well and must not be applied to stiff problems. Generally we find that implicit methods must be used for stiff problems because explicit methods are too expensive. Explicit methods are suitable for equations that are not stiff.

3-2 INTERPOLATION AND QUADRATURE

If we have values of a function at successive times and wish to evaluate the function at some point in between these data points we need an interpolation scheme. Suppose the times are $t_{n-1}, t_n, t_{n+1}, \ldots$ and are equally spaced, and let $y_n = y(t_n)$. Let us define the forward differences

$$\Delta y_n = y_{n+1} - y_n \tag{3-23}$$

$$\Delta^2 y_n = \Delta y_{n+1} - \Delta y_n = y_{n+2} - 2y_{n+1} + y_n \tag{3-24}$$

and then the finite interpolation formula

$$y = y_0 + \alpha \Delta y_0 + \frac{\alpha(\alpha-1)}{2!} \Delta^2 y_0 + \ldots + \frac{\alpha(\alpha-1)\ldots(\alpha-n+1)}{n!} \Delta^n y_0 \tag{3-25}$$

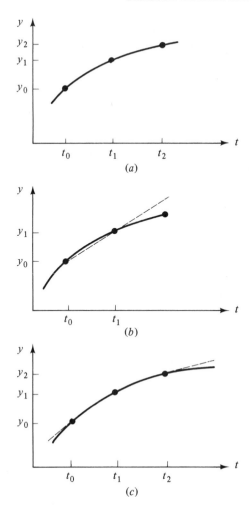

Figure 3-1. Interpolation. (a) Function to be interpolated. (b) Linear interpolation. (c) Quadratic interpolation.

$$\alpha = \frac{t_\alpha - t_0}{h} \qquad h = \Delta t \qquad (3\text{-}26)$$

This is just

$$y_\alpha = y_0 + \alpha(y_1 - y_0) + \frac{\alpha(\alpha-1)}{2!}(y_2 - 2y_1 + y_0) + \ldots \qquad (3\text{-}27)$$

This formula is derived by making an nth-order polynomial in α go through the points $y_0, y_1, \ldots, y_n$. Equation (3-27) provides an interpolation formula to deduce the value of y at any point between t_0 and t_1. If we truncate at the first term the interpolation is linear, as shown in Fig. 3-1. Keeping the second-order terms corresponds to fitting a quadratic polynomial through the points y_0, y_1, and y_2.

Equation (3-27) is a continuous function of α and can be differentiated. Let us differentiate it with respect to t, using

$$\frac{d}{dt} = \frac{d}{d\alpha}\frac{d\alpha}{dt} \qquad \frac{d\alpha}{dt} = \frac{1}{h} \qquad \frac{dy}{dt} = \frac{1}{h}\frac{dy}{d\alpha} \tag{3-28}$$

to give

$$h\frac{dy}{dt} = \frac{dy_\alpha}{d\alpha} = \Delta y_0 + \frac{2\alpha - 1}{2!}\Delta^2 y_0$$

$$+ \frac{(\alpha-1)(\alpha-2) + \alpha(\alpha-2) + \alpha(\alpha-1)}{3!}\Delta^3 y_0 + \dots \tag{3-29}$$

At the point $\alpha = 0$ we get

$$h\left(\frac{dy}{dt}\right)_0 = \Delta y_0 - \tfrac{1}{2}\Delta^2 y_0 + \tfrac{1}{3}\Delta^3 y_0 \tag{3-30}$$

or, since t_0 is arbitrary,

$$hy_n' = (\Delta - \tfrac{1}{2}\Delta^2 + \tfrac{1}{3}\Delta^3 - \dots)y_n \tag{3-31}$$

Expanding this gives

$$hy_n' = y_{n+1} - y_n - \tfrac{1}{2}(y_{n+2} - 2y_{n+1} + y_n) + \dots \tag{3-32}$$

Thus if we know the values of y at times t_n, t_{n+1}, and t_{n+2} we can estimate the first derivative y_n'. Returning to Eq. (3-29), we differentiate it again to get the second derivative

$$h^2 \frac{d^2 y}{dt^2} = \Delta^2 y_0 + (\alpha - 1)\Delta^3 y_0 + \dots \tag{3-33}$$

At $\alpha = 0$

$$h^2 y_n'' = \Delta^2 y_n - \Delta^3 y_n + \dots \tag{3-34}$$

This gives a way to estimate the second derivative. Alternatively, we can say that the second difference $\Delta^2 y_n$ is of order h^2. More generally the nth-order difference is of order h^n.

To obtain an integration formula for

$$I = \int_{t_0}^{t_0+h} y(t)\,dt \tag{3-35}$$

we simply insert y_α and integrate

$$\int_{t_0}^{t_0+h} \left[y_0 + \alpha\Delta y_0 + \frac{\alpha(\alpha-1)}{2!}\Delta^2 y_0 + \dots \right] dt$$

$$= \int_0^1 \left[y_0 + \alpha\Delta y_0 + \frac{\alpha(\alpha-1)}{2!}\Delta^2 y_0 + \dots \right] h\,d\alpha \tag{3-36}$$

and then use the mean-value theorem to include the higher-order terms in the second derivative. Thus we get for some $0 \leqslant \zeta \leqslant 1$

$$\int_{t_0}^{t_0+h} y(t)dt = [y_0 + \tfrac{1}{2}(y_1 - y_0) - \tfrac{1}{12}h^2 y_0''(\zeta)]$$

$$= \frac{h}{2}(y_0 + y_1) + 0(h^3) \tag{3-37}$$

The notation $0(h^3)$ means that $0(h^3)/h^3$ is bounded as $h \rightarrow 0$. Usually this means a term multiplied by h^3. More generally we can write Eq. (3-37) as

$$\int_{t_n}^{t_{n+1}} y(t)dt = \frac{h}{2}(y_n + y_{n+1}) + 0(h^3) \tag{3-38}$$

and add up integrals from a to b, with successive divisions at $a = t_0$, $h = t_1$, $2h = t_2, \ldots, Nh = t_N$, and $(N+1)h = b = t_{N+1}$, giving

$$\int_a^b y(t)dt = \frac{h}{2}(y_0 + 2y_1 + 2y_2 + \ldots + 2y_N + y_{N+1}) + 0(h^3) \tag{3-39}$$

The alert reader will recognize this as the trapezoid rule. It is derived by passing a linear interpolation between the data points and integrating exactly under the piecewise linear interpolant.

Next let us integrate over two intervals and keep the cubic terms to obtain

$$I = \int_{t_0}^{t_0+2h} \left[y_0 + \alpha \Delta y_0 + \frac{\alpha(\alpha-1)}{2!} \Delta^2 y_0 + \frac{\alpha(\alpha-1)(\alpha-2)}{3!} \Delta^3 y_0 + 0(\alpha^4) \right] dt$$

$$= h \int_0^2 \left[y_0 + \alpha \Delta y_0 + \frac{\alpha(\alpha-1)}{2} \Delta^2 y_0 + \frac{\alpha(\alpha-1)(\alpha-2)}{6} \Delta^3 y_0 + 0(\alpha^4) \right] d\alpha \tag{3-40}$$

Carrying out the integration gives the following result (see problem 3-3):

$$I = \frac{h}{3}(y_0 + 4y_1 + y_2) + 0(h^5) \tag{3-41}$$

The term involving $\Delta^3 y_0$ is zero since the α term integrates to zero. More generally, for an arbitrary pair of intervals,

$$\int_{t_n}^{t_{n+2}} y(t)dt = \frac{h}{3}(y_n + 4y_{n+1} + y_{n+2}) + 0(h^5) \tag{3-42}$$

If we add up several pairs of intervals we get Simpson's rule in which

$$\int_a^b y(t)dt = \frac{h}{3}(y_0 + 4y_1 + 2y_2 + 4y_3 + 2y_4 + \ldots$$

$$+ 4y_{N-1} + 2y_N + 4y_{N+2} + y_{N+3}) + 0(h^5) \tag{3-43}$$

N must now be even since the number of intervals must be a multiple of two. This formula corresponds to passing a quadratic polynomial through the three points and integrating exactly under the interpolant. We note in passing that as we go from one subinterval (t_0, t_1, t_2) to the next (t_2, t_3, t_4) the interpolant is continuous, since y_2 is the same in both subintervals, but the first and higher derivatives are not necessarily continuous across the subintervals. The linear interpolant, Eq. (3-39), has an error of $0(h^3)$ and the quadratic interpolant would have an error of $0(h^4)$ except for the fortuitous cancelling of the $\Delta^3 y_0$ term, giving one higher order $0(h^5)$.

Backward difference formulas can also be used

$$\nabla y_n = y_n - y_{n-1} \tag{3-44}$$

$$\nabla^2 y_n = \nabla y_n - \nabla y_{n-1} = y_n - 2y_{n-1} + y_{n-2} \tag{3-45}$$

The interpolation formula is obtained by requiring that a jth-order polynomial in α goes through the points $y_n, y_{n-1}, \ldots, y_{n-j}$. Thus

$$y_{n+\alpha} = y_n + \alpha \nabla y_n + \frac{\alpha(\alpha+1)}{2!} \nabla^2 y_n + \ldots + \frac{\alpha(\alpha+1)\ldots(\alpha+j-1)}{j!} \nabla^j y_n \tag{3-46}$$

Alternatively, we can use the points $y_{n+1}, y_n, \ldots, y_1$. In which case

$$y_{n+\alpha} = y_{n+1} + (\alpha-1)\nabla y_{n+1} + \frac{\alpha(\alpha-1)}{2!} \nabla^2 y_{n+1} + \ldots$$

$$+ \frac{(\alpha-1)(\alpha)(\alpha+1)\ldots(\alpha+j-2)}{j!} \nabla^j y_{n+1} \tag{3-47}$$

These interpolation formulas can be written for the first derivative as well

$$\frac{dy_n(\alpha)}{dt} = y'_{n+\alpha} = y'_n + \alpha \nabla y'_n + \frac{\alpha(\alpha+1)}{2!} \nabla^2 y'_n + \ldots$$

$$+ \frac{\alpha(\alpha+1)\ldots(\alpha+j-1)}{j!} \nabla^j y'_n \tag{3-48a}$$

$$y'_{n+\alpha} = y'_{n+1} + (\alpha-1)\nabla y'_{n+1} + \frac{(\alpha-1)\alpha}{2!} \nabla^2 y'_{n+1} + \ldots$$

$$+ \frac{(\alpha-1)\alpha(\alpha+1)\ldots(\alpha+j-2)}{j!} \nabla^j y_{n+1} \tag{3-48b}$$

If Eq. (3-48a) is differentiated with respect to t and evaluated at $\alpha = 0$ we obtain an estimate of the second derivative

$$h^2 y''_n = \nabla(hy'_n) + \tfrac{1}{2}\nabla^2(hy'_n) + \ldots \tag{3-49}$$

Similarly, the higher derivatives are given by

$$h^{j+1} y_n^{(j+1)} = \nabla^j(hy'_n) + \ldots \tag{3-50}$$

Note that an estimate of a higher derivative can be obtained from values of lower derivatives at successive points. Only values of y_i are needed to obtain y' and only values of y'_i are needed to obtain y''. If only the first terms of Eqs. (3-49) and (3-50) are used the error incurred is one order of h higher, and hence decreases to zero as $h \to 0$.

3-3 EXPLICIT INTEGRATION METHODS

We can use the interpolation formulas to deduce integration methods. If we take the single equation

$$\frac{dy}{dt} = f(y) \tag{3-51}$$

and integrate both sides from t_n to t_{n+1}

$$\int_{t_n}^{t_{n+1}} \frac{dy}{dt} dt = \int_{t_n}^{t_{n+1}} f(y(t))dt \tag{3-52}$$

we get

$$y_{n+1} = y_n + \int_{t_n}^{t_{n+1}} f(y(t))dt = y_n + \int_{t_n}^{t_{n+1}} y'dt \tag{3-53}$$

or

$$y_{n+1} = y_n + h \int_0^1 y'(\alpha)d\alpha \tag{3-54}$$

The integration schemes are generated by inserting various interpolation formulas for $dy/dt(\alpha) = y'(\alpha)$. Substitution of Eq. (3-48a) into Eq. (3-54) gives

$$y_{n+1} = y_n + h \sum_{i=0}^q a_i \nabla^i y'_n \tag{3-55}$$

$$a_i = \int_0^1 \frac{\alpha(\alpha+1)\ldots(\alpha+i-1)}{i!} d\alpha \tag{3-56}$$

$$y_{n+1} = y_n + h(1 + \tfrac{1}{2}\nabla + \tfrac{5}{12}\nabla^2 + \ldots)y'_n \tag{3-57}$$

This can be expanded to give

$$
\begin{aligned}
y_{n+1} &= y_n + hy'_n + \frac{h}{2}(y'_n - y'_{n-1}) + \ldots \\
&= y_n + hy'_n + \frac{h^2}{2} y''_n + \ldots
\end{aligned}
\tag{3-58}
$$

The Euler method is obtained by truncating at $q = 0$ and using $y'_n = f(y_n)$

$$y_{n+1} = y_n + hf(y_n) + 0(h^2) \tag{3-59}$$

The formula is more revealing in the form

$$\frac{y_{n+1} - y_n}{h} = f(y_n) + 0(h) \qquad \text{explicit Euler} \tag{3-60}$$

The left-hand side is a representation of the derivative dy/dt and the derivative is evaluated using the solution at y_n. Graphically this means we evaluate the slope at the nth time level and extend that slope to the next time level to obtain y_{n+1} (see

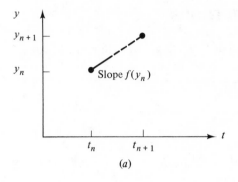

(a)

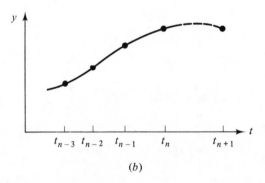

(b)

Figure 3-2 Explicit integration methods. (a) Euler method. (b) Fourth-order Adams–Bashforth method.

Fig. 3-2). Notice also that the linear interpolation gives a method that has accuracy proportional to h or $0(h)$. [Note the difference between Eqs. (3-59) and (3-60).]

The second-order Adams–Bashforth method is obtained by truncating Eq. (3-55) at $q = 1$. Thus

$$y_{n+1} = y_n + h(y_n' + \tfrac{1}{2}\nabla y_n') \tag{3-61}$$

$$= y_n + \frac{h}{2}(3y_n' - y_{n-1}') \tag{3-62}$$

The accuracy of the method is $0(h^2)$ and the appropriate interpolation formula is Eq. (3-46) keeping terms up to second-order differences.

The fourth-order Adams–Bashforth method is obtained by truncating Eq. (3-55) at $q = 3$. Thus

$$y_{n+1} = y_n + h(y_n' + \tfrac{1}{2}\nabla y_n' + \tfrac{5}{12}\nabla^2 y_n' + \tfrac{3}{8}\nabla^3 y_n') \tag{3-63}$$

$$= y_n + \frac{h}{24}(55y_n' - 59y_{n-1}' + 37y_{n-2}' - 9y_{n-3}') + 0(h^5) \tag{3-64}$$

The accuracy of the method is $0(h^4)$, and the method corresponds to passing a third-order polynomial through past values of y_{n-1}', etc. At the beginning of the calculation we know only $y'_0 = f(y_0)$, so we must use another method to get started. After several steps we can then shift to the Adams–Bashforth method. The starting method must be done with a small time step if its accuracy is less than fourth-order, but a low-order method with very small steps is feasible because only a few steps are needed.

3-4 IMPLICIT INTEGRATION METHODS

To obtain an implicit method we use the interpolation formula Eq. (3-48b) and substitute into Eq. (3-54) (see problem 3-5).

$$y_{n+1} = y_n + h(1 - \tfrac{1}{2}\nabla - \tfrac{1}{12}\nabla^2 - \tfrac{1}{24}\nabla^3 - \ldots)y'_{n+1} \tag{3-65}$$

If we truncate this with the first term we get the backward Euler method

$$y_{n+1} = y_n + hy'_{n+1} + 0(h^2) \tag{3-66}$$

$$\frac{y_{n+1} - y_n}{h} = y'_{n+1} = f(y_{n+1}) + 0(h) \qquad \text{implicit Euler} \tag{3-67}$$

The accuracy of this method is only $0(h)$, as in the case of the Euler method, but we see below that this method is more stable. Compare Eqs. (3-60) and (3-67) to illustrate the difference between the explicit and implicit Euler methods.

Truncation of Eq. (3-65) at the second term gives a method

$$y_{n+1} = y_n + h[y'_{n+1} - \tfrac{1}{2}(y'_{n+1} - y'_n)] + 0(h^3) \tag{3-68}$$

$$= y_n + \frac{h}{2}[f(y_{n+1}) + f(y_n)] + 0(h^3) \tag{3-69}$$

which has an accuracy proportional to $0(h^2)$. This method is variously called the modified Euler method, trapezoid rule, or Crank–Nicolson method. Truncation at the fourth term gives the fourth-order Adams–Moulton method

$$y_{n+1} = y_n + \frac{h}{24}(9y'_{n+1} + 19y'_n - 5y'_{n-1} + y'_{n-2}) + 0(h^5) \tag{3-70}$$

How are these equations solved? Since the value of y_{n+1} is unknown, all the equations represent a nonlinear equation to solve for y_{n+1}. If we have several equations instead of just one we get systems of nonlinear equations for y_{n+1}. Chapter 2 describes methods for solving such systems by writing the general implicit methods in the form

$$y_{n+1} = \sum_{i=1}^{k} \alpha_i y_{n+1-i} + h \sum_{i=0}^{k} \beta_i y'_{n+1-i} \tag{3-71}$$

Difference methods have different choices of k, α_i, and β_i. If $\beta_0 = 0$ the method is explicit since the right-hand side can be evaluated. If $\beta_0 \neq 0$ the method is implicit since the right-hand side depends on $f(y_{n+1})$, which is not known. To solve such an equation we write Eq. (3-71) in the form

$$y_{n+1} = h\beta_0 f(y_{n+1}) + w_n \tag{3-72}$$

where w_n represents all the known information. Successive substitution applied to Eq. (3-72) gives

$$y_{n+1}^{(s+1)} = h\beta_0 f(y_{n+1}^{(s)}) + w_n \tag{3-73}$$

and we iterate until an error tolerance is met where

$$|y_{n+1}^{(s+1)} - y_{n+1}^{(s)}| < \varepsilon \tag{3-74}$$

If this tolerance is not met in N iterations we halve the step size and try again. We know that the successive substitution method converges provided that there is a $\mu < 1$ such that

$$h\beta_0 \left| \frac{\partial f}{\partial y} \right| < \mu < 1 \tag{3-75}$$

If $\partial f/\partial y$ is bounded there is always such a μ since we can decrease h to satisfy the inequality. Thus we know that for small enough h the successive substitution methods works.

Newton–Raphson is applied in a similar way with

$$y_{n+1}^{(s+1)} = h\beta_0 \left[f(y_{n+1}^{(s)}) + \frac{\partial f}{\partial y}\bigg|_{y_{n+1}^{(s)}} (y_{n+1}^{(s+1)} - y_{n+1}^{(s)}) \right] + w_n \tag{3-76}$$

Rearrangement gives

$$\left(I - h\beta_0 \frac{\partial f}{\partial y} \right)_{y_{n+1}^{(s)}} (y_{n+1}^{(s+1)} - y_{n+1}^{(s)}) = h\beta_0 f(y_{n+1}^{(s)}) + w_n - y_{n+1}^{(s)} \tag{3-77}$$

If we had multiple equations we would get a system of equations at this point, with $I = \delta_{ij}$ and $\partial f/\partial y = \partial f_i/\partial y_j$ as the jacobian matrix. The Newton–Raphson method also converges provided h is small enough, but it may be more robust than the successive substitution method. It does require calculation of the jacobian matrix, however.

We can conclude that any implicit method is soluble provided the step size is small enough. The strategies described in Secs. 3-6 and 3-9 ensure that this is so.

3-5 PREDICTOR–CORRECTOR AND RUNGE–KUTTA METHODS

An alternative, which is between the explicit and implicit methods, is a predictor–corrector method. In this scheme the predictor is an explicit equation which gives an estimate of y_{n+1}, called $\bar{y}_{n+1}$. This value is then used in the corrector, which is an implicit equation, except that the right-hand side is evaluated using the predicted value $\bar{y}_{n+1}$ rather than y_{n+1}. Combining the Euler method as the predictor and the modified Euler method as the corrector gives the improved Euler method

$$\bar{y}_{n+1} = y_n + hy_n' = y_n + hf(y_n) \tag{3-78}$$

$$y_{n+1} = y_n + \frac{h}{2}(y'_{n+1} + y'_n)$$

$$= y_n + \frac{h}{2}[f(\bar{y}_{n+1}) + f(y_n)] \tag{3-79}$$

Alternatively we can iterate several times with the corrector to give

$$\bar{y}_{n+1}^{(0)} = y_n + hy'_n \tag{3-80}$$

$$\bar{y}_{n+1}^{(s+1)} = y_n + \frac{h}{2}[f(\bar{y}_{n+1}^{(s)}) + f(y_n)] \tag{3-81}$$

The Adams predictor–corrector uses the Adams–Bashforth method to predict

$$\bar{y}_{n+1} = y_n + \frac{h}{24}(55y'_n + \ldots) \tag{3-82}$$

and the Adams–Moulton method to correct

$$y_{n+1} = y_n + \frac{h}{24}(9\bar{y}'_{n+1} + 19y'_n + \ldots) \tag{3-83}$$

The corrector can be applied several times as well. The advantage of these methods is that the stability limitations are less severe than for explicit methods without the necessity of solving the nonlinear equations in the implicit methods.

Runge–Kutta methods are widely used. The explicit schemes involve evaluation of the derivative at points between t_n and t_{n+1}. Let us write the general formula

$$y_{n+1} = y_n + \sum_{i=1}^{v} w_i k_i \tag{3-84}$$

with

$$k_i = hf\left(t_n + c_i h, \, y_n + \sum_{j=1}^{i-1} a_{ij} k_j\right) \tag{3-85}$$

$$c_1 = 0 \tag{3-86}$$

and expand both f and y in a Taylor series

$$y_{n+1} = y_n + y'_n h + \frac{h^2}{2!} y''_n + \ldots \tag{3-87}$$

$$y'_n = f_n \tag{3-88}$$

$$y''_n = \left(\frac{\partial f}{\partial t} + \frac{\partial f}{\partial y} y'\right)_n = (f_t + ff_y)_n \tag{3-89}$$

Putting this into Eq. (3-84) gives

$$y_{n+1} = y_n + hf_n + \frac{h^2}{2}(f_t + ff_y)_n + \ldots \tag{3-90}$$

Now this procedure is repeated for the values of k_i.

$$k_1 = hf(t_n, y_n) = hf_n \tag{3-91}$$

$$k_2 = hf(t_n + c_2 h, y_n + a_{21} k_1)$$
$$= hf_n + h^2[c_2(f_x)_n + a_{21}(f_y)_n f_n] + \cdots \tag{3-92}$$

Substituting this into Eq. (3-84) gives

$$y_{n+1} = y_n + w_1 h f_n + w_2 h f_n + w_2 h^2(c_2 f_t + a_{21} f f_y)_n + \cdots \tag{3-93}$$

Comparison of Eqs. (3-90) to (3-93) shows them to be identical if the following conditions are satisfied:

$$w_1 + w_2 + \ldots + w_v = 1.0$$
$$w_2 a_{21} + \ldots = 0.5 \tag{3-94}$$
$$w_2 c_2 + \ldots = 0.5$$

Examination of the full set of equations reveals that some of the parameters are redundant. For $v = 2$ we have one free parameter at our disposal while still satisfying the equations. For $v = 3$ we have two such free parameters. We obtain a Runge–Kutta method by specifying v and the free parameters. For $v = 2$ we have

$$w_1 + w_2 = 1.0$$
$$w_2 c_2 = 0.5 \tag{3-95}$$
$$c_2 = a_{21}$$

Specification of c_2' then gives a_{21}, w_2, and w_1. With $c_2 = 0.5$ we get the second-order Runge–Kutta scheme

$$y_{n+1} = y_n + hf(t_n + \tfrac{1}{2}h, y_n + \tfrac{1}{2}hf_n) \tag{3-96}$$

or a midpoint scheme. With $c_2 = 1$ we get

$$y_{n+1} = y_n + \frac{h}{2}[f_n + f(t_n + h, y_n + hf_n)] \tag{3-97}$$

which is identical to the Euler predictor–corrector scheme in Eq. (3-79).

A very popular scheme is the Runge–Kutta–Gill method, which is fourth-order and expressed by the algorithm

$$k_1 = hf(t_n, y_n)$$
$$k_2 = hf(t_n + \tfrac{1}{2}h, y_n + \tfrac{1}{2}k_1)$$
$$k_3 = hf(t_n + \tfrac{1}{2}h, y_n + ak_1 + bk_2)$$
$$k_4 = hf(t_n + h, y_n + ck_2 + dk_3)$$
$$y_{n+1} = y_n + \tfrac{1}{6}(k_1 + k_4) + \tfrac{1}{3}(bk_2 + dk_3) \tag{3-98}$$
$$a = \frac{\sqrt{2} - 1}{2} \qquad b = \frac{2 - \sqrt{2}}{2}$$
$$c = -\frac{\sqrt{2}}{2} \qquad d = 1 + \frac{\sqrt{2}}{2}$$

The parameter choices have been made to minimize round-off error. Round-off error occurs in a computer when two n digit numbers are multiplied, giving a $2n$ digit number, but only the first n digits are retained.

It is possible to have implicit Runge–Kutta schemes and here we introduce a semi-implicit scheme due to Caillaud and Padmanabhan.[2] We again write Eq. (3-84) but now allow the summation in Eq. (3-85) to go from 1 to i, making the scheme implicit. Thus

$$k_i = hf\left(y_n + \sum_{j=1}^{i} a_{ij}k_j\right) \tag{3-99}$$

We expand this in a series

$$k_i = hf\left(y_n + \sum_{j=1}^{i-1} a_{ij}k_j\right) + h\frac{\partial f}{\partial y}\left(y_n + \sum_{j=1}^{i-1} a_{ij}k_j\right)a_{ii}k_i \tag{3-100}$$

and generalize to

$$\left[I - ha_{ii}\frac{\partial f}{\partial y}\left(y_n + \sum_{j=1}^{i-1} d_{ij}k_j\right)\right]k_i = hf\left(y_n + \sum_{j=1}^{i-1} a_{ij}k_j\right) \tag{3-101}$$

As before we choose v and expand the equations to the vth order. We choose the parameters so that the same factor multiplies k_i for each i—this minimizes the work of inverting matrices in Eq. (3-102) below—and also so that the method has important stability properties (see Sec. 3-8). The final algorithm for a system of equations is given in the form suggested by Michelsen.[5]

$$\mathbf{k}_1 = h\left[\mathbf{I} - ha_1\frac{\partial \mathbf{f}}{\partial \mathbf{y}}(\mathbf{y}_n)\right]^{-1}\mathbf{f}(\mathbf{y}_n)$$

$$\mathbf{k}_2 = h\left[\mathbf{I} - ha_1\frac{\partial \mathbf{f}}{\partial \mathbf{y}}(\mathbf{y}_n)\right]^{-1}\mathbf{f}(\mathbf{y}_n + b_2\mathbf{k}_1) \tag{3-102}$$

$$\mathbf{k}_3 = h\left[\mathbf{I} - ha_1\frac{\partial \mathbf{f}}{\partial \mathbf{y}}(\mathbf{y}_n)\right]^{-1}(b_{31}\mathbf{k}_1 + b_{32}\mathbf{k}_2)$$

The parameters are

$$a = a_1$$
$$a_1 = 0.43586659 \qquad b_2 = 0.75$$
$$w_1 = \tfrac{11}{27} - b_{31} \qquad w_2 = \tfrac{16}{27} - b_{32} \qquad w_3 = 1.0$$
$$b_{32} = \frac{2}{9a}(6a^2 - 6a + 1) \tag{3-103}$$
$$b_{31} = -\frac{1}{6a_1}(8a^2 - 2a + 1)$$

Notice that the jacobian matrix is evaluated only once per time step, and that the inversion or decomposition of the matrix is needed only once per time step.

We have introduced a variety of methods; many more are known. The possible methods we have discussed are listed in Table 3-1 along with their order of accuracy. Needless to say, to achieve a given overall accuracy with a low-order

Table 3-1 Methods for integrating ordinary differential equations as initial-value problems

Eq.	Method	Truncation error	Need a starting method?	Stability limit p
		Explicit		
(3-59)	Euler	$0(h)$	No	2.0
(3-96)	Second-order Runge–Kutta or mid-point rule	$0(h^2)$	No	2.0
(3-64)	Fourth-order Adams–Bashforth	$0(h^4)$	Yes	0.3
(3-98)	Fourth-order Runge–Kutta–Gill	$0(h^4)$	No	2.8
		Implicit		
(3-67)	Backward Euler	$0(h)$	No	∞
(3-69)	Modified Euler or Trapezoid rule or Crank–Nicolson	$0(h^2)$	No	∞
(3-70)	Fourth-order Adams–Moulton	$0(h^4)$	Yes	3.0
		Predictor–corrector		
(3-78) (3-79) (3-97)	Euler or second-order Runge–Kutta	$0(h^2)$	No	2.0
(3-82) (3-83)	Adams	$0(h^4)$	Yes	1.3

method requires a smaller step size h than with a high-order method. The actual trade-off may be dependent on the problem, however. We have the general categories explicit, implicit, and predictor–corrector, and a selection of order within each category. Before comparing the performance of the methods let us examine the truncation error and stability of the methods.

3-6 EXTRAPOLATION AND STEP-SIZE CONTROL

Once we know the truncation error, or the power n in the formula $0(h^n)$, we can sometimes obtain a more accurate answer by using extrapolation techniques. Suppose we solve the problem with a time step h giving the solution y_1 at time t, and also with a time step $h/2$ giving the solution y_2 at the time t. If a Euler method is used the error in the solution should be proportional to the time step. Let y_0 be the exact solution, and write the error formulas

$$y_1 = y_0 + ch \tag{3-104}$$

$$y_2 = y_0 + \frac{ch}{2} \tag{3-105}$$

Subtraction and rearrangement gives

$$y_0 = 2y_2 - y_1 \tag{3-106}$$

If the error formulas are exact then this procedure gives the exact solution in Eq. (3-106). Usually there is some error in the calculation and the formulas only apply as $h \to 0$, so that Eq. (3-106) is only an approximation to the exact solution. However, it is a more accurate estimate than either y_1 or y_2. The same procedure is used for higher-order methods, except that the error formula Eq. (3-104) must have the correct truncation error. For the trapezoid rule

$$y_1 = y_0 + ch^2 \tag{3-107}$$

Table 3-2 Errors in integrating $y' = -y$ **to** $t = 1$

Number of steps	Total number of steps	Error
	Euler	
2	2	−0.118
4	4	−0.0515
8	8	−0.0243
16	16	−0.0118
32	32	−0.00582
	Extrapolated Euler	
2, 4	6	+0.0149
4, 8	12	+0.00293
8, 16	24	+0.00066
	Trapezoid rule	
1	1	−0.0345
2	2	−0.00788
4	4	−0.00193
8	8	−0.000480
16	16	−0.000120
32	32	−0.0000299
64	64	−0.00000748
	Trapezoid rule, extrapolated	
1, 2	3	+0.00101
2, 4	6	+0.0000543
4, 8	12	+0.00000328
8, 16	24	+0.000000204

$$y_2 = y_0 + c\left(\frac{h}{2}\right)^2 \tag{3-108}$$

$$y_0 = \frac{4y_2 - y_1}{3} \tag{3-109}$$

Let us illustrate the result using a simple problem

$$y' = -y$$
$$y(0) = 1 \tag{3-110}$$

A simple Euler method is used, with a truncation error of $O(h)$. Look at the error at $y(t = 1)$ as a function of h. (See Table 3-2.) The results are plotted in Fig. 3-3. The straight line demonstrates that the error is proportional to the step size h. Next we use the extrapolation formula Eq. (3-106) and obtain the results given in Table 3-2. Clearly the error is much reduced for the same total number of steps. Indeed the extrapolated results based on 8 and 16 steps, or 24 total steps, give results as accurate as using 282 steps without extrapolation. Alternatively, the computation time is only 8 percent of that needed without extrapolation. Results shown in Fig. 3-3 for the trapezoid rule, which has a truncation error of $O(h^2)$, illustrate that the error is proportional to h^2, and extrapolation based on Eq. (3-109) with h^2 is equally successful. The extrapolated results seem to have a truncation error that is the square of the truncation error of the basic method, and indeed the extrapolated results can even be extrapolated to improve the results. Unfortunately the extrapolation is successful only if the step size is small enough for the truncation error formula to be reasonably accurate. In some nonlinear problems this is a very small value and in fact out of reach computationally. It is always a technique worth trying, however.

All the methods discussed so far have used a fixed step size h. This is not necessary provided we have a reasonable way of adjusting the step size while maintaining accuracy. We discuss here three successful methods for doing that.

Bailey[1] has a simple criterion for Eq. (3-1). Letting $y_i^n = y_i(t^n)$ we compute

$$\Delta y_i = |y_i^{n+1} - y_i^n| \tag{3-111}$$

If $\Delta y_i < 0.001$ we ignore that i in the following tests. We take one of the following actions:

1. If all $\Delta y_i / y_i < 0.01$ we double the step size.
2. If any $\Delta y_i / y_i > 0.1$ we halve the step size.
3. Otherwise we keep the same step size.

Bailey applied this scheme to problems involving moving shock fronts and found it worked reasonably well. This method uses no information about the integration method and ignores the information contained in the truncation error formula. The other two schemes do use that information.

Michelsen[5] used a third-order method—a semi-implicit Runge–Kutta scheme, Eqs. (3-102) and (3-103)—and solved the problem twice at each time step, once with time step h and again with two steps of size $h/2$. The error is defined as

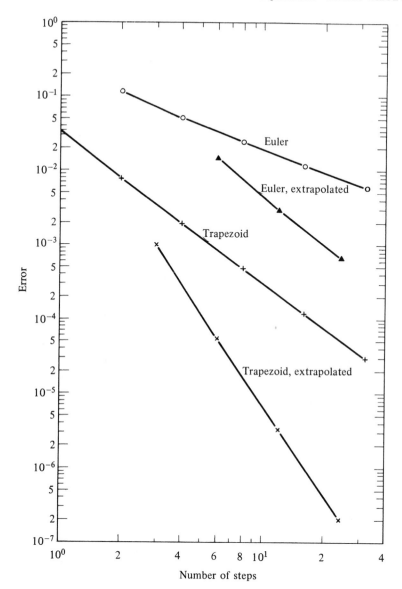

Figure 3-3 Error versus step size, integrating $y' = -y$ to $t = 1$.

$$e_i^{n+1} = y_i^{n+1}\left(\frac{h}{2}\right) - y_i^{n+1}(h) \qquad (3\text{-}112)$$

and the maximum relative error controls whether the step is accepted ($q < 1$) or not

$$q = \max_i \left|\frac{e_i}{\varepsilon_i}\right| \qquad (3\text{-}113)$$

where ε_i is a user-specified tolerance. The actual solution for the $(n+1)$th time step is taken as

$$y_i^{n+1} = y_i^{n+1}\left(\frac{h}{2}\right) + \tfrac{1}{7}e_i^{n+1} = \frac{8y_i^{n+1}(h/2) - y_i^{n+1}(h)}{7} \tag{3-114}$$

where the numbers 8 and 7 come from 2^3 and $2^3 - 1$ for a third-order method. The derivation is identical to that used in Eqs. (3-104) to (3-106) and Eqs. (3-107) to (3-109). With extrapolation the method is fourth-order with error $O(h^4)$. The next step size is taken as

$$h_{n+1} = h_n \min[(4q)^{-1/4}, 3] \tag{3-115}$$

where the $-\tfrac{1}{4}$ comes from the fourth-order method and the 3 is to avoid large increases in step size. If $q > 1$ the result is not accepted, the step size is halved, and reevaluated. In this method the user specifies the desired accuracy in ε_i, and the method tries to achieve it. Even if all the error estimates are exact the error in the solution at t (the global error) is not guaranteed to be less than ε_i, because ε_i controls the error at only one step, whereas the solution at any time is the result of many steps. Even so, such a scheme gives a reasonable control on the step size to make the global error decrease when ε_i decreases. The disadvantage of this method is that it requires three calculations, and three matrix decompositions, to advance one time step.

Gear[3] estimated the local truncation error LTE and compared that to the desired error ε. If the local truncation error has been achieved using a step size h_1,

$$\text{LTE} = ch_1^n \tag{3-116}$$

then we wish to use for the next size one giving

$$\varepsilon = ch_2^n \tag{3-117}$$

or

$$\frac{\text{LTE}}{\varepsilon} = \left(\frac{h_1}{h_2}\right)^n \tag{3-118}$$

This is similar to Eq. (3-115) except it does not require three steps to compute the local truncation error. This is achieved using Eq. (3-50). If we have a first-order method the second derivative is estimated as

$$h^2 y_n'' = \nabla(hy_n') \tag{3-119}$$

Starting the calculation to get $y_0' = f(y_0)$ and stepping forward to t_1, we get $y_1' = f(y_1)$. Then we can estimate

$$h^2 y_1'' = \nabla(hy_1') = h(y_1' - y_0') = h(f_1 - f_0) \tag{3-120}$$

The truncation error of a qth-order method is given by

$$y_{n+1} = y_n + \ldots + C_{q+1}h^{q+1}y_n^{(q+1)} \tag{3-121}$$

or

$$\text{LTE} = C_{q+1}h^{q+1}y_n^{(q+1)} \tag{3-122}$$

We estimate $y_n^{(q+1)}$ and must know C_{q+1} for the method being used. We can then estimate the local truncation error LTE achieved with step size h_1, and then choose the step size h_2 to satisfy Eq. (3-118). The complete integration package is outlined in Sec. 3-8.

3-7 STABILITY

Every numerical analyst has at some time or other seen results of a computer calculation that have a sequence something like the following: 1, 100, -10^5, 10^{24}, $-10^{125}, \ldots$. Indeed the reader of this book, if he or she has not experienced such a result, should reproduce it on a calculator—easily done on a programmable calculator—by applying the Euler method to $y' = -y$ with a time step of 4 and integrating to $t = 100$. The problem could, of course, be a programming error, but it is also possible that the program is correct and the problem is caused by an unstable calculation due to a step size that is too large. We wish to see why this happens, how large is too large, and compare the stability characteristics of the different methods.

We illustrate the phenomenon using the test equation

$$\frac{dy}{dt} = -\lambda y \qquad y(0) = 1 \tag{3-123}$$

where λ is real and positive. Let us write the solution as the sum of the exact solution y_{ex} and an error ε. We put this expression into Eq. (3-123) and note that the exact solution satisfies the differential equation, too. Then the error satisfies

$$\frac{d\varepsilon}{dt} = -\lambda\varepsilon \tag{3-124}$$

We examine the error in successive time steps by looking at $\varepsilon_n = \varepsilon(t_n)$ and ε_{n+1}. An integration method is stable if the error decays in successive time steps. Because of round-off error the computer never solves equations exactly. If the scheme is unstable this round-off error grows with successive time steps and soon swamps the solution.

Applying the Euler method to Eq. (3-124) from t_n to t_{n+1} gives

$$\frac{\varepsilon_{n+1} - \varepsilon_n}{h} = -\lambda\varepsilon_n$$

$$\varepsilon_{n+1} = \varepsilon_n(1 - \lambda h) \tag{3-125}$$

Stability requires that

$$\frac{|\varepsilon_{n+1}|}{|\varepsilon_n|} \leqslant 1 \tag{3-126}$$

and this in turn requires that

$$|1 - \lambda h| \leqslant 1 \qquad 0 \leqslant \lambda h \leqslant 2 \tag{3-127}$$

Thus the Euler method is unstable if the time step is greater than $2/|\lambda|$. Notice that if $h\lambda > 1$ then the errors change sign at each step; if ε_n is positive then ε_{n+1} is negative.

Next we apply the trapezoid rule to Eq. (3-124) and obtain

$$\varepsilon_{n+1} = \varepsilon_n \frac{1 - h\lambda/2}{1 + h\lambda/2} \qquad (3\text{-}128)$$

Equation (3-126) is satisfied for any $h\lambda > 0$. Thus the trapezoid rule is stable for any step size. This is a great advantage, but the disadvantage is that we must solve a system of algebraic equations (if we have more than one differential equation to solve) and the equations may be nonlinear (if the differential equations are nonlinear). Another feature of this method is the oscillatory error. Suppose $h\lambda$ is very big, i.e. $h\lambda \to \infty$. Then

$$\varepsilon_{n+1} = -\varepsilon_n \qquad (3\text{-}129)$$

and the errors are of opposite sign at successive time steps. This causes the numerical solution to oscillate about the exact solution. For some problems these oscillations are noticeable and unacceptable. The point at which the errors are of opposite sign is seen from Eq. (3-128) to be $h\lambda > 2$.

Finally we apply the backward Euler method and obtain

$$\varepsilon_{n+1} = \frac{\varepsilon_n}{1 + h\lambda} \qquad (3\text{-}130)$$

which is stable for all h and does not oscillate.

The results of all three methods are summarized in Table 3-3. We see that the Euler method, simple as it is, requires a small time step for stability. The trapezoid rule requires a small time step to avoid oscillations but is stable for any time step. The backward Euler method does not oscillate at all and is stable for any time step. Both the trapezoid rule and backward Euler method are implicit and require solving sets of algebraic equations. Also the trapezoid rule is second-order, giving a smaller truncation error. The method of choice depends on the difficulty of solving the algebraic equations, whether a time step can be taken small enough that the oscillations are not observable, and whether high accuracy is needed. This

Table 3-3 Comparison of integration methods. Based on $y' = -\lambda y$, $y(0) = 1$

Method	Stable step size, no oscillation in sign of error	Stable step size, oscillation in sign of error	Unstable step size
Euler	$0 < h\lambda < 1$	$1 < h\lambda < 2$	$2 < h\lambda$
Trapezoid	$0 < h\lambda < 2$	$2 < h\lambda < \infty$	None
Backward Euler	$0 < h\lambda < \infty$	None	None

comparison we provide below, after putting the stability theory on a firmer foundation.

The rational approximation to the exponential e^{-z} is defined as

$$r_{mn}(z) \equiv \frac{p_n(z)}{q_m(z)} \simeq e^{-z} \qquad (3\text{-}131)$$

where p_n is a polynomial in z of degree n and q_m is a polynomial in z of degree m. Consider three such approximations

$$e^{-z} = 1 - z \qquad e^{-z} = \frac{1}{1+z} \qquad e^{-z} = \frac{1 - z/2}{1 + z/2} \qquad (3\text{-}132)$$

We now solve the equation

$$\frac{d\mathbf{y}}{dt} = -\mathbf{B}\mathbf{y} + \mathbf{S} \qquad (3\text{-}133)$$

where $\mathbf{B}$ is a constant matrix and $\mathbf{S}$ is a constant vector. The solution is

$$\mathbf{y}(t) = e^{(-t\mathbf{B})}[\mathbf{y}(0) - \mathbf{B}^{-1}\mathbf{S}] + \mathbf{B}^{-1}\mathbf{S} \qquad (3\text{-}134)$$

We need to define the notation for matrix polynomials. The exponential can be expanded to give

$$e^{-z} = 1 - z + \frac{1}{2!}z^2 - \frac{1}{3!}z^3 + \dots \qquad (3\text{-}135)$$

and we define the exponential of a matrix in a similar way

$$e^{-t\mathbf{B}} = 1 - t\mathbf{B} + \frac{1}{2!}t^2\mathbf{B}^2 - \frac{1}{3!}t^3\mathbf{B}^3 + \dots \qquad (3\text{-}136)$$

Of course

$$\mathbf{B}^2 = \mathbf{B}\mathbf{B} \qquad \mathbf{B}^3 = \mathbf{B}\mathbf{B}^2 \qquad (3\text{-}137)$$

and so forth. We want to define rational approximations to the exponential of a matrix argument in a fashion similar to Eq. (3-131). If

$$\mathbf{X} = e^{-t\mathbf{B}} \qquad (3\text{-}138)$$

we define the rational approximation as

$$q_m(t\mathbf{B})\mathbf{X} = p_n(t\mathbf{B}) \qquad (3\text{-}139)$$

where p_n and q_m are matrix polynomials in $t\mathbf{B}$ of degree n and m, respectively. We can differentiate Eq. (3-136) with respect to t to obtain

$$\frac{d}{dt}(e^{-t\mathbf{B}}) = -\mathbf{B} + t\mathbf{B}^2 - \frac{1}{2!}t^2\mathbf{B}^3 + \dots = -\mathbf{B}e^{-t\mathbf{B}} \qquad (3\text{-}140)$$

The reader can thus verify Eq. (3-134).

Next we rearrange Eq. (3-134) by evaluating it from t to $t + \Delta t$ instead of from 0 to t to give

$$\mathbf{y}(t + \Delta t) = e^{-\Delta t\mathbf{B}}[\mathbf{y}(t) - \mathbf{B}^{-1}\mathbf{S}] + \mathbf{B}^{-1}\mathbf{S} \qquad (3\text{-}141)$$

Now if we try to approximate the exponential term using rational approximations we would use

$$q_m(\Delta t \mathbf{B})\mathbf{y}(t + \Delta t) = p_n(\Delta t \mathbf{B})[\mathbf{y}(t) - \mathbf{B}^{-1}\mathbf{S}] + q_m(\Delta t \mathbf{B})\mathbf{B}^{-1}\mathbf{S} \qquad (3\text{-}142)$$

Using in turn the three rational approximations given in Eqs. (3-132) with $q_m = 1$ and $p_n = 1 - z$, we get

$$\mathbf{y}(t + \Delta t) = \mathbf{y}(t) - \mathbf{B}^{-1}\mathbf{S} - \Delta t \mathbf{B}[\mathbf{y}(t) - \mathbf{B}^{-1}\mathbf{S}] + \mathbf{B}^{-1}\mathbf{S} \qquad (3\text{-}143)$$

or

$$\frac{\mathbf{y}(t + \Delta t) - \mathbf{y}(t)}{\Delta t} = -\mathbf{B}\mathbf{y}(t) + \mathbf{S} \qquad (3\text{-}144)$$

Similarly for $q_m = 1 + z$ and $p_n = 1$, we get

$$\mathbf{y}(t + \Delta t) + \Delta t \mathbf{B}\mathbf{y}(t + \Delta t) = \mathbf{y}(t) + \Delta t \mathbf{S}$$

or

$$\frac{\mathbf{y}(t + \Delta t) - \mathbf{y}(t)}{\Delta t} = -\mathbf{B}\mathbf{y}(t + \Delta t) + \mathbf{S} \qquad (3\text{-}145)$$

Using $q_m = 1 + z/2$ and $p_n = 1 - z/2$ we obtain

$$\frac{\mathbf{y}(t + \Delta t) - \mathbf{y}(t)}{\Delta t} = -\tfrac{1}{2}\mathbf{B}[\mathbf{y}(t + \Delta t) + \mathbf{y}(t)] + \Delta t \mathbf{S} \qquad (3\text{-}146)$$

Examination of Eqs. (3-144) to (3-146) reveals that we have applied the Euler method, the backward Euler method, and the trapezoid rule. Indeed, the rational approximations in Eq. (3-132) look very similar to Eqs. (3-125), (3-130), and (3-128) for the three methods. There is a close correspondence between integration schemes and the rational approximations to the exponential. To pursue this relationship more deeply we must solve the equations.

Let λ_i and $\mathbf{x}_i$ be the eigen values and eigen vectors, respectively, of $\mathbf{B}$, i.e. they satisfy

$$\mathbf{B}\mathbf{x}_i = \lambda_i \mathbf{x}_i \qquad (3\text{-}147)$$

We assume

$$\mathbf{x}_i^T \cdot \mathbf{x}_i = 1.0 \qquad (3\text{-}148)$$

and since they are eigen vectors they are orthogonal. Thus

$$\mathbf{x}_i^T \cdot \mathbf{x}_j = 0 \qquad i \neq j \qquad (3\text{-}149)$$

We thus have a set of n eigen values λ_i, and for each one an eigen vector $\mathbf{x}_i$. Let us define the matrices

$$\mathbf{X} = [\mathbf{x}_1, \mathbf{x}_2, \ldots, \mathbf{x}_n] \qquad \mathbf{\Lambda}(\lambda) = \mathrm{diag}(\lambda_i) = \begin{bmatrix} \lambda_1 & & & \\ & \lambda_2 & & 0 \\ & & \ddots & \\ & 0 & & \lambda_n \end{bmatrix} \qquad (3\text{-}150)$$

and compute

$$\mathbf{X}\Lambda(\lambda) = [\lambda_1 \mathbf{x}_1, \lambda_2 \mathbf{x}_2, \dots, \lambda_n \mathbf{x}_n] \tag{3-151}$$

Also the matrix multiplication of $\mathbf{X}$ and its transpose gives

$$\mathbf{X}^T \mathbf{X} = \begin{bmatrix} \cdots \mathbf{x}_1 \cdots \\ \cdots \mathbf{x}_2 \cdots \\ \cdots\cdots\cdots \\ \cdots \mathbf{x}_n \cdots \end{bmatrix} \begin{bmatrix} \vdots & \vdots & \vdots \\ \mathbf{x}_1 \cdots \mathbf{x}_2 \cdots \mathbf{x}_n \\ \vdots & \vdots & \vdots \end{bmatrix} = \begin{bmatrix} 1 & & 0 \\ & 1 & \\ & & \ddots \\ 0 & & 1 \end{bmatrix} \equiv I \tag{3-152}$$

so that

$$\mathbf{X}^T = \mathbf{X}^{-1} \tag{3-153}$$

Next we multiply $\mathbf{X}$ by the matrix $\mathbf{B}$ and use the fact that $\mathbf{X}$ is made up of the eigen vectors [see Eq. (3-150)]

$$\mathbf{B}\mathbf{X} = \mathbf{X}\Lambda(\lambda) \tag{3-154}$$

Let us postmultiply this by $\mathbf{X}^T$ to get

$$\mathbf{B}\mathbf{X}\mathbf{X}^T = \mathbf{X}\Lambda(\lambda)\mathbf{X}^T = \mathbf{B} \tag{3-155}$$

and calculate $\mathbf{B}^2$

$$\mathbf{B}^2 = \mathbf{B}\mathbf{B} = [\mathbf{X}\Lambda(\lambda)\mathbf{X}^T][\mathbf{X}\Lambda(\lambda)\mathbf{X}^T] = \mathbf{X}\Lambda(\lambda)\Lambda(\lambda)\mathbf{X}^T \tag{3-156}$$

Calculation of Λ^2 gives $\Lambda(\lambda^2)$

$$\Lambda(\lambda)\Lambda(\lambda) = \begin{bmatrix} \lambda_1 & & 0 \\ & \lambda_2 & \\ & & \ddots \\ 0 & & \lambda_n \end{bmatrix} \begin{bmatrix} \lambda_1 & & 0 \\ & \lambda_2 & \\ & & \ddots \\ 0 & & \lambda_n \end{bmatrix} = \begin{bmatrix} \lambda_1^2 & & 0 \\ & \lambda_2^2 & \\ & & \ddots \\ 0 & & \lambda_n^2 \end{bmatrix} = \Lambda(\lambda^2) \tag{3-157}$$

so that

$$\mathbf{B}^2 = \mathbf{X}\Lambda(\lambda^2)\mathbf{X}^T \qquad \text{and} \qquad \mathbf{B}^k = \mathbf{X}\Lambda(\lambda^k)\mathbf{X}^T \tag{3-158}$$

and

$$\mathbf{B}^k \mathbf{X} = \mathbf{X}\Lambda(\lambda^k) = [\lambda_1^k \mathbf{x}_1, \lambda_2^k \mathbf{x}_2, \dots, \lambda_n^k \mathbf{x}_n] \tag{3-159}$$

or

$$\mathbf{B}^k \mathbf{x} = \lambda_i^k \mathbf{x} \tag{3-160}$$

With these preliminaries we return to Eq. (3-142) and insert $\mathbf{X}\mathbf{X}^{-1} = \mathbf{I}$ to get

$$q_m(\Delta t \mathbf{B})\mathbf{X}\mathbf{X}^{-1}\mathbf{y}(t+\Delta t) = p_n(\Delta t \mathbf{B})\mathbf{X}\mathbf{X}^{-1}[\mathbf{y}(t) - \mathbf{B}^{-1}\mathbf{S}]$$
$$+ q_m(\Delta t \mathbf{B})\mathbf{X}\mathbf{X}^{-1}\mathbf{B}^{-1}\mathbf{S} \tag{3-161}$$

We simplify the notation

$$\mathbf{v}(t+\Delta t) = \mathbf{X}^{-1}\mathbf{y}(t+\Delta t) \qquad \text{and} \qquad \mathbf{w} = \mathbf{X}^{-1}\mathbf{B}^{-1}\mathbf{S} \tag{3-162}$$

and rewrite Eq. (3-161) giving

$$q_m(\Delta t \mathbf{B}) \mathbf{X} \mathbf{v}(t + \Delta t) = p_n(\Delta t \mathbf{B}) \mathbf{X}[\mathbf{v}(t) - \mathbf{w}] + q_m(\Delta t \mathbf{B}) \mathbf{X} \mathbf{w} \qquad (3\text{-}163)$$

Both the q_m and p_n are matrix polynomials so that we can use Eq. (3-158) to evaluate them. If

$$q_m(z) = \sum_{k=0}^{m} a_k z^k \qquad (3\text{-}164)$$

then the matrix polynomial is

$$q_m(\Delta t \mathbf{B}) \mathbf{X} = \sum_{k=0}^{m} a_k \Delta t^k \mathbf{B}^k \mathbf{X} = \sum_{k=0}^{m} a_k \Delta t^k \mathbf{X} \mathbf{\Lambda}(\lambda^k)$$

$$= \mathbf{X} \sum_{k=0}^{m} a_k \Delta t^k \mathbf{\Lambda}^k = \mathbf{X} q_m(\Delta t \mathbf{\Lambda}) \qquad (3\text{-}165)$$

Thus

$$\mathbf{X} q_m(\Delta t \mathbf{\Lambda}) \mathbf{v}(t + \Delta t) = \mathbf{X} \{ p_n[\Delta t \mathbf{\Lambda}][\mathbf{v}(t) - \mathbf{w}] + q_m(\Delta t \mathbf{\Lambda}) \mathbf{w} \} \qquad (3\text{-}166)$$

Next we multiply by $\mathbf{X}^T$ and use Eq. (3-152). Note that $q_m(\Delta t \mathbf{\Lambda})$ and $p_n(\Delta t \mathbf{\Lambda})$ are diagonal matrices since each $\mathbf{\Lambda}^k$ is diagonal, and we can decouple the equations to write them in the form

$$q_m(\Delta t \lambda_i) \mathbf{v}_i(t + \Delta t) = p_n(\Delta t \lambda_i)[\mathbf{v}_i(t) - \mathbf{w}_i] + q_m(\Delta t \lambda_i) \mathbf{w}_i \qquad (3\text{-}167)$$

We can perform the same operations for Eq. (3-141) to get

$$\mathbf{v}_i(t + \Delta t) = e^{-\Delta t \lambda_i}[\mathbf{v}_i(t) - \mathbf{w}_i] + \mathbf{w}_i \qquad (3\text{-}168)$$

Comparing Eq. (3-167) with (3-168) shows we want the rational approximation

$$r_{mn}(\Delta t \lambda_i) = \frac{p_n(\Delta t \lambda_i)}{q_m(\Delta t \lambda_i)} \qquad (3\text{-}169)$$

to approximate $e^{-\Delta t \lambda_i}$ as well as possible. Based on these results we can examine this relationship for each eigen value individually.

We can now define stability, which has several definitions (see Lambert, p. 233). Dahlquist introduced the term A stability. A numerical method is said to be A stable if its region of absolute stability contains the whole of the left-hand plane (see Fig. 3-4a). Widlund called a numerical method A(α) stable if its region of stability includes the infinite wedge W (see Fig. 3-4b). It is A(0) stable if it is A(α) stable for some small α. Ehle introduced the term L stability. A numerical method is L stable if it is A stable and when applied to $y' = -\lambda y$, $\text{Re}\,\lambda > 0$ yields $y_{n+1} = r(h\lambda)y$ where $|r(h\lambda)| \to 0$ as $\text{Re}(h\lambda) \to \infty$.

Furthermore a rational approximation $r_{mn}(z)$ to e^{-z} is:

1. A acceptable if $|r_{mn}| < 1$ for $\text{Re}\,z > 0$.
2. A(0) acceptable if $|r_{mn}| < 1$ for z real, $z > 0$.
3. L acceptable if it is A acceptable and

$$\lim_{z \to \infty} r_{mn}(z) = 0 \qquad (3\text{-}170)$$

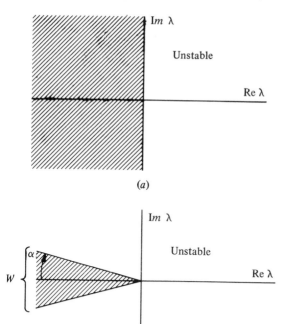

Figure 3-4 Regions of stability.
(a) A stability. (b) A (α) stability.

Consider one-step methods which, when applied to $y' = -\lambda y$, give $y_{n+1} = r_{mn}y_n$, with r_{mn} a polynomial in $h\lambda$. The method is A, A(0), or L stable according to whether the rational polynomial is A, A(0), or L acceptable (Lambert, p. 237). Since each one-step integration method can be related by a rational approximation to e^{-z}, we use that rational approximation to deduce the characteristics of the integration method. We can say:

1. The point where $r_{mn}(\lambda\Delta t) = \pm 1$ gives the $\lambda\Delta t$ for stability limitation.
2. The closer the approximation is to e^{-z} the more accurate it is.
3. If $r_{mn} < 0$ the solutions can oscillate since then the error at the $(n-1)$th step has the opposite sign to the error at the nth step.
4. We can look at a single eigen value λ_i for systems of equations and in fact must concentrate on the largest eigen value since it is for large $|\lambda\Delta t|$ that the integration methods break down.

We now apply these guidelines to the methods already treated. Figure 3-5 shows the rational approximations to the three methods discussed in detail: Euler, trapezoid rule, and backward Euler. The point at which the curve falls below -1 is the limit of stability. The Euler method requires $|\lambda\Delta t| \leqslant 2$ for stability whereas the other methods are always stable. The point at which the approximation falls below zero represents the criterion for the onset of oscillatory errors. For the Euler method and $|\lambda\Delta t| < 1$ the method does not oscillate. The trapezoid rule oscillates

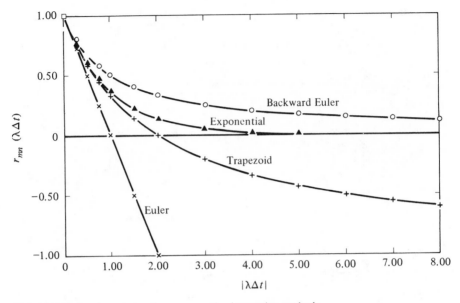

Figure 3-5 Rational approximations representing integration methods.

for $|\lambda\Delta t| > 2$, whereas the backward Euler method does not oscillate at all. The trapezoid rule is $0(h^2)$, whereas the other methods are $0(h)$, and the greater accuracy of the trapezoid rule is evident but only for small $|\lambda\Delta t|$ of less than two. This analysis has to hold for each eigen value, and so it is the largest eigen value that matters.

The other explicit methods are evaluated in a similar way. The rational approximation can be determined by applying the Runge–Kutta methods to Eq. (3-124) (see problem 3-12). The rational approximations are shown in Fig. 3-6. The second-order Runge–Kutta method is stable for

$$|\lambda\Delta t| \leqslant 2.0 \tag{3-171}$$

while the fourth-order Runge–Kutta–Gill method is stable for

$$|\lambda\Delta t| \leqslant 2.8 \tag{3-172}$$

The actual $\lambda\Delta t$ must be kept to about one-third of this limit if accurate results are to be achieved.

For nonlinear equations we can only apply the ideas locally, that is we can consider the system of equations at time t_n. For Eq. (3-1) at time t_n we linearize about the solution $\mathbf{y}_n$, which is known. Thus

$$\frac{dy_i}{dt} = f_i(\mathbf{y}_n) + \sum_{j=1}^{n} \frac{\partial f_i}{\partial y_j}(y_j - y_{jn}) \tag{3-173}$$

Then we examine the stability of a method for the eigen values of the matrix $\partial f/\partial y = \mathbf{A}$, the jacobian. Of course at a later time we have different eigen values, so

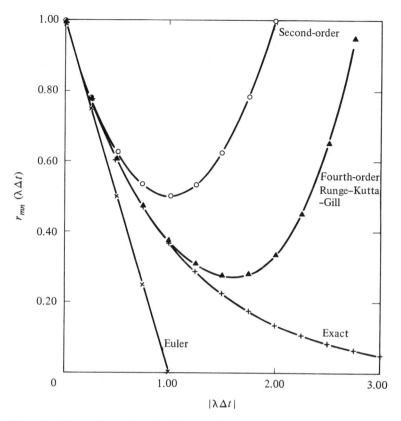

Figure 3-6 Rational approximations for explicit methods.

the method must be stable for whatever eigen values occur during the course of the integration.

We thus reduce the problem of characterizing a method for integrating ordinary differential equations down to an examination of the rational approximation to e^{-z}. The rational approximation gives information about the stability of the method and the tendency of the solution to oscillate, and how these features change with $|\lambda\Delta t|$.

3-8 HIGH-ORDER SCHEMES THAT ARE STABLE AND DO NOT OSCILLATE

For very stiff problems (some λ_i large, some λ_i small) we would like a scheme that is stable, does not oscillate, and is reasonably accurate. The Euler method is not such a scheme since it is not stable for large $|\lambda\Delta t|$. The backward Euler method is stable and does not oscillate, but it is not very accurate, being a first-order method. The trapezoid rule is of higher order (second) but oscillates for large $|\lambda\Delta t|$. Thus

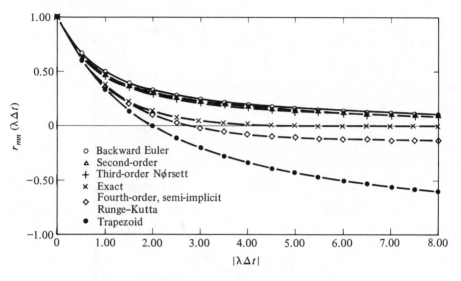

Figure 3-7 Rational approximations for implicit methods.

none of the methods meet our requirements. All the explicit methods fail since they have a stability limitation in terms of $|\lambda\Delta t|$. We want a high-order, implicit method which is A(0) and L stable. There are two methods that have been developed to meet our requirements: semi-implicit Runge–Kutta methods and the Nørsett methods.

The semi-implicit Runge–Kutta method is presented in Eqs. (3-102) and (3-103) and was developed by Caillaud and Padmanabhan[2] to be A and L stable. If we apply the method to Eq. (3-123) and look at y_{n+1}/y_n we get the rational approximation

$$r_{23}(z) = \frac{1 + 0.3075998z - 0.23766072z^2}{(1 + 0.43586659z)^3} \qquad (3\text{-}174)$$

This function, which is plotted in Figs. 3-7 and 3-8, is always between 1 and -1. Thus the integration method is A(0) stable. Since $r(\infty) = 0$ it is also L stable. Furthermore, the method has a truncation error of $O(h^3)$, which can be seen by comparing the polynomial expansion of Eq. (3-174) to the Taylor series for e^{-z}; they agree up to the z^3 term. We must evaluate a jacobian each time step and decompose the matrix only once per time step.

Two other methods are the Nørsett methods of second- and third-order.[6] These are based on Hermite polynomials as rational approximations to e^{-z}. For Eq. (3-133), the method is

$$(I + \alpha\Delta t\mathbf{B}_n)\mathbf{w}_1 = \alpha\Delta t(\mathbf{B}_n\mathbf{y}_n - \mathbf{S}_n)$$
$$(I + \alpha\Delta t\mathbf{B}_n)\mathbf{w}_{i+1} = \alpha\Delta t\mathbf{B}_n\mathbf{w}_i \qquad (3\text{-}175)$$
$$(I + \alpha\Delta t\mathbf{B}_n)y_{n+1} = y_n + \sum_{j=1}^{k-1} L_j\left(\frac{1}{\alpha}\right)\mathbf{w}_j + \alpha\Delta t\mathbf{S}_n$$

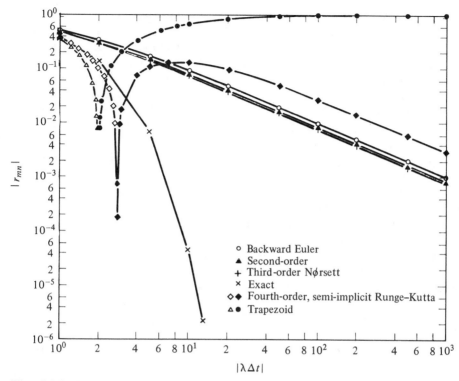

Figure 3-8 Rational approximations for implicit methods for large Δt.

For a second-order method $k = 2$, $\alpha = 1.707106781$ and $L_1(1/\alpha) = 0.4142135623$, while for a third-order method $k = 3$, $\alpha = 2.405147015$, $L_1(1/\alpha) = 0.584225001$, and $L_2(1/\alpha) = 0.254884425$. The order of the truncation error is $0(h^2)$ and $0(h^3)$, respectively, and both methods are A and L acceptable. Once we have inverted the matrix in Eqs. (3-175) we only need to multiply several right-hand sides by the inverse to obtain the solution. Thus we have only one inversion or decomposition of the matrix per time step. The decomposition of the matrix is usually a time-consuming operation, compared to the multiplication by the right-hand side. Thus the work effort is only a little bit greater than that necessary to apply the backward Euler method or the trapezoid rule. In return for the modest increase in work effort we get a second- or third-order method that is L stable as well.

3-9 EQUATION SOLVERS

You are probably already familiar with the Runge–Kutta routine for solving initial-value problems because you may have used the package available at your computer center. These packages usually use a fourth- or fifth-order Runge–Kutta method, such as the Runge–Kutta–Gill method described in Eqs. (3-98), combined with a variable step size. The step size is adjusted to control the local truncation error within a limit set by the user. The local truncation error is estimated using a theory developed for the particular Runge–Kutta method. The method is highly accurate, $0(h^4)$ or $0(h^5)$, requires only function evaluations (no derivatives or matrix decompositions), and is explicit, so it has a stability limitation. While such packages work well for a variety of problems, they will not work well for stiff problems with large $|\lambda_i|$ since the time step is too small for stability reasons, leading to excessive computation time. However, there are other systems, not as widely known, that perform the same function for stiff systems.

Gear's method and the package developed by Hindmarsh[4] is one such system. The system has options of variable step size and variable order of integration. The user can specify either explicit or implicit methods, and the implicit methods can be solved using either successive substitution or Newton–Raphson. There are consequently a variety of choices. The first-order, implicit method is the backward Euler method, but the other-order methods have not been discussed here. We can imagine a program that includes Euler, trapezoid, and Adams–Moulton methods, and the Gear system is similar except that other implicit methods are used. The user specifies a tolerance that is allowed for the local truncation error (the maximum error in one time step). The program then controls the step size to meet this truncation error, as explained in Sec. 3-6. The order of the method is controlled too.

Suppose we are calculating with a kth-order method. The truncation error is determined by the $(k+1)$th derivative. We can estimate this derivative using difference formulas, and we can also estimate the kth derivative and the $(k+2)$th derivative. We thus have a means of estimating the local truncation error of the kth-order method, and the $(k+1)$th- and $(k-1)$th-order methods. We can then determine the step size allowed for each method and choose the method for the next step so as to minimize the work. We generally want the order giving the highest h, but the extra work associated with higher-order methods must be taken into account. We thus can control both the step size and the order of the method.

Hindmarsh's version of Gear's method works as follows, as applied to Eq. (3-1). The user provides a subroutine that evaluates f given y. A tolerance level is specified ε and the user chooses an explicit or implicit approach. If implicit is chosen successive substitution or Newton–Raphson is selected. The integration then proceeds with a small time step and a first-order method. As the solution evolves the program checks the truncation error, adjusts the step size and order, and integrates the problem in as efficient a way as it can. The Newton–Raphson method requires a jacobian, which is evaluated numerically using the function subroutine specified by the user. However, this is an expensive operation, and the matrix decomposition is also expensive, so that the jacobian is not reevaluated at every time step. The old value is used until the iteration (needed to solve the implicit equations) does not converge within three iterations, at which point an updated jacobian is used. If this iteration does not converge within three iterations the time step is decreased and the process begins again. We know that the iterations converge for a small enough time step, so that this scheme always works.

Another system is the fourth-order semi-implicit scheme developed by Michelsen.[5] In this scheme a semi-implicit Runge–Kutta method is used with a fixed-order (third) and a variable time step. The problem is solved twice on each step, once with a step h and once with two steps $h/2$. The truncation error is estimated [see Eqs. (3-112) to (3-114)] and the two solutions are extrapolated to give an even better result that is fourth-order. If the truncation error is within the tolerance ε, the step is accepted, and the next step size is estimated using Eq. (3-115). If the tolerance is not achieved the step size is reduced by a factor of two and the process repeated. This method is implicit, A and L stable, high-order, and thus a candidate method.

In order to illustrate the basic ideas in the Gear system the reader will now be helped to develop his own integration package. To simplify the algebra we use first-order methods, but we can include all the basic features of an integration system that controls the step size to maintain a user-specified accuracy, which is implicit and L acceptable. The step size is controlled by estimating the truncation error and making it less than the user-specified accuracy ε. We obtain the estimated truncation error by using a simple predictor equation. For the first-order predictor we use Eq. (3-57) with $q = 1$. We write here the formula with $q = 2$ so that we can see the error term

$$y_{n+1} = y_n + h_n(y'_n + \tfrac{1}{2}\nabla y'_n) + 0(h_n^3) \tag{3-176}$$

Using Eq. (3-50) the second difference is related to the second derivative and we obtain

$$y_{n+1} = y_n + hy'_n + \tfrac{1}{2}h_n^2 y''_n + 0(h_n^3) \tag{3-177}$$

For the corrector we use the backward Euler method, which is implicit and L acceptable. The formula is Eq. (3-67) with the error term included. Thus

$$y_{n+1} = y_n + h_n(y'_{n+1} - \tfrac{1}{2}\nabla y'_{n+1}) + 0(h_n^3) \tag{3-178}$$
$$= y_n + h_n f(y_{n+1}) - \tfrac{1}{2}h_n^2 y''_{n+1} + 0(h_n^3) \tag{3-179}$$

Now the calculations proceed by using the predictor equation

$$y_{n+1}^p = y_n + h_n f(y_n) \tag{3-180}$$

followed by the corrector equation

$$y_{n+1}^c = y_n + h_n f(y_{n+1}^c) \tag{3-181}$$

It should be noted that the predicted value is not used in the corrector, so the method is not a predictor–corrector method.

The exact solution satisfies Eqs. (3-177) to (3-179). These equations are now written for y_{n+1}^{ex} and subtracted from Eqs. (3-180) and (3-181), respectively. The result is

$$y_{n+1}^p - y_{n+1}^{ex} = -\tfrac{1}{2}h_n^2 y''_n + 0(h_n^3) \tag{3-182}$$

$$y_{n+1}^c - y_{n+1}^{ex} = +\tfrac{1}{2}h_n^2 y''_n + 0(h_n^3) \tag{3-183}$$

We do not know either the exact solution or the second derivative. We can solve these equations for those two quantities and obtain

$$h_n^2 y_n'' = y_{n+1}^c - y_{n+1}^p \tag{3-184}$$

$$y_{n+1}^{ex} = \tfrac{1}{2}(y_{n+1}^c + y_{n+1}^p) \tag{3-185}$$

The truncation error is then

$$d_{n+1} = y_{n+1}^c - y_{n+1}^{ex} = \tfrac{1}{2}(y_{n+1}^c - y_{n+1}^p) \tag{3-186}$$

The calculation proceeds as follows. Equations (3-180) and (3-181) are used to obtain y_{n+1}^p and y_{n+1}^c, and the truncation error is obtained from Eq. (3-186). If the truncation error is less than the specified tolerance ε the step is accepted. If not the step is repeated. In either case the next time step is calculated according to the formula

$$\frac{d_{n+1}}{h_n^2} = \frac{\varepsilon}{h^2} \tag{3-187}$$

$$h = h_n \left(\frac{\varepsilon}{|d_{n+1}|}\right)^{1/2} \tag{3-188}$$

which ensures that the truncation error is equal to ε, within the accuracy of the formula. To avoid small errors we use an h 20 percent smaller than that obtained from Eq. (3-188). The solution of the implicit equations is performed with Newton–Raphson, and the jacobian can be evaluated every time, or old jacobians can be used until convergence is not obtained within N iterations, at which point a new jacobian is evaluated. Also if the iteration does not converge within N iterations the step size is decreased by a factor of two and the calculation resumed.

Since the time step is constantly changing, it is unlikely that results at a given time can be obtained in an integral number of time steps, whose length is unknown. To obtain results at time t between t_n and t_{n+1} we use an interpolation formula, such as Eq. (3-47), of the same accuracy as the calculation to find y_α from y_n and y_{n+1}.

Such an integration system is easy to understand and higher-order schemes can be done as well. For the applications in this book GEARB works quite well and economically. For other specialized applications it may be useful to develop such an integration package.

3-10 COMPARISON

We are now in a position to compare methods using various criteria so as to make an informed and reasonable choice for our problem. First consider methods that have a fixed time step, i.e. those we might program ourselves, and then consider as a separate classification the systems that have a variable step control to maintain the user-specified accuracy (at least locally).

For methods with a fixed step size we must compare accuracy, stability, the tendency of the method to oscillate if large time steps are used, and the work required. We have already provided a quantitative basis for the first three criteria. Let us now provide one for the last.

The work is associated with three operations: function evaluations, jacobian evaluations, and matrix decompositions. We count only the number of multiplications and divisions necessary to perform each of these operations, since additions and subtractions are usually much faster operations on the computer. Let us take a system of equations of the form given by Eq. (3-1) with n unknowns. We assume that each y_i appears in each f_i, i.e. the jacobian is dense, without bands of zeros. We define the following terms:

m_1 = average number of operations to evaluate one term f_j
m_2 = average number of operations to evaluate one jacobian term $\partial f_j / \partial y_i$
M_1 = number of operations to evaluate $\mathbf{f}$, nm_1
M_2 = total number of operations to evaluate the jacobian
  = $n^2 m_2$, or $n^2(m_1 + 1)$ if numerical differentiation is used
M_3 = $(n^3 - n)/3$ = number of operations to decompose the matrix
M_4 = n^2 = number of operations to solve the linear system for one right-hand side
T = total number of time steps needed

An explicit method might require several function evaluations for each time step. Call that number m_5. Then the total number of operations needed to apply the explicit methods is

$$M_{ex} = m_5 m_1 n T_{ex} \qquad (3\text{-}189a)$$

An implicit method will require a jacobian evaluation and assuming it requires one each time step, the total number of operations needed to apply the method (for a numerical jacobian evaluation) is

$$M_{im} = \{[n^2(m_1 + 1) + m_1 n] + \tfrac{1}{3}(n^3 - n) + n^2\} T_{im} \qquad (3\text{-}189b)$$

Equations (3-189a) and (3-189b) may be compared for different conditions to illustrate their implications. Suppose $n = 10$, $m_5 = 5$, and $m_1 = 5$. Then

$$\frac{M_{ex}}{M_{im}} = \frac{T_{ex}}{4.3 T_{im}} \qquad (3\text{-}190)$$

This means that an implicit method must use a time step at least 4.3 times larger than an explicit method if it is to require fewer operations, and presumably less computational cost. If we change n to 100 the number 4.3 increases to 162. If the problem is not stiff, so that using a step size in the range $\lambda \Delta t \simeq 2$ is feasible, then an explicit method of the same order as an implicit method would be preferred. In this situation the two methods would have about equivalent accuracy at the same time step but the explicit method would involve less work. If the problem is stiff, however, so that the maximum eigen value is large, then it may be necessary to use an implicit method with $\lambda \Delta t \simeq 100$. An explicit method limited to something like $\lambda \Delta t = 2$ would be more expensive. Thus we know at the outset that explicit

Table 3-4 Comparison of explicit integration methods

| Method | Eq. | Number of | | Accuracy | Stability limit p |
		Function evaluations	Multiplications		
Euler	(3-59)	1	1	h	2.0
Second-order Runge–Kutta	(3-96)	2	2	h^2	2.0
Euler predictor–corrector	(3-78), (3-79)	2	2	h^2	2.0
Adams–Bashforth	(3-64)	1	5	h^4	0.3
Fourth-order Runge–Kutta–Gill	(3-98)	4	6	h^4	2.8
Fourth-order Adams predictor–corrector	(3-82), (3-83)	2	9	h^4	1.3

methods are likely to be preferred for non-stiff problems, while implicit methods are preferred for stiff problems.

Considering a non-stiff system of equations we examine the explicit schemes. The cost is mainly associated with the function evaluations and the multiplications by step length. Table 3-4 summarizes information gleaned from the formulas for the different methods. Comparing the Euler method with the second-order Runge–Kutta method we see that the Runge–Kutta method requires twice as many function evaluations and multiplications. The Runge–Kutta method is more accurate, however, so that a larger step size should be feasible for the same accuracy. The truncation error for the second-order Runge–Kutta method with a step size h is

$$h^3\left[-\tfrac{1}{12}f^2\frac{d^2f}{dy^2}+\tfrac{1}{6}f\left(\frac{df}{dy}\right)^2\right]_n \tag{3-191}$$

while the error for the Euler method with two steps of size $h/2$ is

$$2\tfrac{1}{2}\left(\frac{h}{2}\right)^2\left(f\frac{df}{dy}\right)_n \tag{3-192}$$

Clearly the error depends on the properties of the function being approximated, but the Runge–Kutta method has the advantage of a higher power of h, which is presumably less than one, giving a smaller error. The actual comparison would depend on numerical experiments, although it is clear that if high accuracy is desired the second-order method would be preferred. The first-order predictor–corrector method is equivalent to the second-order Runge–Kutta method in terms of work requirement, accuracy, and stability limitations. Numerical experiments might show one second-order method preferable to the other, but it probably does not make much difference.

Next let us examine the fourth-order methods. The Adams–Bashforth method requires fewest function evaluations—only one per time step—compared to four for the fourth-order Runge–Kutta method. The Runge–Kutta method has a higher stability limitation, however, in fact nine times as large, so that larger steps are possible. However, if the time step is very small to obtain good accuracy, and is within the limits where the Adams–Bashforth method is stable, then the Adams–Bashforth method would be preferred because of fewer function evaluations per time step. Some finite element methods lead to ordinary differential equations of the form

$$C \frac{d\mathbf{y}}{dt} = \mathbf{f}(\mathbf{y}) \tag{3-193}$$

instead of Eq. (3-1). In principle, these can easily be solved by taking the inverse of C. If C depends on the solution, however, this must be done for each time step. Consequently, explicit methods are not as suitable for equations in the form of Eq. (3-193). Implicit methods handle such equations easily since a matrix is inverted anyway, but the actual equations may need to be rearranged or derived again.

Now let us turn to stiff problems, in which the ratio of maximum to minimum eigen value (in absolute value) is larger than 1,000. In these cases implicit methods must be used because of their stability characteristics, since the maximum eigen value might be very large. The major work in solving these methods is in the formation and decomposition of the jacobian, or the inversion of the matrix. For most cases the major cost is the matrix decomposition (inversion) [see Eq. (3-189b)]. We do not know in advance how many iterations will be required to solve the implicit equations. Without any other guidance we can only assume that the same number of iterations is required by all methods. Since the major work is then the matrix decomposition and since it is assumed that all methods have the same number of decompositions per time step, all methods involve equivalent work. To be more precise we must perform numerical experimentation using the methods for our problems. We can then judge the methods based on the accuracy and their tendency to give solutions that oscillate. The fourth-order Adams–Moulton method can be discarded because it has a stability limitation (see Table 3-1). The other methods are listed in Table 3-5. The trapezoid rule would be eliminated based on its tendency to oscillate, leaving a selection of first-, second-, and third-order methods that are L acceptable. Further selection from the

Table 3-5 Comparison of implicit integration methods

Method	Eq.	Accuracy	Oscillation limit on $\lambda \Delta t$
Backward Euler	(3-67)	h	∞
Trapezoid rule	(3-69)	h^2	2
Semi-implicit Runge–Kutta	(3-102), (3-103)	h^3	∞
Nørsett	(3-175)	h^2, h^3	∞

remaining list in Table 3-5 depends on numerical experimentation and the conclusion may depend on the accuracy desired. If a low-accuracy solution is all that is required even the backward Euler method may be best. For higher accuracy the second- and third-order L acceptable, semi-implicit Runge–Kutta method or the Nørsett methods may be preferred.

Finally, we consider the integration packages with variable step-size control to meet a user-specified local truncation error. We can only compare these packages by calculations using the same problem. Naturally the results depend on the computer program, and the specific numbers may show some variation from one computer installation to another. We used three packages:

1. Runge–Kutta, fourth-order, in the Math. Science Library of Boeing Computer Services. This is a fourth-order scheme with extrapolation to achieve fifth-order accuracy.
2. GEARB, a variable order and time-step method. See Bibliography for availability.
3. The fourth-order, semi-implicit Runge–Kutta method, described by Eqs. (3-102) and (3-103) with the algorithm presented by Michelsen.[5]

First we apply the GEAR and Runge–Kutta packages to the problem

$$\frac{dy_i}{dt} = -\beta_i y_i + y_i^2 \qquad y_i(0) = -1 \tag{3-194}$$

$$\beta_i = (+1{,}000, \ 800, \ -10, \ 0.001) \qquad c_i = -(1+\beta_i)$$

Table 3-6 Comparison of integration methods applied to Eq. (3-194)

t	Time step used	Order of method used	Steps	Function evaluations	Jacobian evaluations	y_4 error
				Number of		
GEAR, $\varepsilon = 10^{-6}$						
0.01	4.7×10^{-4}	5	66	110	9	$-2.2(-12)$
0.1	7.8×10^{-3}	4	99	152	13	$2.4(-8)$
1.0	3.5×10^{-2}	5	154	122	17	$-8.0(-8)$
10	0.45	5	205	279	23	$1.9(-6)$
100	4.69	5	249	329	29	$1.0(-6)$
1,000	92.1	5	285	371	35	$4.4(-7)$
Runge–Kutta, $\varepsilon = 10^{-6}$						
0.01		4–5		221		$-4.9(-12)$
0.1		4–5		460		$-7.7(-12)$
1.0		4–5		2,631		$-1.1(-12)$

with the solution

$$y_i = \frac{\beta_i}{1 + c_i e^{\beta_i t}}$$

At $t = 0$ the eigen values are $-1,002$, -802, 8, -2.001, giving a stiffness ratio of 500. At $t \geqslant 1,000$ values are $-1,000$, -800, -1.0, -0.001, giving a stiffness ratio of 10^6. We expected at the outset that GEAR would be better. Typical results are shown in Table 3-6. Up to time $t = 1$ the two methods used about the same

Table 3-7 Comparison of integration methods applied to Eqs. (1-1) and (1-2)

t	Order of method used	Steps	Function evaluations	Jacobian evaluations	y_1	y_2
			Number of			
			Implicit GEARB, $\varepsilon = 10^{-4}$			
0.4	2	17	57	7	0.985155	3.38178(-5)
1.0	2	19	61	7	0.966345	3.07269(-5)
4.0	3	26	79	8	0.905537	2.24070(-5)
10.0	3	32	94	9	0.841470	1.62419(-5)
			Implicit GEARB, $\varepsilon = 10^{-5}$			
0.4	2	16	51	5	0.985172121	3.38640908(-5)
1.0	3	18	60	6	0.966466087	3.07474038(-5)
4.0	4	28	83	7	0.905542691	2.24076524(-5)
10.0	4	36	101	8	0.841361601	1.62332344(-5)
			Explicit Runge–Kutta, $\varepsilon = 10^{-8}$			
0.4	4		2,127		0.985172	3.38705(-5)
1.0	4		3,258		0.966460	3.07449(-5)
4.0	4		14,685		0.905519	2.24048(-5)
10.0	4		39,926		0.841370	1.62339(-5)
			Semi-implicit Runge–Kutta, $\varepsilon = 10^{-4}$			
0.4	4	8	24	24	0.985172	3.38141(-5)
1.0	4	10	30	30	0.966460	3.07463(-5)
4.0	4	14	42	42	0.905519	2.24053(-5)
10.0	4	18	54	54	0.841370	1.62348(-5)
			Semi-implicit Runge–Kutta, $\varepsilon = 10^{-5}$			
0.4	4	10	30	30	0.9851721	3.386406(-5)
1.0	4	12	36	36	0.9664597	3.074631(-5)
4.0	4	18	54	54	0.9055186	2.240481(-5)
10.0	4	24	72	72	0.8413699	1.623412(-5)

amount of computer time (1.6 sec on a CDC 6400). Notice that the Runge–Kutta method used very many more function evaluations up to $t = 1$, but did not require much more time because it did not have to iterate to solve equations or have to invert matrices. The Runge–Kutta method was not run past $t = 1$ because it would take excessive computation time. The time step could not be increased much beyond what it was in the region up to $t = 1$. The GEAR package, on the other hand, increased the time step drastically; at $t = 1,000$ the time step was about 3,000 times larger than that at $t = 1$. It is this factor of 3,000 that makes implicit methods very powerful when the largest eigen value is large and the equation needs to be integrated for a long time, i.e. for stiff systems.

For the second example we used the problem given in Eqs. (1-1) and (1-2) with

$$k_1 = 0.04 \qquad k_2 = 10^4 \qquad k_3 = 3 \times 10^7 \qquad c_0 = 1.0 \qquad (3\text{-}195)$$

At $t = 0$ the maximum eigen value is 0.04, while at $t = 0.02$ it is 2,450. We applied the explicit Runge–Kutta, implicit GEARB, and semi-implicit Runge–Kutta packages. Typical results are given in Table 3-7. For the semi-implicit Runge–Kutta package the jacobian was evaluated analytically, so that the number of function evaluations does not include function evaluations for the jacobian.

Notice in the GEARB results that going from $\varepsilon = 10^{-4}$ to $\varepsilon = 10^{-5}$ did not increase the cost very much. Another run with $\varepsilon = 10^{-10}$ took about 8 times as long as $\varepsilon = 10^{-5}$. The Runge–Kutta method did not work for $\varepsilon = 10^{-6}$ but did for $\varepsilon = 10^{-8}$. The solution was quite accurate, but at a cost of about 400 times as many function evaluations. The computation time of 15 CPU sec for the Runge–Kutta method was about 40 times larger than the 0.4 CPU sec for the GEARB package, $\varepsilon = 10^{-4}$. Such results are expected for this stiff problem.

The semi-implicit Runge–Kutta method worked very well and took fewer function evaluations. It used more jacobians than GEAR, but for $\varepsilon = 10^{-4}$ proved to take the same overall time of 0.4 CPU sec and to be about as accurate.

It may be concluded that we generally use implicit methods for stiff problems and explicit methods for the others. Which method within these classifications depends on accuracy, stability, tendency to oscillate, and work effort. The eventual decision depends on the user's goals and the problem to be solved, but the material in this chapter, if properly applied, will lead the reader to a suitable, and possibly the best, method.

STUDY QUESTIONS

1. Interpolation—how to apply it
2. Terminology
   a. Explicit versus implicit
   b. Stiff equations
   c. Stiffness ratio
3. Explicit schemes
   a. Advantages and characteristics

    b. Order
    c. Stability limitations
        i Euler methods
        ii Adams–Bashforth methods
        iii Runge–Kutta methods
4. Implicit schemes
    a. How to solve equations
    b. Application of the theory of algebraic equations to the difficulty in solving equations
    c. Advantages and characteristics
    d. Order
    e. Stability limitations
        i Backward Euler methods
        ii Crank–Nicolson methods
        iii Adams–Moulton methods
5. Predictor–corrector methods—difference from implicit methods
6. Extrapolation techniques—Richardson extrapolation
7. Step-size control—information needed to control the size
8. Stability
    a. Relationships to rational polynomials
    b. How to interpret rational polynomials
    c. Determination of A and L stability
9. Integration of ordinary differential equations
    a. Information needed to choose a method
    b. How information is obtained

PROBLEMS

3-1 Interpolate the following sequence to obtain $y_{1/2}$ using first-, second-, and third-order interpolation orders:

$$y_0 = 1.0 \qquad y_1 = 0.3678794411 \qquad y_2 = 0.1353352832 \qquad y_3 = 0.0497870683$$

3-2 Develop first-, second-, and third-order extrapolation formulas to give y_5 from known values of y_1, y_2, y_3, and y_4. Apply the formulas and compare results for the functions:

    (a) $y_i = e^{-(i-1)0.05}$
    (b) $y_i = \sin(i+1)$ (in radians).

3-3 *(a)* Derive Eq. (3-41). Why is there no $0(h^4)$ term?

    (b) Derive the fourth-order Adams–Bashforth method, Eq. (3-63), from the general formulas, Eqs. (3-55), (3-54), and (3-48*a*).

3-4 Derive the derivative estimate, Eq. (3-49), from the general formula, Eq. (3-48*a*).

3-5 Derive the implicit interpolation formula, Eq. (3-65), from the interpolation formulas, Eqs. (3-48*b*) and (3-54).

3-6 Solve the following problem from $t = 0$ to $t = 1$ as a test of your computer programming. Use a Runge–Kutta method, Gear's method, and a scheme you choose with a fixed step size. The Runge–Kutta method and Gear's method should be available as packages.

$$y_1' = -y_1 \qquad y_2' = -10^2 y_2 \qquad y_3' = -y_3$$
$$y_1(0) = 1 \qquad y_2(0) = 1 \qquad y_3(0) = 1$$

3-7 Solve Eqs. (1-1) and (1-2) with the constants given in Eqs. (3-195) using the same three integration schemes after testing them in problem 3-6.

3-8 Use the best available method to integrate the following equations from $t = 0$ to $t = 1.0$:

$$\text{Le}\frac{dT_1}{dt} = -10.5(T_1 - T_2) + \beta R_1$$

$$\frac{dc_1}{dt} = -10.5(c_1 - c_2) - R_1$$

$$c_2 = 1.0 \qquad T_2 = 1.0 \qquad R_1 = \phi^2 c_1 e^{\gamma(1 - 1/T_1)}$$

$$T_1(0) = 1.0 \qquad c_1(0) = 0.73$$

$$\beta = 0.15 \qquad \phi^2 = 1.21 \qquad \gamma = 30 \qquad \text{Le} = 0.1$$

3-9 Develop the equations to solve the problem

$$\frac{1}{\text{Pe}}\frac{d^2 c}{dx^2} - \frac{dc}{dx} - R(c) = 0$$

$$\frac{dc}{dx}(1) = 0 \qquad c(1) = a$$

from $x = 1$ to $x = 0$. The function $R(c)$ is a general reaction rate expression.

In Chapter 4 we will use this method to solve two-point boundary-value problems, choosing the constant a such that $c(0)$ satisfies a boundary condition at $x = 0$.

3-10 Derive the extrapolation formula, Eq. (3-114). Write the solution obtained with step size h as $y(h)$ and that obtained with step size $h/2$ as $y(h/2)$. The error formulas are

$$y(h) = y^e + ah^m$$

$$y\left(\frac{h}{2}\right) = y^e + a\left(\frac{h}{2}\right)^m$$

Solve these for y^e to get Eq. (3-114).

3-11 Consider the general nonlinear problem of Eq. (3-1) to be solved with a backward Euler method with a variable step size. The nonlinear equations are to be solved using either successive substitution or Newton–Raphson. Using the theorems stated in Chapter 2 determine the conditions under which these iteration schemes will converge. What happens for $\Delta t \to 0$? Apply your theorem to the case $dy/dt = -y^2$ and $y(0) = 1$.

3-12 Apply the second-order Runge–Kutta method of Eq. (3-96) to Eq. (3-123) to go from y_n to y_{n+1}. Compute y_{n+1}/y_n, the characteristic polynomial or rational approximation. Compare this polynomial to the Taylor series for $e^{-\lambda \Delta t}$. Find the conditions under which the rational approximation is greater than one.

3-13 Answer the same questions as in problem 3-12 but for the semi-implicit Runge–Kutta method, Eqs. (3-102) and (3-103). Derive the rational approximation, Eq. (3-174).

BIBLIOGRAPHY

Only the highlights of methods for integrating ordinary differential equations have been considered. The reader desiring additional information should consult books dealing entirely with that topic. A good summary of basic information is by

    Lapidus, L., and J. H. Seinfeld: *Numerical Solution of Ordinary Differential Equations*, Academic Press, New York, 1971.

A more mathematical treatment is by

    Lambert, J. D.: *Computational Methods in Ordinary Differential Equations*, John Wiley & Sons, Inc., New York, 1973.

The author found the following report extremely useful for clarifying some of the ideas of this chapter

    Smith, I. M., J. L. Siemienivich, and I. Gladwell: "A Comparison of Old and New Methods for

Large Systems of Ordinary Differential Equations Arising from Parabolic Partial Differential Equations," *Num. Anal. Rep.*, Department of Engineering, University of Manchester, England, no. 13, 1975.

The GEAR package was presented here as a general purpose implicit (and explicit) equations solver. The GEAR program is available as either GEAR (for a dense $n \times n$ jacobian) or GEARB (for a banded jacobian) from the Argonne Code Center, Applied Mathematics Division, Building 221, Argonne National Laboratory, 9700 South Cass Avenue, Argonne, Ill. 60439, U.S.A. Other packages are available and some are compared in the following articles. Explicit codes are compared by

Hull, T. E., W. H. Enright, B. M. Fellen, and A. E. Sedgwick: "Comparing Numerical Methods for Ordinary Differential Equations," *SIAM J. Num. Anal.*, vol. 9, pp. 603–637, 1972.

Shampine, L. F., H. A. Watts, and S. M. Davenport: "Solving Nonstiff Ordinary Differential Equations—The State of the Art," *SIAM Review*, vol. 18, pp. 376–411, 1976.

Implicit codes are compared by

Byrne, G., A. Hindmarsh, K. R. Jackson, and H. G. Brown: "A Comparison of Two ODE Codes: GEAR and EPISODE," *Comp. Chem. Eng.*, vol. 1, pp. 133–147, 1977.

GEAR though not always the best package is quite a good one.

The Runge–Kutta package is given by

Forsythe, G. E., M. A. Malcolm, and C. B. Moler: *Computer Methods for Mathematical Computations*, Prentice-Hall, Englewood Cliffs, 1977.

REFERENCES

1. Bailey, H. E.: "Numerical Integration of the Equations Governing the One-Dimensional Flow of a Chemically Reactive Gas," *Phys. Fluids*, vol. 12, pp. 2292–2300, 1969.
2. Caillaud, J. B., and L. Padmanabhan: "An Improved Semi-implicit Runge–Kutta Method for Stiff Systems," *Chem. Eng. J.*, vol. 2, pp. 227–232, 1971.
3. Gear, C. W.: *Numerical Initial-Value Problems in Ordinary Differential Equations*, Prentice-Hall, Englewood Cliffs, 1971.
4. Hindmarsh, A. C.: "GEARB: Solution of Ordinary Differential Equations Having Banded Jacobian," *UCID-30059, Rev. 1 Computer Documentation*, Lawrence Livermore Laboratory, University of California, 1975.
5. Michelsen, M. L.: "An Efficient General Purpose Method for the Integration of Stiff Ordinary Differential Equations," *A.I.Ch.E. J.*, vol. 22, pp. 594–597, 1976.
6. Nørsett, S. P.: "One-Step Methods of Hermite-Type for Numerical Integration of Stiff Systems," *BIT*, vol. 14, pp. 63–77, 1974.

FOUR

ORDINARY DIFFERENTIAL EQUATIONS— BOUNDARY-VALUE PROBLEMS

Diffusion problems in one space dimension lead to boundary-value problems. These problems are characterized by ordinary differential equations, usually of second order, with two boundary conditions, which are applied at two different locations in space. This means that initial-value methods cannot be applied in a straightforward fashion (but see Sec. 4-11) and also leads to the other nomenclature: two-point boundary-value problems. In this chapter we examine a variety of techniques applicable to these problems and see what information is best gleaned from which method.

4-1 METHOD OF WEIGHTED RESIDUALS

The first example considered is steady-state heat conduction in a slab, as illustrated in Fig. 1-2. We allow the thermal conductivity to depend on temperature in a linear fashion, which makes the problem nonlinear. We take

$$k = k_0 + k_0'(T - T_0) \tag{4-1}$$

The equation is the analog of Eq. (1-6) for heat transfer instead of mass transfer, and with no heat generation in the slab

$$\frac{d}{dx}\left(k\frac{dT}{dx}\right) = 0 \tag{4-2}$$

We assume the temperature of one side is maintained at T_0 while the other side is kept at temperature T_1. The nondimensional problem can be written as

$$\frac{d}{dx}\left[(1+a\theta)\frac{d\theta}{dx}\right] = 0$$

$$\theta(0) = 0 \qquad \theta(1) = 1$$

or

$$(1+a\theta)\frac{d^2\theta}{dx^2} + a\left(\frac{d\theta}{dx}\right)^2 = 0 \tag{4-3}$$

We expand the solution in a series of known functions with unknown coefficients. These coefficients are determined to satisfy the differential equation in some best sense. The criterion used to choose this "best" sense determines the method. The approximate solution is taken as a polynomial because of its simplicity.

$$\theta_N = \sum_{i=0}^{N} c_i' x^i \tag{4-4}$$

This function can be made to satisfy the boundary conditions by requiring

$$c_0' = 0 \qquad \sum_{i=0}^{N} c_i' = 1 \tag{4-5}$$

Thus Eq. (4-4) can be written as

$$\theta_N = x + \sum_{j=1}^{N} c_j(x^{j+1} - x) \tag{4-6}$$

We note that this function satisfies the boundary conditions for any value of the unknown constants $\{c_j\}$. Alternatively, the first term satisfies the boundary conditions of the problem, and each of the additional terms satisfy the homogeneous boundary conditions, i.e. the same boundary conditions but with the right-hand side zero. Making our trial function satisfy the boundary conditions means we have already satisfied part of the problem.

The next step is to form the residual. We substitute the trial function, Eq. (4-6), into Eq. (4-3) to form the residual

$$R(x, \theta_N) = \frac{d}{dx}\left[(1+a\theta_N)\frac{d\theta_N}{dx}\right] \tag{4-7}$$

The weighted residual is required to be zero. Thus

$$\int_0^1 W_k R(x, \theta_N)dx = 0 \tag{4-8}$$

Finally we choose a criterion or a weighted function. If we take the weighting functions to be the dirac delta function we have the collocation method

$$W_k = \delta(x - x_k) \tag{4-9}$$

$$\int_0^1 W_k R(x, \theta_N)dx = R(x_k, \theta_N) = 0 \tag{4-10}$$

This corresponds to satisfying the differential equation at the collocation points, but not necessarily for x in between. In the method of moments we choose $W_k = x^{k-1}$. In the subdomain method we choose W_k to be one on a subdomain $x_{k-1} < x < x_{k+1}$ and zero elsewhere. For the Galerkin method we choose the weighting function to be $\partial \theta_N / \partial c_k$, which in this case is

$$W_k = x^{k+1} - x \tag{4-11}$$

The least squares method uses

$$W_k = \frac{\partial R}{\partial c_k} \tag{4-12}$$

so that the interpretation is that the mean square residual

$$I = \int_0^1 R^2(x, \theta_N) dx \tag{4-13}$$

is being minimized. It is clear to see that as the number of collocation points increases we satisfy the differential equation at more and more points, and presumably we force the approximation to become the exact solution, which has a zero residual for all points. Similarly, in the subdomain method as the intervals get smaller and smaller, the residual approaches zero on average in smaller and smaller subdomains. In the least squares method the mean square residual is zero for the exact solution, so that as more and more parameters are allowed, and the mean square residual gets smaller, the approximate solution approaches the exact solution. The rationale behind the moments and Galerkin methods is more abstract and uses functional analysis. The key theorem states that if a function is orthogonal to each member of a complete set of functions then that function can only be zero. Two functions f_1 and f_2 are orthogonal if the integral of their product is zero

$$\int_0^1 f_1 f_2 dx = 0 \tag{4-14}$$

In this case one function is the residual and the complete set of functions are the weighting functions. The Galerkin and the moments methods then make the residual orthogonal to the weighting functions in Eq. (4-8), thus making the residual approach zero as $N \to \infty$. Further details and historical remarks about the Method of Weighted Residuals are given elsewhere.[3]

Let us apply several methods in the first approximation

$$\theta_1 = x + c_1(x^2 - x) \qquad \theta'_1 = 1 + c_1(2x - 1) \qquad \theta''_1 = 2c_1 \tag{4-15}$$

First we apply the collocation method. We choose as the collocation point $x = \frac{1}{2}$, since this is the midpoint of the interval. The residual evaluated at this point is then

$$\left(1 + a\frac{1 - c_1/2}{2}\right)2c_1 + a = 0 \tag{4-16}$$

which determines c_1. We choose to calculate numerical results only for the case $a = 1$, when

$$-\frac{c_1^2}{2} + 3c_1 + 1 = 0 \qquad \text{or} \qquad c_1 = -0.317 \tag{4-17}$$

The other solution to the quadratic is rejected as being physically unrealistic, since it gives the heat flux in the wrong direction at $x = 1$. The approximate solution is then

$$\theta_1 = x - 0.317(x^2 - x) \tag{4-18}$$

We can put this approximate solution into the differential equation and look at the residual. It is zero at $x = \frac{1}{2}$, but nonzero elsewhere. When the residual is zero everywhere we have the exact solution. Indeed in some cases the size of the residual can be related to the error in the solution (see Finlayson, p. 338). Without going into details we can apply some tests to the approximate solution. How good a heat balance does it give? Since Eq. (4-3) governs heat transfer across a slab under steady conditions the heat flux at both sides should be the same. Indeed the heat flux at all x should be the same. We find that at $x = 0$

$$(1 + a\theta_1)\frac{d\theta_1}{dx} = 1 - c_1 = 1.317 \tag{4-19}$$

and at $x = 1$

$$(1 + a\theta_1)\frac{d\theta_1}{dx} = 2(1 + c_1) = 1.366 \tag{4-20}$$

Thus there is a 4 percent difference in the two fluxes.

The next step is to compute the next approximation by taking $N = 2$. The trial function is then

$$\theta_2 = x + c_1(x^2 - x) + c_2(x^3 - x) \tag{4-21}$$

We substitute this trial function into the differential equation to form the residual and now make the residual zero at two points (since we have two constants to find). We again choose the equispaced points of $x = \frac{1}{3}$ and $x = \frac{2}{3}$. For $a = 1$ we get two nonlinear algebraic equations

$$2(\tfrac{4}{3} - \tfrac{2}{9}c_1 - \tfrac{8}{27}c_2)(c_1 + c_2) + (1 - \tfrac{1}{3}c_1 - \tfrac{2}{3}c_2)^2 = 0 \tag{4-22}$$

$$2(\tfrac{5}{3} - \tfrac{2}{9}c_1 - \tfrac{10}{27}c_2)(c_1 + 2c_2) + (1 + \tfrac{1}{3}c_1 + \tfrac{1}{3}c_2)^2 = 0 \tag{4-23}$$

The methods of Chapter 2 can be used to solve these equations, giving $c_1 = -0.5992$ and $c_2 = 0.1916$. The fluxes at the two sides are 1.4076 and 1.568, with the average 1.49. The exact answer is 1.5, so we have a better result than the first approximation. Figure 4-1 shows that the residual is smaller on average and is zero at the collocation points. Thus the second approximation is an improvement over the first approximation. We could continue in this fashion, but find it easier to use orthogonal collocation as described in Sec. 4-4.

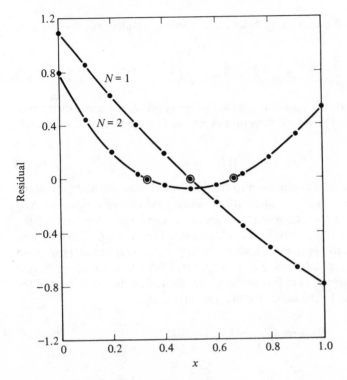

Figure 4-1 Residuals for collocation solutions.

Returning to the first approximation we use the Galerkin method

$$\int_0^1 x(1-x)R(x,\theta_1)dx = 0 \tag{4-24}$$

This gives the following equation to solve for c_1:

$$\int_0^1 x(1-x)\{[1+x+c_1(x^2-x)]2c_1+[1+c_1(2x-1)]^2\}dx = 0 \tag{4-25}$$

Solving gives $c_1 = -0.326$, with the fluxes at $x = 0$ and $x = 1$ being 1.326 and 1.348, respectively. Notice that in the Galerkin method we must calculate the integrals appearing in Eq. (4-25), whereas in the collocation method we merely needed to evaluate the residual at specific points. Thus the collocation method is easier to apply.

Next let us apply the method of moments in the first approximation

$$\int_0^1 R(x,\theta_1)dx = 0 \tag{4-26}$$

This results in the solution $c_1 = -0.333$. The second approximation requires that

$$\int_0^1 R(x,\theta_2)dx = 0 \qquad \int_0^1 xR(x,\theta_2)dx = 0 \tag{4-27}$$

Table 4-1 Approximate solution to heat conduction Eq. (4-3)

x	Collocation Eq. (4-18)	Eq. (4-21)	Galerkin $c_1 = -0.326$	Moments $c_1 = -0.333$	Eq. (4-28)	Finite difference	Exact
0.10	0.129	0.135	0.129	0.130	0.143	0.129	0.140
0.25	0.309	0.317	0.311	0.313	0.332	0.310	0.323
0.50	0.579	0.578	0.582	0.583	0.594	0.580	0.581
0.75	0.809	0.800	0.811	0.813	0.809	0.810	0.803
0.90	0.929	0.921	0.929	0.930	0.925	0.929	0.924
$\theta'_N(0)$	1.317	1.408	1.326	1.333	1.500	1.320	1.500
$2\theta'_N(1)$	1.367	1.568	1.348	1.333	1.500	1.360	1.500
Average flux	1.34	1.49	1.34	1.33	1.50	1.34	1.500

which gives two nonlinear equations that have the solutions $c_1 = -\frac{3}{4}$ and $c_2 = \frac{1}{4}$. The approximation is then

$$\theta_2 = \tfrac{3}{2}x - \tfrac{3}{4}x^2 + \tfrac{1}{4}x^3 \tag{4-28}$$

and the fluxes at the two boundaries are both 1.5. This result suggests that the answer is perhaps the best of all the methods, since the integral energy balance is satisfied.

Another way to test the methods is to compare them to the exact solution, which in this case is easily found to be $\theta = -1 + (1 + 3x)^{1/2}$, as shown in Table 4-1. All the methods give results within about 10 percent, and the first approximation is easy to derive. If this accuracy is acceptable we can stop with the first approximation. The accuracy is not guaranteed, however! If the accuracy is unacceptable, we must compute higher approximations to obtain more accurate answers as well as to assess the accuracy of the results. We see that the accuracy given by the different criteria is about the same, so that the choice of criteria can be based on other considerations, such as the ease of setting up the problem.

The advantages of the Method of Weighted Residuals are that the first approximation is easy to do, often contains the main features of the result, and may even be quite accurate. Higher approximations are more accurate, but only a few terms are necessary in any case. The disadvantage, which is shared by many numerical methods for boundary-value problems, is that the accuracy of the approximate solution is difficult to determine. We see below that the approach outlined here works well when the solution is relatively smooth, without sharp gradients or derivatives. Solutions with sharp gradients require so large an N that other methods are preferable, as outlined below for finite element methods.

4-2 FINITE DIFFERENCE METHOD

The finite difference method also is concerned with specific points in the domain called grid points. The domain is divided up into equidistant intervals, as shown in Fig. 4-2a, although the assumption of equal intervals is not necessary. Let us

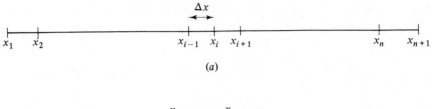

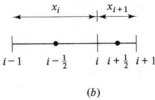

Figure 4-2 (*a*) Finite difference grid. (*b*) Variable grid spacing.

suppose we have a continuous function $\theta(x)$ and use Taylor series expansions to deduce difference formulas for first and second derivatives involving only the values at x_{i-1}, x_i, and x_{i+1}. We let $\theta_i = \theta(x_i)$ and write a Taylor series for θ_{i+1} and θ_{i-1} giving

$$\theta_{i+1} = \theta_i + \theta'_i \Delta x + \theta''_i \frac{\Delta x^2}{2!} + \theta'''_i \frac{\Delta x^3}{3!} + \theta''''_i \frac{\Delta x^4}{4!} + \dots \tag{4-29}$$

$$\theta_{i-1} = \theta_i - \theta'_i \Delta x + \theta''_i \frac{\Delta x^2}{2!} - \theta'''_i \frac{\Delta x^3}{3!} + \theta''''_i \frac{\Delta x^4}{4!} + \dots \tag{4-30}$$

These formulas are rearranged and divided by Δx to obtain two expressions for the first derivative

$$\frac{\theta_{i+1} - \theta_i}{\Delta x} = \theta'_i + \frac{\Delta x}{2} \theta''_i + \dots \tag{4-31}$$

$$\frac{\theta_i - \theta_{i-1}}{\Delta x} = \theta'_i - \frac{\Delta x}{2} \theta''_i + \dots \tag{4-32}$$

Each formula is correct to $0(\Delta x)$. Alternatively, we can subtract Eqs. (4-29) and (4-30), rearrange, and divide by Δx to obtain

$$\frac{\theta_{i+1} - \theta_{i-1}}{2\Delta x} = \theta'_i + \frac{1}{3!} \theta'''_i \Delta x^2 + \dots \tag{4-33}$$

which is correct to $0(\Delta x^2)$. We add Eqs. (4-29) and (4-30), rearrange, and divide by Δx^2 to obtain an expression for the second derivative

$$\frac{\theta_{i+1} - 2\theta_i + \theta_{i-1}}{\Delta x^2} = \theta''_i + \frac{2}{4!} \theta''''_i \Delta x^2 + \dots \tag{4-34}$$

This is correct to $0(\Delta x^2)$.

Next let us consider Eq. (4-3). In the case of $a = 0$ we can write the differential equation at the ith grid point using the difference formulas just derived

$$\frac{d^2\theta}{dx^2} = 0 \qquad \frac{\theta_{i+1} - 2\theta_i + \theta_{i-1}}{\Delta x^2} = 0 \tag{4-35}$$

If this equation is repeated for each interior point, and the values $\theta_1 = 0$ and $\theta_{N+1} = 1$ are taken to satisfy the boundary conditions, we have enough equations to solve for the θ_i. We thus have a representation of the solution.

Next we consider when the thermal conductivity varies with temperature. We write the equation as

$$-\frac{dq}{dx} = 0 \qquad q = -k\frac{d\theta}{dx} \tag{4-36}$$

and apply the second-order-correct Eq. (4-33) to get

$$-\frac{q_{i+1/2} - q_{i-1/2}}{\Delta x} = 0 \tag{4-37}$$

Equation (4-36) can in turn be applied using the second-order-correct Eq. (4-33) giving

$$q_{i+1/2} = -k(\theta_{i+1/2})\frac{\theta_{i+1} - \theta_i}{\Delta x} + 0(\Delta x^2)$$

$$q_{i-1/2} = -k(\theta_{i-1/2})\frac{\theta_i - \theta_{i-1}}{\Delta x} + 0(\Delta x^2) \tag{4-38}$$

Combining the formulas gives

$$\frac{k(\theta_{i+1/2})(\theta_{i+1} - \theta_i) - k(\theta_{i-1/2})(\theta_i - \theta_{i-1})}{\Delta x^2} = 0 \tag{4-39}$$

to solve at each grid point. To do this we must find a way to evaluate $k_{i+1/2}$ and $k_{i-1/2}$. There are several ways this can be done.

The first method is to use the interpolation formula developed in Chapter 3. Here we use Eq. (3-47) for θ instead of y

$$\theta_{i+\alpha} = \theta_{i+1} + (\alpha - 1)\nabla\theta_{i+1} + \frac{\alpha(\alpha - 1)}{2!}\nabla^2\theta_{i+1} \tag{4-40}$$

Applying this formula at $\alpha = \frac{1}{2}$ and $\alpha = -\frac{1}{2}$ gives

$$\theta_{i+1/2} = \theta_{i+1} - \tfrac{1}{2}(\theta_{i+1} - \theta_i) - \tfrac{1}{8}(\theta_{i+1} - 2\theta_i + \theta_{i-1}) \tag{4-41}$$

$$\theta_{i-1/2} = \theta_{i+1} - \tfrac{3}{2}(\theta_{i+1} - \theta_i) + \tfrac{3}{8}(\theta_{i+1} - 2\theta_i + \theta_{i-1}) \tag{4-42}$$

We must solve Eq. (4-39) combined with Eqs. (4-41) and (4-42). Let us do that for $\Delta x = \frac{1}{2}$ and define the variables:

$$\theta_1 = \theta(x = 0) = 0$$
$$\theta_{1.5} = \theta(x = \tfrac{1}{4})$$
$$\theta_2 = \theta(x = \tfrac{1}{2})$$
$$\theta_{2.5} = \theta(x = \tfrac{3}{4})$$
$$\theta_3 = \theta(x = 1) = 1 \tag{4-43}$$

$$(1+\theta_{2.5})(1-\theta_2)-(1+\theta_{1.5})(\theta_2-0)=0$$
$$\theta_{2.5}=\tfrac{3}{8}+\tfrac{3}{4}\theta_2 \qquad \theta_{1.5}=-\tfrac{1}{8}+\tfrac{3}{4}\theta_2 \tag{4-44}$$

Simplifying gives

$$-\tfrac{3}{2}\theta_2^2-\tfrac{3}{2}\theta_2+\tfrac{11}{8}=0 \tag{4-45}$$

which has the solution $\theta_2 = 0.580$. We can use the interpolation formula to obtain $\theta_{1.5} = 0.310$ and $\theta_{2.5} = 0.810$. To find $\theta(x = 0.1)$ we need to interpolate 80 percent of the way from θ_2 back to $\theta_{1.2}$ or use Eq. (4-40) with $i = 2$ and $\alpha = -0.8$. It is found that $\theta_{1.2} = 0.129$; likewise $\theta_{2.8} = 0.929$. These values are comparable to others derived by the Method of Weighted Residuals, as shown in Table 4-1.

Again we apply a test of the results by looking at the heat flux at the two sides. We cannot apply Eq. (4-33) at the point $x = 0$ since we have no value of $\theta_0 = \theta(-\Delta x)$. Applying Eq. (4-31) in terms of only θ_1 and θ_2 gives fluxes at the two sides

$$-q_1 = (1+\theta_1)\frac{\theta_2-\theta_1}{\Delta x} = 1.16$$
$$-q_3 = (1+\theta_3)\frac{\theta_3-\theta_2}{\Delta x} = 1.68 \tag{4-46}$$

These fluxes clearly differ greatly. Such inaccurate results are achieved when we use a formula correct only to $0(\Delta x)$.

To improve on this result we develop a one-sided derivative that is correct to $0(\Delta x^2)$. The Taylor series for θ_{i+2} is

$$\theta_{i+2} = \theta_i+2\theta_i'\Delta x+\theta_i''\frac{4}{2!}\Delta x^2+\theta_i'''\frac{8}{3!}\Delta x^3 \tag{4-47}$$

Four times Eq. (4-29) minus Eq. (4-47) gives after rearrangement the desired result

$$\theta_i' = \frac{-3\theta_i+4\theta_{i+1}-\theta_{i+2}}{2\Delta x}+0(\Delta x^2) \tag{4-48}$$

This one-sided difference formula is correct to $0(\Delta x^2)$. The analogous formula in the other direction is

$$\theta_i' = \frac{\theta_{i-2}-4\theta_{i-1}+3\theta_i}{2\Delta x}+0(\Delta x^2) \tag{4-49}$$

Using these difference formulas to evaluate the derivatives then gives the fluxes at the sides

$$-q_1 = -3\theta_1+4\theta_2-\theta_3 = 1.32$$
$$-q_3 = 2(\theta_1-4\theta_2+3\theta_3) = 1.36 \tag{4-50}$$

These are comparable to the results obtained with the Method of Weighted Residuals. Clearly we must evaluate fluxes to $0(\Delta x^2)$.

The second method for treating the nonlinear thermal conductivity is to average the k at the two grid points.

$$k(\theta_{i+1/2}) = \tfrac{1}{2}(k_i + k_{i+1}) \tag{4-51}$$

Now $k_1 = 1$, $k_2 = 1 + \theta_2$, and $k_3 = 2$. Applying this method to the three-node solution gives

$$\tfrac{1}{2}(2 + 1 + \theta_2)(\theta_3 - \theta_2) - \tfrac{1}{2}(1 + 1 + \theta_2)(\theta_2 - \theta_1) = 0 \tag{4-52}$$

which is

$$-2\theta_2^2 - 4\theta_2 + 3 = 0 \tag{4-53}$$

The result is $\theta_2 = 0.581$, giving a result very close to that obtained using the interpolation formula for θ.

The second approximation uses $\Delta x = \tfrac{1}{3}$ and the points $\theta_1 = \theta(x = 0)$, $\theta_2 = \theta(\Delta x)$, $\theta_3 = \theta(2\Delta x)$, and $\theta_4 = \theta(x = 1)$. Using the averaged thermal conductivities gives the equations

$$
\begin{aligned}
(k_3 + k_2)(\theta_3 - \theta_2) - (k_2 + 1)(\theta_2) &= 0 \\
(k_3 + 2)(1 - \theta_3) - (k_3 + k_2)(\theta_3 - \theta_2) &= 0
\end{aligned}
\tag{4-54}
$$

These are solved using an iterative scheme in which the thermal conductivities are evaluated at the old iteration and the set of linear equations solved. Thus

$$
\begin{aligned}
(k_3^s + k_2^s)(\theta_3^{s+1} - \theta_2^{s+1}) - (k_2^s + 1)\theta_2^{s+1} &= 0 \\
(k_3^s + 2)(1 - \theta_3^{s+1}) - (k_3^s + k_2^s)(\theta_3^{s+1} - \theta_2^{s+1}) &= 0
\end{aligned}
\tag{4-55}
$$

The solution is $\theta_2 = 0.414$ and $\theta_3 = 0.732$. The fluxes evaluated with the three-sided second-order expressions of Eqs. (4-48) and (4-49) are 1.39 and 1.46, respectively. The finite difference second approximation is not as accurate as the second approximation found with the Method of Weighted Residuals.

When the grid spacing is not uniform the same procedures can be applied. For the variable grid shown in Fig. 4-2b we want to write the difference equation for Eq. (4-36). The first equation is

$$-\frac{q_{i+1/2} - q_{i-1/2}}{\tfrac{1}{2}(\Delta x_i + \Delta x_{i+1})} = 0 \tag{4-56}$$

while the constitutive equations are

$$q_{i+1/2} = -k_{i+1/2}\frac{\theta_{i+1} - \theta_i}{\Delta x_{i+1}}$$

$$q_{i-1/2} = -k_{i-1/2}\frac{\theta_i - \theta_{i-1}}{\Delta x_i} \tag{4-57}$$

With the approximation of Eq. (4-51) we get the difference formula

$$\frac{[(k_{i+1} + k_i)(\theta_{i+1} - \theta_i)]/\Delta x_{i+1} - [(k_i + k_{i-1})(\theta_i - \theta_{i-1})]/\Delta x_i}{\Delta x_i + \Delta x_{i+1}} = 0 \tag{4-58}$$

The finite difference method has the advantage that the method is easy to formulate, although it may need a large number of grid points for high accuracy. Derivatives must be carefully evaluated in order not to destroy the accuracy that has been achieved. The solution must be interpolated at points between the grid points, using a formula that is at least as accurate as the error in the difference formulation.

4-3 REGULAR PERTURBATION

Perturbation methods are useful when a parameter in the problem is either very small or very large. Consider Eq. (4-3). The exact solution is derived by integrating the equation twice by separating variables

$$(1+a\theta)\frac{d\theta}{dx} = c_1 \tag{4-59}$$

$$\int(1+a\theta)d\theta = c_1 \int dx \tag{4-60}$$

$$\theta + \frac{a}{2}\theta^2 = c_1 x + c_2 \tag{4-61}$$

Application of the boundary conditions gives $c_2 = 0$, $c_1 = 1 + a/2$, and

$$\theta = -\frac{1}{a} + \frac{1}{a}[1 + (2a + a^2)x]^{1/2} \tag{4-62}$$

The solution is a function of both x, position, and the parameter a. Usually this parameter is given and the problem is solved for that specified a. However, let us consider the exact solution as a function of both x and a, and expand in a Taylor series (actually a Maclaurin series) about $a = 0$ to give

$$\theta(x, a) = \theta(x, 0) + \frac{\partial\theta}{\partial a}(x, 0)a + \frac{\partial^2\theta}{\partial a^2}(x, 0)\frac{a^2}{2!} + \dots \tag{4-63}$$

We can combine terms and write this simply as

$$\theta(x, a) = \theta_0(x) + \theta_1(x)a + \theta_2(x)a^2 + \dots \tag{4-64}$$

The various terms can be evaluated using l'Hospital's rule

$$\lim_{a \to 0} \frac{f(a)}{g(a)} = \frac{df/da(0)}{dg/da(0)} \quad \text{when} \quad f(0) = g(0) = 0 \tag{4-65}$$

$$\theta_0(x) = x \qquad \theta_1(x) = \tfrac{1}{2}x(1 - x) \tag{4-66}$$

Thus the first two terms are

$$\theta(x, a) = x + \frac{a}{2}x(1 - x) \tag{4-67}$$

For $a = 1$ the approximate solution gives the value $\theta(0.5) = 0.625$, with an 8 percent error. However, for smaller values of a we get more accurate answers, as shown in Table 4-2. What we would like is a method for obtaining these results without having to solve the problem first!

Table 4-2 Perturbation solution to Eq. (4-3)

a	$\theta(0.5), 0(a)$ Eq. (4-67)	$\theta(0.5), 0(a^2)$ Eq. (4-74)	Exact $\theta(0.5)$ Eq. (4-62)	Flux (0), $0(a)$	Flux (1), $0(a)$	Flux (0), $0(a^2)$	Flux (1), $0(a^2)$	Exact flux
0.01	0.503	0.503	0.501	1.005	1.005	1.005	1.005	1.005
0.1	0.525	0.524	0.512	1.050	1.045	1.050	1.045	1.050
1.0	0.625	0.609	0.581	1.500	1.000	1.500	2.000	1.500

In the perturbation method we expand the solution in the series of Eq. (4-64). This expansion is substituted into the differential equation to obtain, in this case

$$(1 + a\theta_0 + a^2\theta_1 + \dots)(\theta_0'' + a\theta_1'' + a^2\theta_2'') + a(\theta_0' + a\theta_1' + \dots)^2 = 0 \qquad (4\text{-}68)$$

We next collect terms multiplied by like powers of a.

$$a^0(\theta_0'') + a[\theta_1'' + \theta_0\theta_0'' + (\theta_0')^2] + a^2(\theta_2'' + \theta_0\theta_1'' + \theta_1\theta_0'' + 2\theta_0'\theta_1') + \dots = 0 \quad (4\text{-}69)$$

We do the same thing for the boundary conditions to give

$$a^0\theta_0(0) + a\theta_1(0) + a^2\theta_2(0) + \dots = 0$$
$$a^0\theta_0(1) + a\theta_1(1) + a^2\theta_2(1) + \dots = 1 \qquad (4\text{-}70)$$

Now if these equations are to be satisfied for all a they must be satisfied for $a = 0$. This gives the first problem to solve:

$$\theta_0'' = 0 \qquad \theta_0(0) = 0 \qquad \theta_0(1) = 1 \qquad (4\text{-}71)$$

If Eqs. (4-71) are satisfied, then these terms drop out of Eqs. (4-69) and (4-70), leaving only powers of a, a^2, etc. We then divide the result by a and apply the same argument. If the equation is true for $a = 0$ the next collection of terms must be zero. In a similar way we can obtain the result that the coefficient of each power of a must individually be zero. Thus we have the separate problems for a'

$$\theta_1'' + \theta_0\theta_0'' + (\theta_0')^2 = 0 \qquad \theta_1(0) = \theta_1(1) = 0 \qquad (4\text{-}72)$$

and for a^2

$$\theta_2'' + \theta_0\theta_1'' + \theta_1\theta_0'' + 2\theta_0'\theta_1' = 0 \qquad \theta_2(0) = \theta_2(1) = 0 \qquad (4\text{-}73)$$

We solve these in turn to obtain the perturbation solution. Putting the results back into Eq. (4-64) gives the solution

$$\theta(x, a) = x + \frac{a}{2}x(1 - x) + \frac{a^2}{2}x^2(x - 1) \qquad (4\text{-}74)$$

This approach is a regular perturbation method. In this case it is easier to solve the perturbation equations than it is to evaluate the derivatives of the exact solution, as in Eq. (4-65). With the regular perturbation method we get the expansion of the exact solution without knowing what it is. The algebra and the difficulty of solving the equation may increase tremendously as we solve for higher approximations. Also, the larger a is the more terms are needed for a good solution. Thus the method is best for small values of the parameter. The fact that the solution is made more accurate by including more terms is demonstrated in Table 4-2, where the second approximation is also recorded. For small a only a first approximation is needed, but for large a the second approximation improves the accuracy greatly. We need a method to provide the solution for large parameters when the perturbation methods are not accurate. Techniques that do this are the Method of Weighted Residuals and the finite difference method. In the next section the best of the methods using weighted residuals—the orthogonal collocation method—is described.

4-4 ORTHOGONAL COLLOCATION

The orthogonal collocation method has several advantages over the collocation method presented in Sec. 4-1. Namely, the collocation points are picked automatically, thus avoiding the arbitrary choice (and a possible poor one) by the user, and the error decreases much faster as the number of terms increases. There are three differences in the orthogonal collocation method: the trial function is taken as a series of orthogonal polynomials, the collocation points are taken as the roots to one of those polynomials, and the dependent variables are the solution values at the collocation points rather than the coefficients in the expansion.

First we examine the advantage of solving for the solution at the collocation points rather than the coefficients. (Note that the same approach can be used with the Method of Weighted Residuals.) We expand the solution in the form

$$y(x) = \sum_{i=1}^{N} a_i y_i(x) \tag{4-75}$$

where $\{y_i(x)\}$ are known functions of position. Usually we express the solution by providing the set $\{a_i\}$. Then we evaluate Eq. (4-75) at a set of N points to give

$$y(x_j) = \sum_{i=1}^{N} a_i y_i(x_j) \tag{4-76}$$

Remember that for all problems the $y_i(x_j)$ are known numbers. Thus using Eq. (4-76) gives $y(x_j)$ if the coefficients $\{a_i\}$ are known. Conversely, rearranging Eq. (4-76) and solving for $\{a_i\}$ we obtain

$$a_i = \sum_{i=1}^{N} [y_i(x_j)]^{-1} [y(x_j)] \tag{4-77}$$

This means that if the value of the solution is known at N points then the coefficients $\{a_i\}$ are determined. Consequently, we can solve a problem using as unknowns either the coefficients $\{a_i\}$ or the set of solution values at the collocation points $\{y(x_j)\}$.

To solve a differential equation that includes derivatives of y as well we differentiate Eq. (4-75) once or twice, for example, and evaluate the result at the collocation points

$$y'(x_j) = \sum_{i=1}^{N} a_i y_i'(x_j) \tag{4-78}$$

$$y''(x_j) = \sum_{i=1}^{N} a_i y_i''(x_j) \tag{4-79}$$

Since the coefficients $\{a_i\}$ can be expressed in terms of the solution values at the collocation points $\{y(x_j)\}$, the derivatives can also. We simply substitute Eq. (4-77) into Eqs. (4-78) and (4-79). Then the derivative at a particular collocation point, which is needed for the residual, is expressed in terms of the solution at all the collocation points

$$y'(x_j) = \sum_{i,k=1}^{N} [y_i(x_k)]^{-1}[y(x_k)]y_i'(x_j) \qquad (4\text{-}80)$$

We write the result as

$$y'(x_j) = \sum_{i=1}^{N} A_{jk}y(x_k) \qquad y''(x_j) = \sum_{i=1}^{N} B_{jk}y(x_k) \qquad (4\text{-}81)$$

To illustrate the idea let us take a function of $0 \leqslant x \leqslant 1$ and apply Eq. (4-75) with $N = 3$ and the functions

$$y_1 = 2(x-1)(x-\tfrac{1}{2}) \qquad y_2 = 4x(1-x) \qquad y_3 = 2x(x-\tfrac{1}{2}) \qquad (4\text{-}82)$$

The series Eq. (4-75) is just linear combinations of 1, x, and x^2, but so chosen that $y_i(x_j) = \delta_{ij}$ and

$$a_i = y(x_i) \qquad (4\text{-}83)$$

We let $x_1 = 0$, $x_2 = \tfrac{1}{2}$, and $x_3 = 1$. Next we differentiate Eq. (4-75) once and twice, and then evaluate the results at the midpoint

$$y'(\tfrac{1}{2}) = -a_1 + a_3$$
$$y''(\tfrac{1}{2}) = 4a_1 - 8a_2 + 4a_3 \qquad (4\text{-}84)$$

Thus the first and second derivatives at the midpoint are given in terms of the values of the function at the collocation points

$$y'(\tfrac{1}{2}) = -y(x_1) + y(x_3)$$
$$y''(\tfrac{1}{2}) = 4y(x_1) - 8y(x_2) + 4y(x_3)$$

In this respect the method is similar to the finite difference method, which would write in place of Eq. (4-84)

$$\frac{dy}{dx}(x_2) = \frac{y(x_3) - y(x_1)}{1} \qquad \frac{d^2y}{dx^2}(x_2) = \frac{y(x_3) - 2y(x_2) + y(x_1)}{(\tfrac{1}{2})^2} \qquad (4\text{-}85)$$

These are in fact identical in this case. When more collocation points are used, however, the derivatives are expressed in terms of the solution at all the collocation points, whereas in the finite difference method the derivatives are expressed only in terms of the solution at the grid points immediately adjacent. Similarly we get

$$y_1' = 4x - 3 \qquad y_2' = 4 - 8x \qquad y_3' = 4x - 1$$
$$y_1'(0) = -3 \qquad y_2'(0) = 4 \qquad y_3'(0) = -1$$
$$y_1'(1) = 1 \qquad y_2'(1) = -4 \qquad y_3'(1) = 3$$
$$y'(0) = -3a_1 + 4a_2 - a_3 = -3y_1 + 4y_2 - y_3$$
$$y'(1) = a_1 - 4a_2 + 3a_3 = y_1 - 4y_2 + 3y_3$$

These are the same as the finite difference formulas of Eqs. (4-48) and (4-49), with $\Delta x = 1$.

The next improvement to be introduced into the collocation method is to choose orthogonal polynomials for trial functions. We define the polynomial $P_m(x)$ as

$$P_m(x) = \sum_{j=0}^{m} c_j x^j \tag{4-86}$$

and we say that the polynomial has degree m and order $m+1$. The coefficients in Eq. (4-86) are defined by requiring that P_1 be orthogonal to P_0, P_2 be orthogonal to both P_1 and P_0, and P_m be orthogonal to each P_k, where $k \leqslant m-1$. The orthogonality condition can include a weighting function $W(x) \geqslant 0$. Thus

$$\int_a^b W(x)P_k(x)P_m(x)dx = 0 \qquad k = 0, 1, 2, \ldots, m-1 \tag{4-87}$$

This procedure specifies the polynomials to within a multiplicative constant, which we determine by requiring the first coefficient to be one. For illustration, let us use $W(x) = 1$, $a = 0$, and $b = 1$. The polynomials are

$$P_0 = 1 \qquad P_1 = 1+bx \qquad P_2 = 1+cx+dx^2 \tag{4-88}$$

The first one is already known: $P_0 = 1$. The second one is found by requiring

$$\int_0^1 P_0 P_1 dx = 0 \qquad \text{or} \qquad \int_0^1 (1+bx)dx = 0 \tag{4-89}$$

which makes $b = -2$. The third one P_2 is found from

$$\int_0^1 P_0 P_2 dx = 0 \qquad \int_0^1 P_1 P_2 dx = 0 \tag{4-90}$$

and so forth. The results are

$$P_0 = 1 \tag{4-91}$$

$$P_1 = 1-2x \qquad P_1(x) = 0 \text{ at } x = \tfrac{1}{2} \tag{4-92}$$

and

$$P_2 = 1-6x+6x^2 \qquad P_2(x) = 0 \text{ at } x = \tfrac{1}{2}(1 \pm \sqrt{3}/3) \tag{4-93}$$

The polynomial $P_m(x)$ has m roots in the interval a to b, and these serve as convenient choices of the collocation points. Thus if the expansion involves P_0 and P_1, such that

$$y = a_1 P_0(x) + a_2 P_1(x) \tag{4-94}$$

we need two collocation points to evaluate two residuals to determine the two constants a_1 and a_2, and we choose the two roots to $P_2(x) = 0$. We see that the whole procedure is automatic once the weighting function $W(x)$ is chosen. The user thus has fewer arbitrary choices as to trial functions and collocation points, although the weighting function must be specified.

We next apply orthogonal collocation to boundary-value problems. We expand in orthogonal polynomials, Eqs. (4-91) to (4-93), but we wish to have a first term that satisfies the boundary conditions followed by a series that has unknown coefficients, with each term satisfying the homogeneous boundary conditions. Let us take

$$y = x + x(1-x) \sum_{i=1}^{N} a_i P_{i-1}(x) \tag{4-95}$$

We can easily write this in the form

$$y = \sum_{i=1}^{N+2} b_i P_{i-1}(x) \tag{4-96}$$

and identify coefficients so that the two series are identical. For simplicity in deriving the derivative matrices we can also write the series as

$$y = \sum_{i=1}^{N+2} d_i x^{i-1} \tag{4-97}$$

In subroutine PLANAR in the appendix a more sophisticated way of finding the matrices is used based on Eq. (4-95). Taking the first and second derivatives of Eq. (4-97) we evaluate them at the collocation points. We take the collocation points as the N roots to $P_N(x) = 0$; these roots are between zero and one. The collocation points are then $x_1 = 0.0$, $x_2, \ldots, x_{N+1}$ are the interior roots, and $x_{N+2} = 1.0$, as shown in Fig. 4-3. The derivatives at the $N+2$ collocation points are

$$y(x_j) = \sum_{i=1}^{N+2} d_i x_j^{i-1} \tag{4-98}$$

$$\frac{dy}{dx}(x_j) = \sum_{i=1}^{N+2} d_i(i-1)x_j^{i-2} \tag{4-99}$$

$$\frac{d^2y}{dx^2}(x_j) = \sum_{i=1}^{N+2} d_i(i-1)(i-2)x_j^{i-3} \tag{4-100}$$

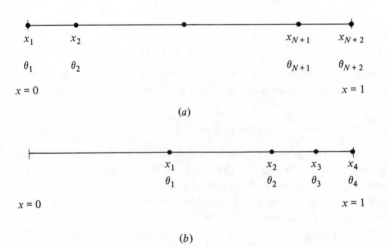

Figure 4-3 (a) Location of collocation points for $N = 3$. (b) Location of collocation points for $N = 3$ in symmetric problems.

We can write these equations in matrix notation, where $\mathbf{Q}$, $\mathbf{C}$, and $\mathbf{D}$ are $N+2 \times N+2$ matrices,

$$\mathbf{y} = \mathbf{Qd} \qquad \frac{d\mathbf{y}}{dx} = \mathbf{Cd} \qquad \frac{d^2\mathbf{y}}{dx^2} = \mathbf{Dd} \qquad (4\text{-}101)$$

$$Q_{ji} = x_j^{i-1} \qquad C_{ji} = (i-1)x_j^{i-2} \qquad D_{ji} = (i-1)(i-2)x_j^{i-3} \qquad (4\text{-}102)$$

Solving the first equation for $\mathbf{d}$ we can rewrite the first and second derivatives as

$$\frac{d\mathbf{y}}{dx} = \mathbf{CQ}^{-1}\mathbf{y} \equiv \mathbf{Ay} \qquad \frac{d^2\mathbf{y}}{dx^2} = \mathbf{DQ}^{-1}\mathbf{y} \equiv \mathbf{By} \qquad (4\text{-}103)$$

Thus the derivative at any collocation point is expressed in terms of the value of the function at the collocation points.

To evaluate integrals accurately we use the quadrature formula

$$\int_0^1 f(x)dx = \sum_{j=1}^{N+2} W_j f(x_j) \qquad (4\text{-}104)$$

Table 4-3 Polynomial roots and weighting functions defined by Eq. (4-87)

N	x_j	W_j
1	0.50000 00000	0.66666 66667
2	0.21132 48654	0.50000 00000
	0.78867 51346	0.50000 00000
3	0.11270 16654	0.27777 77778
	0.50000 00000	0.44444 44444
	0.88729 83346	0.27777 77778
4	0.06943 18442	0.17392 74226
	0.33000 94783	0.32607 25774
	0.66999 05218	0.32607 25774
	0.93056 81558	0.17392 74226
5	0.04691 00771	0.11846 34425
	0.23076 53450	0.23931 43353
	0.50000 00000	0.28444 44444
	0.76923 46551	0.23931 43353
	0.95308 99230	0.11846 34425
6	0.03376 52429	0.08566 22462
	0.16939 53068	0.18038 07865
	0.38069 04070	0.23395 69678
	0.61930 95931	0.23395 69678
	0.83060 46933	0.18038 07865
	0.96623 47571	0.08566 22462

In Eq. (4-87) $W = 1$, $a = 0$, and $b = 1$. For a given N the collocation points $x_2,\ldots,x_{N+1}$ are given above; $x_1 = 0$ and $x_{N+2} = 1.0$. For $N = 1$, $W_1 = W_3 = \frac{1}{6}$ and for $N \geqslant 2$, $W_1 = W_{N+2} = 0$.

Table 4-4 Matrices for orthogonal collocation for polynomials with roots in Table 4-3

N	W	A	B
1	$\begin{pmatrix} \frac{1}{6} \\ \frac{2}{3} \\ \frac{1}{6} \end{pmatrix}$	$\begin{pmatrix} -3 & 4 & -1 \\ -1 & 0 & 1 \\ 1 & -4 & 3 \end{pmatrix}$	$\begin{pmatrix} 4 & -8 & 4 \\ 4 & -8 & 4 \\ 4 & -8 & 4 \end{pmatrix}$
2	$\begin{pmatrix} 0 \\ \frac{1}{2} \\ \frac{1}{2} \\ 0 \end{pmatrix}$	$\begin{pmatrix} -7 & 8.196 & -2.196 & +1 \\ -2.732 & 1.732 & 1.732 & -0.7321 \\ 0.7321 & -1.732 & -1.732 & 2.732 \\ -1 & 2.196 & -8.196 & 7 \end{pmatrix}$	$\begin{pmatrix} 24 & -37.18 & 25.18 & -12 \\ 16.39 & -24 & 12 & -4.392 \\ -4.392 & 12 & -24 & 16.39 \\ -12 & 25.18 & -37.18 & 24 \end{pmatrix}$

To determine W_j we evaluate Eq. (4-104) for $f_i = x^{i-1}$. Thus

$$\int_0^1 x^{i-1} dx = \sum_{j=1}^{N+2} W_j x_j^{i-1} = \frac{1}{i} \tag{4-105}$$

$$WQ = f \qquad W = fQ^{-1} \tag{4-106}$$

The quadrature formula is exact if $f(x)$ is a polynomial of degree $2N-1$ in x. The needed collocation points are listed in Table 4-3 and the matrices for the first two approximations are given in Table 4-4. Note that for $N = 1$ we get the second-order finite difference derivatives, Eqs. (4-48) and (4-49).

Let us next apply the method to solve Eq. (4-3). The boundary conditions require that

$$\theta_1 = 0 \qquad \theta_{N+2} = 1 \tag{4-107}$$

and the residual is evaluated at the N interior collocation points $\{x_2, x_3, \dots, x_{N+1}\}$.

$$(1+\theta_i) \sum_{j=1}^{N+2} B_{ij}\theta_j + \left(\sum_{j=1}^{N+2} A_{ij}\theta_j\right)^2 = 0 \tag{4-108}$$

This can also be written as

$$\sum_{k=1}^{N+2} A_{ik}(1+\theta_k) \sum_{j=1}^{N+2} A_{kj}\theta_j = 0 \tag{4-109}$$

These equations are solved for $\{\theta_2, \theta_3, \dots, \theta_{N+1}\}$. The derivative at the two sides is given by the A matrix for the appropriate row. The fluxes at the two sides are then

$$\text{flux}(0) = (1+\theta_1) \sum_{j=1}^{N+2} A_{1j}\theta_j$$

$$\text{flux}(1) = (1+\theta_{N+2}) \sum_{j=1}^{N+2} A_{N+2,j}\theta_j \tag{4-110}$$

For the first approximation we have for Eq. (4-108)

$$(1+\theta_2)(4\theta_1 - 8\theta_2 + 4\theta_3) + (-\theta_1 + \theta_3)^2 = 0 \tag{4-111}$$

where the B_{2j} and A_{2j} matrices are found in Table 4-4. With $\theta_1 = 0$ and $\theta_3 = 1.0$ we get

$$(1+\theta_2)(-8\theta_2+4)+1 = 0 \tag{4-112}$$

which gives $\theta_2 = 0.579$. The fluxes at the boundaries are given by

$$\text{flux }(0) = (1+\theta_1)(-3\theta_1 + 4\theta_2 - \theta_3) = 4\theta_2 - 1 = 1.317$$
$$\text{flux }(1) = (1+\theta_3)(\theta_1 - 4\theta_2 + 3\theta_2) = -8\theta_2 + 6 = 1.367 \tag{4-113}$$

We notice that this solution is the same as that derived by collocation in Sec. 4-1. This is because for $N = 1$ the trial function is a second-order polynomial in x in both cases, and the collocation point for orthogonal collocation is $x = \frac{1}{2}$ for $N = 1$, as is used in Sec. 4-1. Thus the solutions must be the same.

For the second approximation $N = 2$ we again use the matrices listed in Table 4-4 to evaluate the residuals at the two collocation points 0.21132 and 0.78868. Since these are different collocation points from those used in Sec. 4-1 we expect slightly different solutions. The solutions are $\theta_2 = 0.2844$ and $\theta_3 = 0.8392$. The fluxes at the two sides are

$$\text{flux }(0) = 1.488$$
$$\text{flux }(1) = 1.493 \tag{4-114}$$

The two fluxes agree very well and are within 0.3 percent, lending confidence to the accuracy of the solution. The values of $\theta(x)$ at x that are not collocation points are found from any of the expansions, Eqs. (4-95) to (4-97). Equation (4-95) gives

$$\theta = x + x(1-x)[a_1 + a_2(1-2x)] \tag{4-115}$$

which can be evaluated at x_2 and x_3, since we now know θ_2 and θ_3 there. The resulting set of equations can then be solved for a_1 and a_2. In this case we get $a_1 = 0.3709$ and $a_2 = 0.1173$. The values of $\theta(x)$ are very close to those from the second approximation by the method of moments, Eq. (4-28).

A detailed comparison of the methods is provided in Sec. 4-13. We note here that the second approximation using orthogonal collocation is very accurate, giving the fluxes within 0.8 percent of the exact solution, whereas the collocation method using equispaced collocation points gives an accuracy of only 6 percent. The finite difference method with two interior grid points ($\Delta x = \frac{1}{3}$) has the same number of terms as the two-term orthogonal collocation method and takes the same work to solve, but provides an accuracy in flux of only 8 percent. Thus the orthogonal collocation method provides the highest accuracy for this example.

4-5 DIFFUSION AND REACTION—EXACT RESULTS

An important problem in chemical engineering is to predict the diffusion and reaction in a porous catalyst pellet. Diffusion is expressed by Eq. (1-6). Also the reaction rate can depend on concentration and temperature. Of course more general cases are possible. For example, the reaction rate may depend on several concentrations or on the activity of the catalyst, which may depend on position. We consider here the reaction of $A \to B$, with the reaction rate depending on the nth power of concentration of A, denoted by c'. The goal is to predict the overall reaction rate, or the mass transfer into and out of the catalyst pellet. Conservation of mass in a spherical domain gives

$$\frac{1}{(r')^2} \frac{d}{dr'} \left[D_e (r')^2 \frac{dc'}{dr'} \right] - k_0 R'(c, T) = 0 \tag{4-116}$$

while conservation of energy gives

$$\frac{1}{(r')^2} \frac{d}{dr'} \left[(r')^2 k_e \frac{dT'}{dr'} \right] + (-\Delta H_R) k_0 R'(c, T) = 0 \tag{4-117}$$

Here D_e is the effective diffusivity of the porous medium, k_0 is the rate constant, k_e is the effective thermal conductivity of the porous medium, and $-\Delta H_R$ is the heat of reaction. The rate of removal of A is $k_0 R'$ in units of concentration per time, or mass or moles per volume per time. We use boundary conditions at the center to have no flux through the center, making the problem symmetric about the origin. At $r' = 0$

$$\frac{dc'}{dr'} = \frac{dT'}{dr'} = 0 \tag{4-118}$$

At the boundary of the pellet we use the boundary conditions of the third kind. Thus at $r' = R$

$$-D_e \frac{dc'}{dr'} = k_g(c' - c_0) \qquad -k_e \frac{dT'}{dr'} = h_p(T' - T_0) \tag{4-119}$$

where k_g and h_p are mass and heat transfer coefficients of the transfer from the porous pellet to the surrounding medium. The concentration and temperature in the surrounding medium are c_0 and T_0, respectively, while R is the pellet radius.

The dimensionless equations are derived from these boundary conditions, here for the limiting case of constant physical properties, thus D_e, k_e, k_0, $-\Delta H_R$, k_g, and h_p are constants. Let $r = r'/R$, $c = c'/c_0$, $T = T'/T_0$, and $R_1 = R'/c_0^n$. Thus

$$\frac{1}{r^2} \frac{d}{dr} \left(r^2 \frac{dc}{dr} \right) = \phi^2 R_1(c, T) \tag{4-120}$$

and

$$\frac{1}{r^2} \frac{d}{dr} \left(r^2 \frac{dT}{dr} \right) = -\beta \phi^2 R_1(c, T) \tag{4-121}$$

at $r = 0$

$$\frac{dc}{dr} = 0 \qquad \frac{dT}{dr} = 0 \tag{4-122}$$

while at $r = 1$

$$-\frac{dc}{dr} = \text{Bi}_m(c - 1) \qquad -\frac{dT}{dr} = \text{Bi}(T - 1) \tag{4-123}$$

The dimensionless groups are:

$$\phi^2 = \frac{k_0 R^2 c_0^{n-1}}{D_e} \qquad \text{Thiele modulus squared}$$

$$\text{Bi}_m = \frac{k_g R}{D_e} \qquad \text{Biot number for mass transfer}$$

$$\text{Bi} = \frac{h_p R}{k_e} \qquad \text{Biot number for heat transfer}$$

(4-124)

$$\beta = \frac{(-\Delta H_R)c_0 D_e}{k_e T_0} \qquad \text{Dimensionless heat of reaction}$$

The corresponding equations for cylindrical and planar geometry are obtained by replacing r^2 by r and 1, respectively, in Eqs. (4-120) and (4-121).

The dimensionless groups have physical meaning. The most important parameter is the Thiele modulus ϕ. The group $1/(k_0 c_0^{n-1})$ is a characteristic time for reaction, while R^2/D_e is a characteristic time for diffusion. The Thiele modulus squared is thus the ratio of two characteristic times, diffusion to reaction. If the reaction is very fast its characteristic time is small and the Thiele modulus is large. Likewise, if the diffusion is very fast its characteristic time is small and the Thiele modulus is small. The Thiele modulus thus measures the relative importance of the diffusion and reaction phenomena.

The Biot number for mass transfer Bi_m is the ratio of two characteristic times: R^2/D_e for diffusion or mass transfer across the inside of the pellet and R/k_g for mass transfer across the boundary layer outside the pellet. For large Bi_m the characteristic time for internal diffusion is large compared with external diffusion, and internal diffusion dominates. The Biot number for heat transfer Bi is likewise the ratio of the characteristic time $R^2 \rho C_p T_0/k_e$ for diffusion of heat internal to the pellet to the characteristic time $R\rho C_p T_0/h_p$ for heat transport across the thermal boundary layer external to the catalyst pellet. Typically the Biot number for mass is large (say 100 or larger), making internal diffusion important. The Biot number for heat is smaller (about five) making external heat transport important. The effect of temperature is also related to the dimensionless heat of reaction β and the dimensionless activation energy

$$\gamma = \frac{E\hat{R}}{T_0}$$

(4-125)

where E is the activation energy and $\hat{R}$ is the gas constant. Clearly, if β or γ is small the temperature effect is less.

These equations can be reduced to a single equation in the following steps. Firstly we multiply the first equation by β and add it to the second giving

$$\frac{1}{r^2}\frac{d}{dr}\left[r^2\left(\beta\frac{dc}{dr} + \frac{dT}{dr}\right)\right] = 0$$

(4-126)

and then integrate this once to obtain

$$r^2\left(\beta\frac{dc}{dr}+\frac{dT}{dr}\right) = K_1 = 0 \tag{4-127}$$

The boundary conditions at $r = 0$ make both derivatives zero so that the constant is zero. We integrate once more to give

$$\beta c(r)+T(r) = K_2 \tag{4-128}$$

and evaluate at $r = 1$

$$\beta c(1)+T(1) = K_2 \tag{4-129}$$

We also have from Eq. (4-127)

$$\beta\frac{dc}{dr}(1) + \frac{dT}{dr}(1) = 0 \tag{4-130}$$

Multiplying the first boundary condition, Eq. (4-123), by β and adding it to the second we obtain

$$-\left(\beta\frac{dc}{dr}+\frac{dT}{dr}\right)\Bigg|_{r=1} = \mathrm{Bi}_m\beta[c(1)-1]+\mathrm{Bi}[T(1)-1] \tag{4-131}$$

The left-hand side according to Eq. (4-130) is zero, however, so that we get

$$T(1) = 1+\beta\delta[1-c(1)] \tag{4-132}$$

with

$$\delta = \frac{\mathrm{Bi}_m}{\mathrm{Bi}} \tag{4-133}$$

Now the constant K_2 can be evaluated solely in terms of $c(1)$

$$K_2 = \beta c(1)+1+\beta\delta[1-c(1)] \tag{4-134}$$

The temperature is then given by

$$T(r) = -\beta c(r)+1+\beta c(1)+\beta\delta[1-c(1)] \tag{4-135}$$

The original problem can be rewritten as

$$\frac{1}{r^2}\frac{d}{dr}\left(r^2\frac{dc}{dr}\right) = \phi^2 f(c(r),c(1),\beta,\delta) \tag{4-136}$$

$$\frac{dc}{dr}(0) = 0 \tag{4-137a}$$

$$-\frac{dc}{dr}(1) = \mathrm{Bi}_m[c(1)-1] \tag{4-137b}$$

If the reaction is nth-order and irreversible the reaction rate expression is

$$R_1(c,T) = c^n e^{\gamma-\gamma/T} \tag{4-138}$$

Then

$$f = c^n(r)e^{\gamma - \gamma/T} \tag{4-139}$$

where $T(r)$ is given by Eq. (4-135).

One special case is important, namely when the Biot numbers for heat and mass transfer become large. Then the boundary conditions are simply

$$c(1) = 1 \qquad T(1) = 1 \tag{4-140}$$

and the relation between temperature and concentration is

$$T(r) = 1 + \beta - \beta c(r) \tag{4-141}$$

Equation (4-141) also holds when $\delta = 1$ or $\text{Bi} = \text{Bi}_m$.

Before solving these equations consider their implications. If the reaction is external mass transfer-limited (i.e. small Bi_m) the concentration change occurs primarily from the bulk-stream value (here one in nondimensional form) to the value on the surface of the pellet $c(1)$. If the reaction is irreversible and very fast, the concentration on the surface of the pellet is very small. Letting $c(1) \to 0$ in Eq. (4-132) gives $T(1) = 1 + \beta\delta$ and $T(0) = 1 + \beta\delta$. Also $c(0) = 0$ if $c(1) = 0$. Thus the maximum temperature rise is $1 + \beta\delta$ and the pellet temperature is constant. If the reaction is very slow, however, then $c(r)$ is very close to the bulk-stream value, and setting $c(r) = c(1) = 1$ in Eq. (4-135) gives $T(r) = 1$, or the bulk-stream value. This case is isothermal and can also be brought about by having a small heat of reaction β.

We define the effectiveness factor as the average reaction rate with diffusion divided by the average reaction rate if the rate of reaction is evaluated at the bulk-stream (or boundary condition) values. This last quantity is the average reaction rate if diffusion is very fast, presenting no limitation to the mass transfer. The effectiveness factor is thus

$$\eta = \frac{\phi^2 \displaystyle\int_0^1 R_1(c(r), T(r))r^{a-1}dr}{\phi^2 \displaystyle\int_0^1 R_1(1, 1)r^{a-1}dr} \tag{4-142}$$

The parameter $a = 1, 2,$ or 3 respectively, for planar, cylindrical, or spherical geometry. We can integrate Eq. (4-120) over the domain $r = 0$ to 1 to obtain

$$\int_0^1 \frac{1}{r^2} \frac{d}{dr}\left(r^2 \frac{dc}{dr}\right)r^2 dr = \phi^2 \int_0^1 R_1(c, T)r^{a-1}dr = \frac{dc}{dr}(1) \tag{4-143}$$

Hence we can rewrite Eq. (4-142) as

$$\eta = \frac{a}{\phi^2}\frac{dc/dr(1)}{R(1, 1)} \tag{4-144}$$

Exact solutions give the same result in Eqs. (4-142) to (4-144), but numerical or perturbation solutions may not.

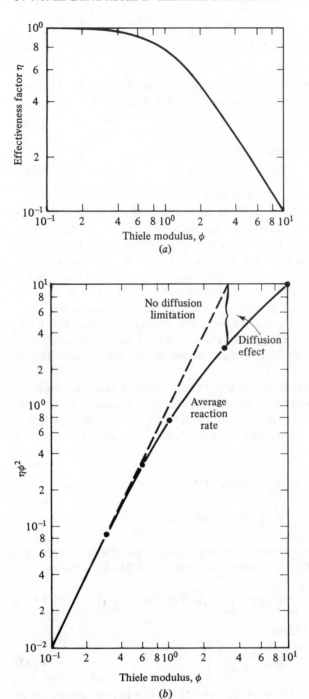

Figure 4-4 First-order, irreversible reaction. (*a*) Effectiveness factor. (*b*) Reaction rate. (*c*) Concentration profiles.

Next we consider Eq. (4-126) for an isothermal reaction that is nth-order and irreversible in planar geometry

$$\frac{d^2c}{dr^2} - \phi^2 c^n = 0 \tag{4-145}$$

$$\frac{dc}{dr}(0) = 0 \qquad c(1) = 1$$

$$\eta = \frac{1}{\phi^2}\frac{dc}{dr}(1) = \int_0^1 c^n(r)\,dr$$

We first solve Eq. (4-145) for the still simpler case of a first-order equation. Since the equation is linear with constant coefficients we try a solution of the form e^{kr} and find that we need certain k to satisfy the equation

$$k^2 e^{kr} - \phi^2 e^{kr} = 0 \qquad \text{or} \qquad k^2 = \phi^2 \tag{4-146}$$

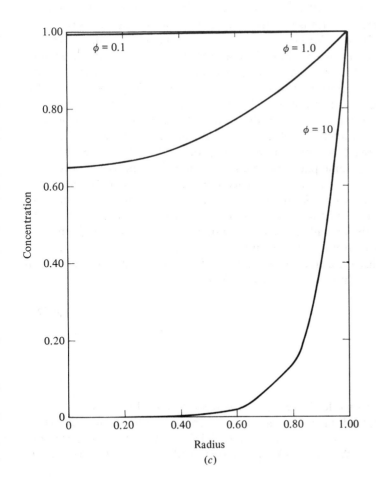

(c)

We thus write the solution as

$$c(r) = A'e^{\phi r} + B'e^{-\phi r} = A \sinh \phi r + B \cosh \phi r \qquad (4\text{-}147)$$

where

$$\sinh \phi r = \frac{e^{\phi r} - e^{-\phi r}}{2} \qquad \cosh \phi r = \frac{e^{\phi r} + e^{-\phi r}}{2} \qquad (4\text{-}148)$$

Application of the boundary conditions gives the solution $A = 0$, $B = 1/\cosh \phi$, and

$$c(r) = \frac{\cosh \phi r}{\cosh \phi} \qquad (4\text{-}149)$$

The effectiveness factor is then

$$\eta = \frac{1}{\phi} \tanh \phi \qquad (4\text{-}150)$$

The effectiveness factor is plotted as a function of Thiele modulus ϕ in Fig. 4-4a. At small ϕ the effectiveness factor is one, meaning that the rate of reaction is relatively uninfluenced by diffusion. For large ϕ the effectiveness factor is smaller than one, meaning that the average reaction rate is reduced below what it would be without diffusion limitations. The reaction rate as a function of ϕ is shown in Fig. 4-4b. We see that the rate is proportional to ϕ^2 for small ϕ but is proportional to ϕ for larger ϕ values. Since ϕ^2 is proportional to the reaction rate constant this means that the actual reaction rate is lowered due to the influence of diffusion. This effect must be correctly modeled by the chemical engineer in the design and operation of catalytic chemical reactors. The concentration profiles inside the pellet shown in Fig. 4-4c illustrate the same phenomenon. For small ϕ the concentration remains at the boundary value and diffusion effects are minimal. For larger ϕ, the concentration decreases away from the pellet surface due to diffusion, and since the reaction rate is less when the concentration is less, the inner part of the pellet contributes less to the overall reaction rate. For the largest ϕ shown the mass is confined to a narrow layer near the boundary.

We next simplify the exact solution for the effectiveness factor in preparation for the perturbation solution. For small $\phi \to 0$ we get

$$\eta = 1 - \frac{\phi^2}{3} \qquad (4\text{-}151)$$

while for large ϕ

$$\eta = \frac{1}{\phi} \qquad (4\text{-}152)$$

The dependence of η on ϕ for large ϕ is clearly represented in Fig. 4-4a.

The nonisothermal problems are very interesting subjects for numerical analysis. Consider Eqs. (4-136), (4-137a) and (4-137b) with a first-order reaction, $n = 1$ in Eq. (4-139). We first look at situations with the boundary conditions

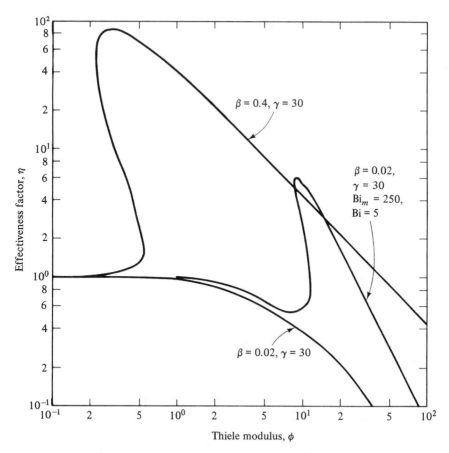

Figure 4-5 Effectiveness factor for first-order, irreversible reaction in spherical catalyst pellet.

given by Eq. (4-140) in the case of large Biot numbers for mass and heat transfer. We take the specific numbers $\beta = 0.4$ and $\gamma = 30$, and solve the problem for a variety of ϕ^2. The solution methods are the ones described below. A typical curve of η versus ϕ is shown in Fig. 4-5. Notice that a vertical line through $\phi = 0.4$ passes three times through the curve. This means that for a given reaction rate condition, set by ϕ, the problem has three solutions, each with a different η and different $c(r)$. We say the problem has multiple solutions. This problem has multiple solutions for $0.21 < \phi < 0.56$. In this range of ϕ the numerical problem to find the solution is formidable.

Values of β (the dimensionless heat of reaction) are not typically so large as 0.4. Values of 0.02 are common, and with $\gamma = 30$ and $\delta = 1$ the curve of the effectiveness factor versus Thiele modulus is shown in Fig. 4-5. Clearly, no multiple solutions are possible. If we use realistic values of Bi_m (say 250) and Bi (say 5) the curve takes the shape shown in Fig. 4-5. Multiple steady states are then possible for a range of ϕ. In this case the multiple steady states come about because the external heat transfer resistance is so great that the heat of reaction liberated in the pellet cannot escape, thus raising the temperature. The net reaction rate is increased due to the higher temperature even though the effect of concentration diffusion is to decrease the reaction rate.

A similar phenomenon occurs for certain isothermal reactions. Here we use the form appropriate to carbon monoxide oxidation on a platinum catalyst

$$k_0 R' = \frac{k_0 c'}{(1 + K_A c')^2} \tag{4-153}$$

We thus must solve Eqs. (4-136), (4-137a) and (4-137b) when

$$f = \frac{c}{(1 + \alpha c)^2} \qquad \alpha = K_A c_0 \tag{4-154}$$

For large values of α this problem has multiple solutions. Here they are caused by the adsorption of carbon monoxide onto the catalyst. The reaction rate is inversely proportional to the carbon monoxide concentration except at very small concentrations. Thus at any point in the catalyst, if the reactant is used up the reaction rate is actually larger than it would be at zero concentration.

We can now turn to the task of predicting the results shown in Fig. 4-5. First we apply the perturbation method, and then the orthogonal collocation and finite difference methods.

4-6 PERTURBATION METHOD FOR DIFFUSION AND REACTION

We see in the heat transfer example that the perturbation method can simplify a nonlinear problem to a succession of linear problems. The solution to these problems gives results that reflect the exact results for small values of the perturbation parameter. The diffusion and reaction problems have many parameters, and we next derive perturbation solutions by using ϕ^2 as a perturbation parameter.

We apply the perturbation method first for small ϕ. The series

$$c(r, \phi) = \phi^0 c_0(r) + \phi^2 c_1(r) + \phi^4 c_2(r) + \dots \tag{4-155}$$

is substituted into Eq. (4-145) with $n = 1$ to obtain the perturbation problems

$$\phi^0: \qquad c_0'' = 0 \qquad c_0'(0) = 0 \qquad c_0(1) = 1 \tag{4-156}$$

$$\phi^2: \qquad c_1'' = c_0 \qquad c_i'(0) = 0 \qquad c_i(1) = 0 \qquad i \geq 1 \tag{4-157}$$

$$\phi^4: \qquad c_2'' = c_1 \tag{4-158}$$

These are solved to obtain

$$c_0(r) = 1 \qquad c_1(r) = -\frac{1 - r^2}{2} \qquad c_2(r) = \frac{+5 - 6r^2 + r^4}{24} \tag{4-159}$$

We get the effectiveness factor from Eq. (4-145) for $n = 1$

$$\eta = 1 - \tfrac{1}{3}\phi^2 + \tfrac{2}{15}\phi^4 \tag{4-160}$$

Notice that if we use the derivative expression, Eq. (4-145), to evaluate the effectiveness factor we must find the c_2 term to get the ϕ^2 term in the expression, but if we use the integrated expression, Eq. (4-145), we get the ϕ^2 term from the c_1 term and ϕ^4 from the c_2 term. We don't know how good the solution is, although we know we need $\phi \to 0$. If we somewhat arbitrarily say the first approximation is acceptable if the additional term in the second approximation is only 10 percent of the first approximation, the solution is good for $\phi < 0.5$. A 1 percent criterion leads to the condition $\phi < 0.16$.

For a large ϕ we try a perturbation series in $1/\phi$. Letting $a = 1/\phi^2$ we rewrite Eq. (4-145) as

$$a \frac{d^2 c}{dr^2} - c = 0 \tag{4-161}$$

The perturbation method gives the simpler problems

$$a^0: \qquad -c_0(r) = 0 \qquad c_0(1) = 1 \qquad c_0'(0) = 0 \tag{4-162}$$

$$a^1: \qquad c_0'' - c_1(r) = 0 \qquad c_1(1) = 0 \qquad c_1'(0) = 0 \tag{4-163}$$

If we try to solve Eq. (4-162) for c_0 we see that the conditions are mutually exclusive; c_0 cannot be zero throughout but be one at the boundary. Thus the method does not work. Notice that for $a = 0$ the type of equation is changed. Indeed the equation is no longer a differential one. This feature is a clue that the regular perturbation method will not work: if the coefficient of the highest derivative goes to zero on application of the regular perturbation method, then it does not work.

The solution to the dilemma is to apply asymptotic expansions, or inner and outer expansions. We know that for large ϕ the solution is confined to a region near the boundary $r = 1$, so we use a coordinate system which expands that region. Near $r = 1$ we derive the inner solution c_1 as a function of $s = \phi(1-r)$. The equations are then

$$ds = -\phi dr \tag{4-164}$$

$$\frac{d^2 c_1}{ds^2} - c_1 = 0 \tag{4-165}$$

$$c_1(s = 0) = 1 \tag{4-166}$$

We do not apply a boundary condition at the other end point, thus we get the solution

$$c_1(s) = A \sinh s + \cosh s \tag{4-167}$$

Near $r = 0$ we derive the outer solution c_2 as a function of r (we could expand this region, too, if necessary) and apply the boundary condition at $r = 0$ but not at $r = 1$

$$\frac{d^2 c_2}{dr^2} - \phi^2 c_2 = 0 \tag{4-168}$$

$$\frac{dc_2}{dr}(0) = 0 \tag{4-169}$$

The solution is thus

$$c_2(r) = C \cosh \phi r \tag{4-170}$$

We would like these solutions to be the same where they meet, at least for large ϕ. Thus we want at large ϕ, $c_2(r)$ as r increases $= c_1(s)$ as s increases. This gives

$$\lim_{\phi \to \infty} c_2(r) = \frac{C}{2} e^{\phi r} \qquad \lim_{\phi \to \infty} c_1(r) = \frac{A+1}{2} e^{\phi(1-r)} \tag{4-171}$$

The matching condition at large ϕ requires

$$\frac{C}{2} e^{\phi r} = \frac{A+1}{2} e^{\phi(1-r)} \tag{4-172}$$

The only way this can be true is if $C = 0$ and $A + 1 = 0$. Thus the inner and outer solutions are

$$c_1 = \cosh s - \sinh s \qquad c_2 = 0 \tag{4-173}$$

The effectiveness factor at large ϕ is then

$$\eta = \frac{1}{\phi^2} \frac{dc_1}{dr}\bigg|_{r=1} = -\frac{1}{\phi} \frac{dc_1}{ds}\bigg|_{s=0} = \frac{1}{\phi} \tag{4-174}$$

$$= \int_0^1 c(r)dr = \frac{1}{\phi} \int_0^\phi (\cosh s - \sinh s)ds = \frac{1}{\phi} - \frac{e^{-\phi}}{\phi} \tag{4-175}$$

The solution for small ϕ is Eq. (4-155) together with Eqs. (4-159). For large ϕ we have Eqs. (4-173) and (4-174). These are compared to the exact solution in Fig. 4-6. The solution is good for small ϕ (<0.6) or large ϕ (>3).

Next we treat the nonlinear problem

$$\frac{d^2 c}{dr^2} - \phi^2 f(c) = 0$$

$$\frac{dc}{dr}(0) = 0 \qquad c(1) = 1 \tag{4-176}$$

The outer solution is valid near $r = 0$ and satisfies

$$\frac{d^2 c_2}{dr^2} - \phi^2 f(c_2) = 0 \qquad \frac{dc_2}{dr}(0) = 0 \tag{4-177}$$

Clearly for large ϕ it is necessary that the reaction rate term be zero, making c_2 the equilibrium concentration. For an irreversible reaction $c_2 = 0$. The inner solution must satisfy

$$\frac{d^2 c_1}{ds^2} - f(c_1) = 0 \qquad c_1(1) = 1 \tag{4-178}$$

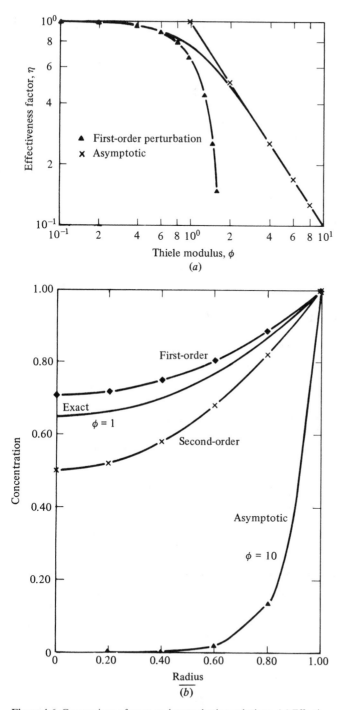

Figure 4-6 Comparison of exact and perturbation solutions. (a) Effectiveness factor. (b) Comparison of first and second-order perturbation solutions with exact solution.

This equation is still difficult to solve, but the following technique provides an exact solution. Letting

$$p = \frac{dc}{ds} \tag{4-179}$$

we calculate

$$p\frac{dp}{dc} = \frac{dc}{ds}\frac{d}{dc}\left(\frac{dc}{ds}\right) = \frac{dc}{ds}\frac{d^2c}{ds^2}\frac{ds}{dc} = \frac{d^2c}{ds^2} \tag{4-180}$$

Thus Eq. (4-178) becomes

$$p\frac{dp}{dc_1} = f(c_1) \tag{4-181}$$

This we integrate to obtain

$$\frac{p^2}{2}\bigg|_{p(0)}^{p(\infty)} = \int_0^1 f(c_1)dc_1 \tag{4-182}$$

To match with the inner solution, which has $c_2 = 0$ for all r, it is clear that $p(\infty)$ must be zero, i.e. a zero slope condition holds where the inner and outer solutions are matched. We thus obtain

$$-\frac{1}{2}\left[\frac{dc_1}{ds}(0)\right]^2 = \int_0^1 f(c)dc \tag{4-183}$$

and this gives the effectiveness factor as

$$\eta = \frac{1}{\phi^2}\frac{dc}{dr}\bigg|_{r=1} = -\frac{1}{\phi}\frac{dc}{ds}\bigg|_{s=0} = \frac{1}{\phi}\left[2\int_0^1 f(c)dc\right]^{1/2} \tag{4-184}$$

For other geometries and the more general boundary condition, Eq. (4-123), the corresponding result is

$$\eta = \frac{\sqrt{2}}{\hat{\phi}}\left[\int_0^{c(1)} f(c)dc\right]^{1/2} \qquad \hat{\phi} = \frac{V_p}{A_pR}\phi \tag{4-185}$$

$$Bi_m[c(1)-1] = -\sqrt{2}\phi\left[\int_0^{c(1)} f(c)dc\right]^{1/2} \tag{4-186}$$

In these formulas it is assumed that $f(1) = 1$. V_p and A_p are the volume and external surface area of the catalyst pellet. For regular pellets $\hat{\phi} = \phi/a$.

Next we apply the perturbation method to the nonisothermal problem, Eqs. (4-136), (4-137a) and (4-137b). We take the case when the Biot numbers for heat and mass transfer are large so that the problem reduces to

$$\frac{1}{r^2}\frac{d}{dr}\left(r^2\frac{dc}{dr}\right) = \phi^2 f(c)$$

$$f(c) = c\exp\frac{\gamma\beta(1-c)}{1+\beta(1-c)} \tag{4-187}$$

$$c'(0) = 0 \qquad c(1) = 1$$

For small ϕ we use the regular perturbation method but we must expand the reaction rate expression

$$c = c_0 + \phi^2 c_1 + \phi^4 c_2 + \ldots$$

$$f(c) = f(c_0) + \left(\frac{df}{dc}\right)\Bigg|_{c_0} \phi^2 c_1 + \ldots \tag{4-188}$$

We get the simpler problems

$$\frac{1}{r^2}\frac{d}{dr}\left(r^2\frac{dc_0}{dr}\right) = 0 \qquad \frac{dc_0}{dr}(0) = 0 \qquad c_0(1) = 1$$

$$\frac{1}{r^2}\frac{d}{dr}\left(r^2\frac{dc_1}{dr}\right) = f(c_0) \qquad \frac{dc_1}{dr}(0) = 0 \qquad c_1(1) = 0 \tag{4-189}$$

We can easily solve for $c_0 = 1$ and $c_1 = \frac{1}{6}(r^2 - 1)$. The effectiveness factor for spherical geometry is

$$\eta = \frac{\phi^2 \displaystyle\int_0^1 c(r)r^2\,dr}{\phi^2 \displaystyle\int_0^1 r^2\,dr} = 3\int_0^1 c(r)r^2\,dr = \frac{3}{\phi^2}\frac{dc}{dr}\Bigg|_{r=1} \tag{4-190}$$

The integral expression gives

$$\eta = 1 - \frac{\phi^2}{15} \tag{4-191}$$

The next approximation is very difficult to calculate so we obtain useful information only for small ϕ. We might choose $\phi < 1.2$, as the region of validity of Eq. (4-191), since then the second, correction term is 10 percent or less of the first one. Unfortunately, the approximation is not valid at all. As shown in Fig. 4-5, η increases with ϕ at small ϕ.

For large ϕ we use the singular perturbation method with Eqs. (4-185) giving the result

$$\eta = \frac{A}{\phi} \qquad \phi \to \infty \tag{4-192}$$

where

$$A = 3\sqrt{2}\left[\int_0^1 c\exp\frac{\gamma\beta(1-c)}{1+\beta-\beta c}\,dc\right]^{1/2} \tag{4-193}$$

This requires numerical quadrature to evaluate, but does provide the exact constant in Eq. (4-192). Results given below when $\beta = 0.4$ and $\gamma = 30$ suggest that Eq. (4-192) is adequate for $\phi > 0.6$. Thus we have no reasonable perturbation solution for $\phi < 0.6$.

Usually the regular and singular perturbation methods provide useful information in limited regions of parameter space. It may be possible to find the first approximation, but more laborious to find the second approximation, and when the work is done there is still a significant region of parameter space for which no solution has been found. We need a method to provide the solution for intermediate values of ϕ, when neither perturbation solution is valid. Section 4-7 applies a very good method for doing this—the orthogonal collocation method.

4-7 ORTHOGONAL COLLOCATION FOR DIFFUSION AND REACTION

The orthogonal collocation method has proved to be a useful method for problems of diffusion and reaction. Frequently, a first approximation gives accurate results, and it also gives insight into the solution. If desired the higher approximations can be calculated to provide more accurate answers, and the method is suitable for bridging the gap between the regions of validity of the perturbation solutions.

We next turn to problems of the type given in Eq. (4-136). In many of these problems it is possible to prove that the solution is a symmetric function of x, i.e. a function of only even powers of x and excluding all the odd powers. In such a case it is our prerogative to include that information in the choice of trial functions. To do this we construct orthogonal polynomials that are functions of x^2. One choice is

$$y(x^2) = y(1) + (1 - x^2) \sum_{i=1}^{N} a_i P_{i-1}(x^2) \tag{4-194}$$

Equivalent choices are

$$y(x^2) = \sum_{i=1}^{N} b_i P_{i-1}(x^2) \tag{4-195a}$$

$$= \sum_{i=1}^{N+1} d_i x^{2i-2} \tag{4-195b}$$

We define the polynomials to be orthogonal with the condition

$$\int_0^1 W(x^2) P_k(x^2) P_m(x^2) x^{a-1} dx = 0 \qquad k \leqslant m-1 \tag{4-196}$$

where we use $a = 1$, 2, or 3 for planar, cylindrical, or spherical geometry, respectively. Again we take the first coefficient of the polynomial as one, so that the choice of the weighting function $W(x^2)$ completely determines the polynomial, and hence the trial function and the collocation points.

We differentiate Eq. (4-195b) once and take the laplacian of it, too, where

$$\nabla^2 y = \frac{1}{x^{a-1}} \frac{d}{dx} \left(x^{a-1} \frac{dy}{dx} \right) \tag{4-197}$$

for the three geometries

$$\frac{dy}{dx} = \sum_{i=1}^{N+1} d_i(2i - 2) x^{2i-3} \tag{4-198}$$

thus

$$\nabla^2 y = \sum_{i=1}^{N+1} d_i(2i-2)[(2i-3)+a-1]x^{2i-4} \tag{4-199}$$

Now the collocation points are N interior points $0 < x_j < 1$ and one boundary point $x_{N+1} = 1$. The point $x = 0$ is not included because the symmetry condition requires that the first derivative be zero at $x = 0$ and that condition is already built into the trial function. The location of the collocation points is shown in Fig. 4-3b. The derivatives are evaluated at the collocation points to give

$$y(x_j) = \sum_{i=1}^{N+1} x_j^{2i-2}d_i \tag{4-200}$$

$$\frac{dy}{dx}(x_j) = \sum_{i=1}^{N+1} x_j^{2i-3}(2i-2)d_i \tag{4-201}$$

$$\nabla^2 y(x_j) = \sum_{i=1}^{N+1} \nabla^2(x^{2i-2})|_{x_j}d_i \tag{4-202}$$

and in matrix notation we have

$$\mathbf{y} = \mathbf{Qd} \qquad \frac{d\mathbf{y}}{d\mathbf{x}} = \mathbf{Cd} \qquad \nabla^2\mathbf{y} = \mathbf{Dd} \tag{4-203}$$

$$Q_{ji} = x_j^{2i-2} \qquad C_{ji} = (2i-2)x_j^{2i-3} \qquad D_{ji} = \nabla^2(x^{2i-2})|_{x_j} \tag{4-204}$$

Solving for $\mathbf{d}$ gives, as before,

$$\frac{d\mathbf{y}}{d\mathbf{x}} = \mathbf{CQ}^{-1}\mathbf{y} \equiv \mathbf{Ay} \qquad \nabla^2\mathbf{y} = \mathbf{DQy} \equiv \mathbf{By} \tag{4-205}$$

Quadrature formulas are

$$\int_0^1 f(x^2)x^{a-1}dx = \sum_{j=1}^{N+1} W_j f(x_j^2) \tag{4-206}$$

and W_j found by using Eq. (4-206) for $f_i = x^{2i-2}$

$$\int_0^1 x^{2i-2}x^{a-1}dx = \sum_{j=1}^{N+1} W_j x_j^{2i-2} = \frac{1}{2i-2+a} \equiv f_i \tag{4-207}$$

$$\mathbf{WQ} = \mathbf{f} \qquad \mathbf{W} = \mathbf{fQ}^{-1}$$

The integration is exact for functions f that are polynomials of degree $2N$ in x^2, provided the interior collocation points are the roots to $P_N(x^2)$ defined by Eq. (4-196) with $W = 1-x^2$. If $W = 1$ then the integration is exact for polynomials of degree $2N-1$. The collocation points for the different geometries and weighting functions are given in Table 4-5. The matrices for $W = 1-x^2$ and $N = 1$ or 2 are given in Table 4-6.

Table 4-5 Roots of polynomials defined by Eq. (4-196) with $W = 1$

	Geometry		
N	Planar $a = 1$	Cylindrical $a = 2$	Spherical $a = 3$
1	0.57735 02692	0.70710 67812	0.77459 66692
2	0.33998 10436	0.45970 08434	0.53846 93101
	0.86113 63116	0.88807 38340	0.90617 93459
3	0.23861 91861	0.33571 06870	0.40584 51514
	0.66120 93865	0.70710 67812	0.74153 11856
	0.93246 95142	0.94196 51451	0.94910 79123
4	0.18343 46425	0.26349 92300	0.32425 34234
	0.52553 24099	0.57446 45143	0.61337 14327
	0.79666 64774	0.81852 94874	0.83603 11073
	0.96028 98565	0.96465 96062	0.96816 02395
5	0.14887 43390	0.21658 73427	0.26954 31560
	0.43339 53941	0.48038 04169	0.51909 61292
	0.67940 95683	0.70710 67812	0.73015 20056
	0.86506 33667	0.87706 02346	0.88706 25998
	0.97390 65285	0.97626 32447	0.97822 86581
6	0.12523 34085	0.18375 32119	0.23045 83160
	0.36783 14990	0.41157 66111	0.44849 27510
	0.58731 79543	0.61700 11402	0.64234 93394
	0.76990 26742	0.78696 22564	0.80157 80907
	0.90411 72564	0.91137 51660	0.91759 83992
	0.98156 06342	0.98297 24091	0.98418 30547

For a given N the collocation points $x_1, \ldots, x_N$ are listed above. $x_{N+1} = 1.0$.

The orthogonal collocation method is first applied to the diffusion–reaction problem, Eq. (4-145) with $n = 2$. We initially test whether the solution is symmetric in x. To do this we derive a power series solution using the expansion

$$c(x) = \sum_{i=0}^{\infty} a_i x^i \qquad (4\text{-}208)$$

This form is substituted into Eq. (4-145) and the coefficients of successive powers of x are set to zero

$$x^0: \qquad 2a_2 - \phi^2 a_0^2 = 0 \qquad (4\text{-}209)$$

$$x^1: \qquad 6a_3 - \phi^2(2a_0 a_1) = 0 \qquad (4\text{-}210)$$

$$x^2: \qquad 12a_4 - \phi^2(a_1^2 + 2a_0 a_2) = 0 \qquad (4\text{-}211)$$

$$x^3: \qquad 20a_5 - \phi^2(2a_0 a_3 + 2a_1 a_2) = 0 \qquad (4\text{-}212)$$

Application of the boundary condition at $x = 0$ gives $a_1 = 0$. If a_1 is zero then by Eq. (4-210) a_3 is zero, too and Eq. (4-212) then says that $a_5 = 0$. This can be

Table 4-6 Matrices for orthogonal collocation for polynomials in Eq. (4-196) with $W = 1 - x^2$

N	x	W	A			B		
			Planar geometry $a = 1$					
1	$\begin{pmatrix} 0.4472 \\ 1.0000 \end{pmatrix}$	$\begin{pmatrix} 0.8333 \\ 0.1667 \end{pmatrix}$	$\begin{pmatrix} -1.118 & 1.118 \\ -2.500 & 2.500 \end{pmatrix}$			$\begin{pmatrix} -2.5 & 2.5 \\ -2.5 & 2.5 \end{pmatrix}$		
2	$\begin{pmatrix} 0.2852 \\ 0.7651 \\ 1.0000 \end{pmatrix}$	$\begin{pmatrix} 0.5549 \\ 0.3785 \\ 0.0667 \end{pmatrix}$	$\begin{pmatrix} -1.753 & 2.508 & -0.7547 \\ -1.371 & -0.6535 & 2.024 \\ 1.792 & -8.791 & 7 \end{pmatrix}$			$\begin{pmatrix} -4.740 & 5.677 & -0.9373 \\ 8.323 & -23.26 & 14.94 \\ 19.07 & -47.07 & 28 \end{pmatrix}$		
			Cylindrical geometry $a = 2$					
1	$\begin{pmatrix} 0.5774 \\ 1.0000 \end{pmatrix}$	$\begin{pmatrix} 0.375 \\ 0.125 \end{pmatrix}$	$\begin{pmatrix} -1.732 & 1.732 \\ -3 & 3 \end{pmatrix}$			$\begin{pmatrix} -6 & 6 \\ -6 & 6 \end{pmatrix}$		
2	$\begin{pmatrix} 0.3938 \\ 0.8031 \\ 1.0000 \end{pmatrix}$	$\begin{pmatrix} 0.1882 \\ 0.2562 \\ 0.0556 \end{pmatrix}$	$\begin{pmatrix} -2.540 & 3.826 & -1.286 \\ -1.378 & -1.245 & 2.623 \\ 1.715 & -9.715 & 8 \end{pmatrix}$			$\begin{pmatrix} -9.902 & 12.30 & -2.397 \\ 9.034 & -32.76 & 23.73 \\ 22.76 & -65.42 & 42.67 \end{pmatrix}$		
			Spherical geometry $a = 3$					
1	$\begin{pmatrix} 0.6547 \\ 1.0000 \end{pmatrix}$	$\begin{pmatrix} 0.2333 \\ 0.1 \end{pmatrix}$	$\begin{pmatrix} -2.291 & 2.291 \\ -3.5 & 3.5 \end{pmatrix}$			$\begin{pmatrix} -10.5 & 10.5 \\ -10.5 & 10.5 \end{pmatrix}$		
2	$\begin{pmatrix} 0.4688 \\ 0.8302 \\ 1.0000 \end{pmatrix}$	$\begin{pmatrix} 0.0949 \\ 0.1908 \\ 0.0476 \end{pmatrix}$	$\begin{pmatrix} -3.199 & 5.015 & -1.816 \\ -1.409 & -1.807 & 3.215 \\ 1.697 & -10.70 & 9 \end{pmatrix}$			$\begin{pmatrix} -15.67 & 20.03 & -4.365 \\ 9.965 & -44.33 & 34.36 \\ 26.93 & -86.93 & 60 \end{pmatrix}$		

continued by induction to prove that all the odd powers of x are absent from the series. Thus it is appropriate to use the matrices in Table 4-6. The residuals at the N interior collocation points are

$$\sum_{i=1}^{N+1} B_{ji} c_i - \phi^2 c_j^2 = 0 \qquad j = 1, 2, \dots, N \tag{4-213}$$

while the boundary condition is

$$c_{N+1} = 1 \tag{4-214}$$

After solution the effectiveness factor, Eq. (4-145), is given by

$$\eta = \frac{\displaystyle\sum_{j=1}^{N+1} W_j c_j^2}{\displaystyle\sum_{j=1}^{N+1} W_j} \tag{4-215}$$

The first approximation is taken to give

$$-2.5 c_1 + 2.5 c_2 - \phi^2 c_1^2 = 0 \qquad c_2 = 1 \tag{4-216}$$

The solution is

$$c_1 = \frac{-2.5 + (6.25 + 10\phi^2)^{1/2}}{2\phi^2} \tag{4-217}$$

and the effectiveness factor is

$$\eta = \tfrac{1}{6} + \tfrac{5}{6}c_1^2 \tag{4-218}$$

The effectiveness factor is plotted versus the ϕ in Fig. 4-7 and gives an accurate approximation for $\phi < 2$. For large ϕ a higher approximation is required. The reason for the poor approximation at large ϕ can be deduced from the approximate profile. When put into the form

$$c = 1 + a(1 - x^2) \tag{4-219}$$

the coefficient a is determined from c_1, which applies at x_1. Here $1 - x_1^2 = \tfrac{4}{5}$. Hence

$$a = 1.25(c_1 - 1) \tag{4-220}$$

The concentration at the center of the pellet ($x = 0$) is then

$$c(0) = 1.25c_1 - 0.25 \tag{4-221}$$

and this value becomes negative for $\phi^2 \geqslant 50$. This is clearly unrealistic and higher approximations are necessary.

The asymptotic analysis gives another reason why Eq. (4-218) is not good for large ϕ. Equation (4-184) gives the exact result for large ϕ

$$\eta = \frac{(2/3)^{1/2}}{\phi} \tag{4-222}$$

This is shown in Fig. 4-7 and the one-term orthogonal collocation OC solution

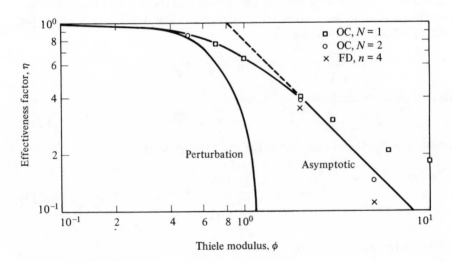

Figure 4-7 Effectiveness factor for a second-order reaction in slab.

does not approach this asymptote. Conversely the orthogonal collocation solution does approach the exact solution for small ϕ and does a considerably better job than the perturbation solution for intermediate ϕ. To improve the results we apply Eq. (4-213) with $N = 2$. The results are shown in Fig. 4-7 and provide a very good approximation for $\phi < 3$.

The diffusion–reaction problem, Eq. (4-145), is solved for a linear reaction ($n = 1$) in a slab using orthogonal collocation with different levels of approximation (N). For three different ϕ the errors in the effectiveness factors are given in Fig. 4-8. For small ϕ the concentration distribution is well approximated by a quadratic or quartic function of position (see Fig. 4-4c), and the effectiveness factors are well determined with a two-term solution (which corresponds to a quartic function of x). For the larger ϕ ($\phi = 10$) the concentration profile is much

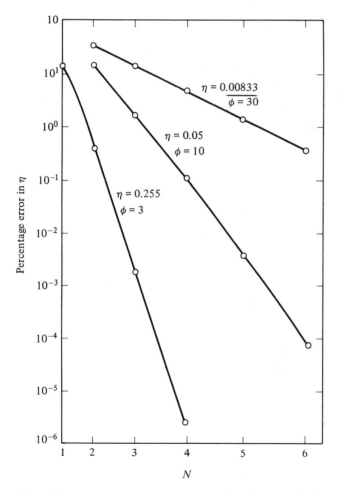

Figure 4-8 Accuracy versus N in orthogonal collocation method for linear reaction in a slab.

steeper and is very flat over much of the region. A much higher degree of polynomial is then necessary to approximate the concentration, and the errors in ϕ are consequently larger for a given N. Conversely higher N must be used to preserve the same accuracy as ϕ increases.

These examples of the orthogonal collocation method illustrate one of its advantages: a first-term solution is often quite accurate and easy to derive. It contains many of the qualitative features of the solution and is thus useful for analysis and study. Higher approximations can be obtained to improve the accuracy, and the accuracy of the orthogonal collocation method is generally higher than that of the straight collocation method as well as usually being more accurate than the Method of Weighted Residuals. It is also easy to apply and has the further advantage of expressing the equations in terms of the solution at the collocation points rather than the coefficients. It suffers from the difficulty that a high-degree polynomial (large N) is necessary if the solution has sharp gradients. In such cases other approaches are necessary.

The orthogonal collocation method is next applied to the problem of nonisothermal diffusion and reaction in a spherical catalyst pellet expressed by Eqs. (4-136), (4-137a) and (4-137b). We take the case of a first-order, irreversible reaction so that the reaction rate term is given by Eq. (4-138) with the temperature given by Eq. (4-135). We look only for solutions that are symmetric about $r = 0$ so that we can use the collocation points and matrices in Tables 4-5 and 4-6, respectively. The residuals are

$$\sum_{i=1}^{N+1} B_{ji}c_i - \phi^2 f(c_j, c_{N+1}, \beta, \delta) = 0 \tag{4-223}$$

and the boundary condition gives

$$-\sum_{i=1}^{N+1} A_{N+1,i}c_i = \text{Bi}_m(c_{N+1} - 1) \tag{4-224}$$

with

$$f(c_j, c_{N+1}, \beta, \delta) = c_j e^{\gamma - \gamma/T_j} \tag{4-225}$$

$$T_j = -\beta c_j + 1 + \beta c_{N+1} + \beta\delta(1 - c_{N+1}) \tag{4-226}$$

The effectiveness factor is

$$\eta = \frac{3\dfrac{dc}{dr}\Big|_{r=1}}{\phi^2} = \frac{3\displaystyle\sum_{i=1}^{N+1} A_{N+1,i}c_i}{\phi^2} \tag{4-227}$$

and this can be evaluated using the average reaction rate as

$$\eta = \frac{\displaystyle\sum_{i=1}^{N+1} W_i f(c_i, c_{N+1}, \beta, \delta)}{\displaystyle\sum_{i=1}^{N+1} W_i f(1, 1, \beta, \delta)} \tag{4-228}$$

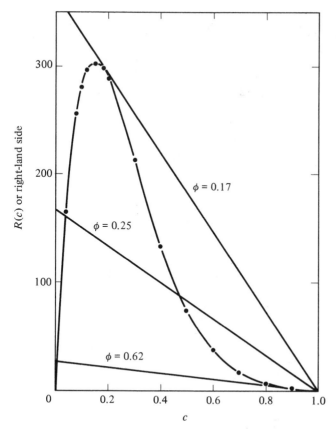

Figure 4-9 Multiple steady states in catalyst, Eqs. (4-229) and (4-230).

The set of nonlinear equations is solved using Newton–Raphson, since the successive substitution method would only converge for small ϕ. The program OCRXN does this.

As example computations, we consider the case with $\text{Bi}_m \to \infty$ so that $c(1) = 1$, $\gamma = 30$, and $\beta = 0.4$. The first approximation gives the equation

$$\frac{-10.5c_1 + 10.5}{\phi^2} = c_1 e^{\gamma - \gamma/T_1} = R(c_1) \tag{4-229}$$

$$T_1 = 1 + \beta - \beta c_1 \tag{4-230}$$

This equation can have more than one solution. This can be seen by plotting the right-hand side versus c_1 in Fig. 4-9. The left-hand side then depends on ϕ. For large particles, hence large ϕ, the two curves intersect only once, as is the case for $\phi = 1.0$. This corresponds to a diffusion-controlled situation and gives a unique steady state. For small particles, only one intersection occurs ($\phi = 0.62$), which corresponds to the case of fast diffusion and for which the concentration gradients are small. For intermediate values of ϕ, for example $\phi = 0.25$, however, the two curves intersect at more than one place. Graphically we can obtain more than one steady state solution for $0.17 < \phi < 0.62$. More accurate finite difference computations (using 100 grid points) give the values $0.22 < \phi < 0.54$, so that the collocation method gives reasonable results very easily in this case. The essential difficulty in this problem is the large changes in the reaction rate. We recall that

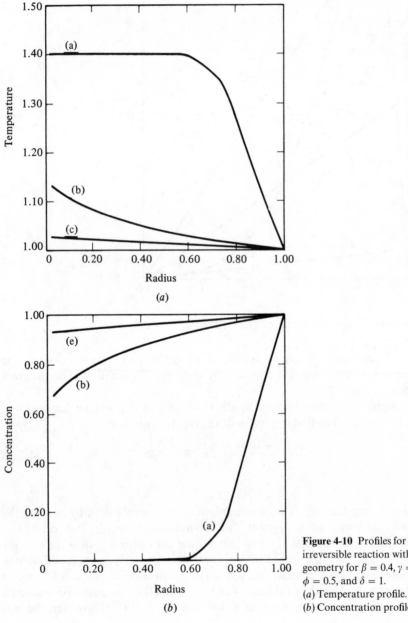

Figure 4-10 Profiles for first-order, irreversible reaction with spherical geometry for $\beta = 0.4$, $\gamma = 30$, $\phi = 0.5$, and $\delta = 1$.
(a) Temperature profile.
(b) Concentration profile.

the convergence properties of an iterative solution method are dependent on the derivative equation, in this case on $dR(c_1)/dc_1$. Figure 4-9 reveals that this quantity goes through large changes as the concentration c_1 goes from zero to one.

Higher approximations can be calculated as shown above, and the iterations will usually converge to one of the steady-state solutions. Which solution is chosen depends on the initial guess, and some experimentation is necessary to obtain them all. The lower steady-state solution (lower temperature) is usually obtained by starting with an initial guess $c(r) = 1$, while the upper solution is usually obtained by the guess $c(r) = 0$. The intermediate solution is more difficult to obtain, but the first-order approximation gives a good first guess when calculating the higher approximations. Typical temperature and concentration profiles are shown in Fig. 4-10 for $\phi = 0.5$, with $N = 10$ and $W = 1 - r^2$. The upper steady state is very flat over much of the region, necessitating a large number of terms in the series. These solutions were obtained with the program OCRXN, which is listed in the appendix. The Newton–Raphson method does not always converge, and the convergence may depend critically on the initial guess. This is an unfortunate feature of this problem. Solutions on the lower leg of the curve are readily found. The intermediate steady states ($\phi = 0.22$–0.55, $\eta = 2$–48) and the upper steady states (highest η) are much harder to obtain.

The case studied is for an extreme value of β, which is usually much smaller ($\beta < 0.1$). For boundary conditions of the first kind (i.e. $\text{Bi}_m \to \infty$) and with β this small, the solution is unique. However, if the complete boundary condition of the third kind is used as in Eq. (4-123) multiple solutions are possible. For example with $\text{Bi}_m = 250$, $\text{Bi} = 5$, $\beta = 0.02$, $\gamma = 20$, and $\delta = 50$ we get

$$K_2 = 1 + \beta\delta + \beta(1-\delta)c(1) \tag{4-231}$$

For a case with $\phi = 14.44$ one solution has $c(1) = 0.16$ and $c(0) = 0$, which gives $K_2 = 1.84$. Then

$$T(r) = K_2 - \beta c(r) \tag{4-232}$$

giving $T(1) = 1.836$ and $T(0) = 1.84$, while the external value of temperature is $T = 1$. The corresponding values of $e^{\gamma - \gamma/T}$ for $T = 1.0$, 1.836, and 1.84 are 1.0, 9005, and 9232. Thus the reaction rate is extremely large. The solution for these parameters is confined near the boundary. The concentration is essentially nonzero only between $r = 0.998$ and $r = 1$, and is zero in the inner core $0 < r < 0.998$. If the orthogonal collocation method is applied to such a case it is necessary to use at least 40 terms before a collocation point is contained within this region. In this case the concentration has a sharp gradient near the boundary, and some other method, which allows steep gradients, must be used.

Before proceeding to alternative methods let us return to the question of the choice of weighting function in the defining equation for the orthogonal collocation method, Eq. (4-87) or (4-196), since that one choice determines the collocation points and all the matrices, etc. For symmetric polynomials we provided two choices, $W = 1$ or $W = 1 - x^2$, while for polynomials in x we used $W = 1$. Other choices are possible. The author has found from experience when solving diffusion and reaction problems that the first approximation is more accurate if the matrices corresponding to $W = 1 - x^2$ are used. For higher approximations it is not too crucial, although for boundary conditions of the third kind the choice $W = 1$ seems better. This may be because the choice $W = 1$ makes the collocation points closer to the boundary ($x = 1$) and the solution is unknown there. Hence collocation points are needed there. By contrast, if the boundary condition is of the first kind the solution is known at $x = 1$, and the choice $W = 1 - x^2$ moves the collocation points away from $x = 1$ to regions where they are presumably needed more. The author's philosophy, however, is not to optimize the choice of polynomial—through the choice of $W(x)$—for a given problem, but to depend on the higher approximations to give the needed accuracy. Improvements achieved by changing polynomials are rarely significant compared to those obtained by adding more terms to the expansion. The only exception to this philosophy is for problems with symmetry, and then only for the first approximation.

When solving a problem it is usually necessary to solve it for several choices of N, the number of interior collocation points, in order to assess the accuracy. The question then arises as to the way in which the errors decrease as N is increased. Theoretical results are discussed in Sec. 4-13, but the following extrapolation technique may prove useful. We denote the answer obtained with N as S_N. An improved result is frequently given by the Shanks' formula[8]

$$S = \frac{S_N^2 - S_{N-1}S_{N+1}}{2S_N - S_{N-1} - S_{N+1}} \tag{4-233}$$

4-8 LOWER–UPPER DECOMPOSITION OF MATRICES

Three methods—finite difference, collocation finite element, and Galerkin finite element—give rise to large sets of equations, and their solution inevitably requires inverting a matrix, or at least solving a large set of equations. These equations have a special property in that a great many of the elements are zero, and indeed there is a pattern of zero and nonzero elements. Such matrices are called sparse. For example, a finite difference method with 1,000 grid points would yield a $1,000 \times 1,000$ matrix with 10^6 elements. Only about 3,000 of these are nonzero, however. The work to solve the system of equations without taking advantage of the zeros would be about 3×10^8 multiplications. Even with a fast computer such a calculation would be lengthy. For example, if one multiplication takes 10^{-6} sec then the calculation would take 300 sec or 5 minutes. If the pattern of zeros is taken into account we would be able to solve the system with about 5,000 multiplications, a reduction by a factor of 60,000. Consequently for an efficient solution we must take into account the pattern of zeros or the structure of the matrix.

The standard method of solving a linear system is to do a lower–upper (LU) decomposition on the matrix or a gaussian elimination. We illustrate the LU decomposition first on a dense $n \times n$ matrix (all elements nonzero) before considering matrices with special structure.

In the $n \times n$ linear system

$$\mathbf{Ax} = \mathbf{f} \tag{4-234}$$

$$\begin{bmatrix} a_{11} & a_{12} & a_{13}\ldots \\ a_{21} & a_{22} & a_{23}\ldots \\ a_{31} & a_{32} & a_{33}\ldots \\ \ldots\ldots\ldots\ldots\ldots \\ \ldots\ldots\ldots\ldots\ldots \\ \ldots\ldots\ldots\ldots\ldots \end{bmatrix} \begin{bmatrix} x_1 \\ x_2 \\ x_3 \\ . \\ . \\ . \end{bmatrix} = \begin{bmatrix} f_1 \\ f_2 \\ f_3 \\ . \\ . \\ . \end{bmatrix} \tag{4-235}$$

The 21 element can be made to be zero by multiplying the first row by $-a_{21}/a_{11}$ and adding it to the second row. The same operation of multiplying the first row by $-a_{31}/a_{11}$ and adding it to the third row leads to a zero in the first element of the third row. By doing this for each row we can end with a column of zeros in the first column, except for the diagonal position

$$\mathbf{A}^{(1)}\mathbf{x} \equiv \begin{bmatrix} a_{11} & a_{12} & a_{13} & \cdots \\ 0 & a_{22} - \dfrac{a_{21}}{a_{11}}a_{12} & a_{23} - \dfrac{a_{21}}{a_{11}}a_{13}\ldots \\ 0 & a_{32} - \dfrac{a_{31}}{a_{11}}a_{12} & a_{33} - \dfrac{a_{31}}{a_{11}}a_{13}\ldots \\ \ldots\ldots\ldots\ldots\ldots\ldots\ldots\ldots \\ \ldots\ldots\ldots\ldots\ldots\ldots\ldots\ldots \\ \ldots\ldots\ldots\ldots\ldots\ldots\ldots\ldots \end{bmatrix} \begin{bmatrix} x_1 \\ x_2 \\ x_3 \\ . \\ . \\ . \end{bmatrix} = \begin{bmatrix} f_1 \\ f_2 - \dfrac{a_{21}}{a_{11}}f_1 \\ f_3 - \dfrac{a_{31}}{a_{11}}f_1 \\ . \\ . \\ . \end{bmatrix} \equiv \mathbf{f}^{(1)}$$

$$\tag{4-236}$$

In the sequel we define

$$a_{ij}^{(k)} = a_{ij}^{(k-1)} - \frac{a_{i,k-1}^{(k-1)}}{a_{k-1,k-1}^{(k-1)}} a_{k-1,j}^{(k-1)} \tag{4-237}$$

$$f_i^{(k)} = f_i^{(k-1)} - \frac{a_{i,k-1}^{(k-1)}}{a_{k-1,k-1}^{(k-1)}} f_{k-1} \tag{4-238}$$

We now want to do the same thing on the second column, to make it a column of zeros below the diagonal

$$\mathbf{A}^{(2)}\mathbf{x} = \begin{bmatrix} a_{11} & a_{12} & a_{13} & \cdots \\ 0 & a_{22}^{(2)} & a_{23}^{(2)} & \cdots \\ 0 & 0 & a_{33}^{(2)} & \cdots \\ \cdots & \cdots & a_{43}^{(2)} & \cdots \\ \ldots\ldots\ldots\ldots\ldots \\ \ldots\ldots\ldots\ldots\ldots \end{bmatrix} \begin{bmatrix} x_1 \\ x_2 \\ x_3 \\ . \\ . \\ . \end{bmatrix} = \begin{bmatrix} f_1 \\ f_2^{(2)} \\ f_3^{(2)} \\ f_4^{(2)} \\ . \\ . \end{bmatrix} \tag{4-239}$$

We continue doing this in sequence until the whole lower triangle is filled with zeros. The result is upper triangular and we set $\mathbf{A}^{(n)} = \mathbf{U}$.

$$\mathbf{A}^{(n)}\mathbf{x} = \begin{bmatrix} a_{11} & a_{12} & a_{13} & \cdots & \cdots \\ 0 & a_{22}^{(2)} & a_{23}^{(2)} & \cdots & \cdots \\ 0 & 0 & a_{33}^{(3)} & a_{34}^{(3)} & \cdots \\ 0 & 0 & 0 & a_{44}^{(4)} & \cdots \\ \cdots\cdots\cdots\cdots\cdots\cdots\cdots \\ \cdots & \cdots & \cdots & \cdots & a_{nn}^{(n)} \end{bmatrix} \begin{bmatrix} x_1 \\ x_2 \\ x_3 \\ x_4 \\ \vdots \\ x_n \end{bmatrix} = \begin{bmatrix} f_1 \\ f_2^{(2)} \\ f_3^{(3)} \\ f_4^{(4)} \\ \vdots \\ f_n^{(n)} \end{bmatrix} \qquad (4\text{-}240)$$

We define $\mathbf{L}$ as the matrix with zeros in the upper triangle, one on the diagonal, and the scalar multiples we just used in the lower triangle

$$\mathbf{L} = \begin{bmatrix} 1 & 0 & 0 & \cdots & 0 \\ -\dfrac{a_{21}}{a_{11}} & 1 & 0 & \cdots & 0 \\ -\dfrac{a_{31}}{a_{11}} & -\dfrac{a_{32}^{(2)}}{a_{22}^{(2)}} & 1 & & \vdots \\ \vdots & \vdots & & \ddots & 0 \\ -\dfrac{a_{n1}}{a_{11}} & -\dfrac{a_{n2}^{(2)}}{a_{22}^{(2)}} & \cdots & -\dfrac{a_{n,n-1}^{(n-1)}}{a_{n-1,n-1}^{(n-1)}} & 1 \end{bmatrix} \qquad (2\text{-}241)$$

The unit diagonal can be understood and then $\mathbf{L}$ and $\mathbf{U}$ can be stored in the same space as $\mathbf{A}$. We see that the solution is easily obtained now, because we can simply solve for

$$x_n = \frac{f_n^{(n)}}{a_{nn}^{(n)}}$$

$$x_{n-1} = \frac{f_{n-1}^{(n-1)} - a_{n-1,n}^{(n-1)}x_n}{a_{n-1,n-1}^{(n-1)}} \qquad (2\text{-}242)$$

$$x_i = \frac{f_i^{(i)} - \displaystyle\sum_{j=i+1}^{n} a_{i,j}^{(i)}x_j}{a_{i,i}^{(i)}}$$

in reverse sequence.

It is possible to show (see Forsythe and Moler,[4] p. 28) that $\mathbf{A} = \mathbf{LU}$. Then we can represent the equations to be solved as

$$\mathbf{Ax} = \mathbf{LUx} = \mathbf{f} \qquad (2\text{-}243)$$

This represents two triangular systems that are easily solved

$$\mathbf{Ly} = \mathbf{f} \qquad \mathbf{Ux} = \mathbf{y} \qquad (2\text{-}244)$$

Once $\mathbf{L}$ and $\mathbf{U}$ have been found additional problems can be solved that have the same matrix and different right-hand sides. The triangular systems of Eq. (2-244) are easily evaluated; solving the first one is called the forward sweep and solving the second one is called the aft sweep. The combined process of a fore-and-aft sweep takes fewer multiplications than the original decomposition. The subroutines DECOMP and SOLVE in the appendix do this.

The number of multiplications and divisions needed to do one LU decomposition and m fore-and-aft sweeps for a dense matrix is

$$\text{Operation count} = \tfrac{1}{3}n^3 - \tfrac{1}{3}n + mn^2 \qquad (4\text{-}245)$$

This is fewer operations than it takes to calculate an inverse, so the decomposition is usually the method of choice. Notice that the decomposition is proportional to n^3 whereas the fore-and-aft sweep is proportional to n^2. For large n the decomposition is a significant cost. This was the reason why the integration packages, such as GEAR, described in Chapter 3 do not always re-evaluate the jacobian at each time step. Instead these packages use the old jacobian for several time steps to avoid the continual decomposition cost.

Before seeing how the LU decomposition works for matrices with a significant number of zeros, let us apply the finite difference method to the diffusion and reaction problem, Eqs. (4-135) to (4-139). The grid spacing is taken as shown in Fig. 4-2a, and the equation is written at each grid point, including $r_1 = 0$ and $r_{n+1} = 1$,

$$\frac{c_{i+1} - 2c_i + c_{i-1}}{\Delta r^2} + \frac{a-1}{r_i}\frac{c_{i+1} - c_{i-1}}{2\Delta r} = \phi^2 R(c_i) \qquad (4\text{-}246)$$

For $i = 1$ and $i = n+1$ the above equation involves c_0 and c_{n+2}, which are undefined. To define them we introduce a false boundary and apply the boundary condition. At $r = 0$, we define y_0 as the value of y at $r = -\Delta r$. Then the boundary condition, Eq. (4-137a), is

$$\frac{c_2 - c_0}{2\Delta r} = 0 \qquad (4\text{-}247)$$

and is correct to $0(\Delta r^2)$. Also at $r = 0$ the value of the second term in the differential equation is evaluated using l'Hospital's rule

$$\lim_{r \to 0} \frac{dc/dr}{r} = \frac{d^2 c/dr^2|_{r=0}}{1} \qquad (4\text{-}248)$$

Combining Eq. (4-246) for $i = 1$ with Eqs. (4-247) and (4-248) gives

$$a\frac{2(c_2 - c_1)}{\Delta r^2} = \phi^2 R(c_1) \qquad (4\text{-}249)$$

At $r = 1$ we introduce a false boundary and let c_{n+2} represent the solution at $r = 1 + \Delta r$. We apply the boundary condition of Eq. (4-137b) in a manner correct to second order to give

$$-\frac{c_{n+2} - c_n}{2\Delta r} = \text{Bi}_m(c_{n+1} - 1) \qquad (4\text{-}250)$$

If this is substituted into Eq. (4-246) for $n+1$ we get

$$\frac{2c_n - \text{Bi}_m 2\Delta r(c_{n+1} - 1) - 2c_{n+1}}{\Delta r^2} - (a-1)\text{Bi}_m(c_{n+1} - 1) = \phi^2 R(c_{n+1}) \qquad (4\text{-}251)$$

Next we collect the equations, multiply each of them by r^2, and solve with a successive substitution method. The resulting matrix problem is

$$
\begin{bmatrix}
-2a & 2a & & & & \\
1-b_2 & -2 & 1+b_2 & & \mathbf{0} & \\
& 1-b_3 & -2 & 1+b_3 & & \\
& & \ddots & \ddots & \ddots & \\
& \mathbf{0} & & 1-b_{n-1} & -2 & 1+b_{n-1} \\
& & & 1-b_n & -2 & 1+b_n \\
& & & & 2 & -2-\mathrm{Bi}_m\Delta r[2+\Delta r(a-1)]
\end{bmatrix}
\begin{bmatrix}
c_1^{s+1} \\
c_2^{s+1} \\
c_3^{s+1} \\
\vdots \\
c_{n-1}^{s+1} \\
c_n^{s+1} \\
c_{n+1}^{s+1}
\end{bmatrix}
=
\begin{bmatrix}
\phi^2\Delta r^2 R(c_1^s) \\
\phi^2\Delta r^2 R(c_2^s) \\
\phi^2\Delta r^2 R(c_3^s) \\
\vdots \\
\phi^2\Delta r^2 R(c_{n-1}^s) \\
\phi^2\Delta r^2 R(c_n^s) \\
\phi^2\Delta r^2 R(c_{n+1}^s) \\
-\mathrm{Bi}_m\Delta r[2+\Delta r(a-1)]
\end{bmatrix}
$$

(4-252)

where we have let

$$b_i = \frac{a-1}{2} \frac{\Delta r}{r_i} \tag{4-253}$$

We say the Eq. (4-252) is tridiagonal, and we want to learn how to do an LU decomposition for such a matrix.

The LU decomposition of a tridiagonal matrix is done using gaussian elimination and is sometimes called the Thomas algorithm. Using a standard form of the tridiagonal matrix as

$$\begin{bmatrix} b_1 & c_1 & & & & \\ a_2 & b_2 & c_2 & & \mathbf{0} & \\ & a_3 & b_3 & c_3 & & \\ & & & \ddots & & \\ \mathbf{0} & & a_n & b_n & c_n & \\ & & & a_{n+1} & b_{n+1} \end{bmatrix} \begin{bmatrix} x_1 \\ x_2 \\ x_3 \\ \vdots \\ x_n \\ x_{n+1} \end{bmatrix} = \begin{bmatrix} d_1 \\ d_2 \\ d_3 \\ \vdots \\ d_n \\ d_{n+1} \end{bmatrix} \tag{4-254}$$

we calculate in succession

$$c_1' = \frac{c_1}{b_1} \qquad d_1' = \frac{d_1}{b_1}$$

$$c_{k+1}' = \frac{c_{k+1}}{b_{k+1} - a_{k+1}c_k'} \qquad d_{k+1}' = \frac{d_{k+1} - a_{k+1}d_k'}{b_{k+1} - a_{k+1}c_k'} \tag{4-255}$$

$$c_{n+1} = 0 \qquad x_{n+1} = d_n' \qquad x_k = d_k' - c_k'x_{k+1}$$

If these steps are performed on the Eq. (4-254) the reader will see that we are just doing a gaussian elimination. We can rearrange the computation slightly to make it perform an LU decomposition. The important point is that there is no fill outside the tridiagonal matrix, in other words the structure of the matrix remains the same. This is an important advantage when we want to reduce the amount of work. This algorithm is contained in subroutines INVTRI and SWEEP in the appendix. The number of operation counts to solve m such systems in a tridiagonal matrix, size $n+1$, is

$$\text{Operation count} = 2n + m(3n+1) \tag{4-256}$$

which is a significant saving over $\frac{1}{3}n^3$ of Eq. (4-245).

While applying the finite difference method to the problem of diffusion and reaction in a pellet the successive substitution method might not be a very good method to solve the nonlinear equations. It is clear from the structure of Eq. (4-252) that if Δr is taken small enough, the successive substitution method converges. It is better, however, to use a more robust method, like Newton–Raphson. In that case the structure of the zero and nonzero elements in Eq. (4-252) remains the same, although the nonzero diagonal entries are different, so that the same economy results. If the reaction rate expression also depends on $c(1)$, the algorithm must be adjusted to allow operation on the last column, which has a nonzero entry for each row. This is easily done, however.

When Newton–Raphson is applied to Eq. (4-246) we get

$$R(c_i^{s+1}) \simeq R(c_i^s) + \frac{dR}{dc}\bigg|_{c_i^s} (c_i^{s+1} - c_i^s) \tag{4-257}$$

This means that the diagonal term in Eq. (2-252) is changed to

$$-2 - \phi^2 \Delta r \frac{dR}{dc}\bigg|_{c_i^s} \qquad i = 2,\ldots,n \tag{4-258}$$

while the right-hand side is changed to

$$\phi^2 \Delta r^2 \left[R(c_i^s) - \frac{dR}{dc}\bigg|_{c_i^s} c_i^s \right] \tag{4-259}$$

We then use these definitions and the tridiagonal algorithm to solve the equations. This algorithm is in program FDRXN in the appendix. Reaction problems can then be solved with the finite difference method.

Solving Eqs. (4-136), (4-137a), and (4-137b) with $\beta = 0.4$, $\gamma = 30$, and $\text{Bi}_m = \infty$ for $\phi = 1$ gives the results listed in Table 4-7. These results indicate that the error is proportional to Δx^2, which is because the derivative boundary conditions are evaluated to $0(\Delta x^2)$. Problem 4-13 illustrates that only $0(\Delta x)$ is achieved if the boundary conditions are evaluated to $0(\Delta x)$.

We can also apply extrapolation techniques to these results as we do in Sec. 3-6. Let η_0 be the exact solution for the effectiveness factor, η_1 the solution with grid size Δx, and η_2 the solution with grid size $\Delta x/2$. For a second-order scheme the error obeys

$$\eta_1 = \eta_0 + c\Delta x^2$$
$$\eta_2 = \eta_0 + c\left(\frac{\Delta x}{2}\right)^2 \tag{4-260}$$

with higher-order terms neglected. Solving this for η_0 we obtain

$$\eta_0 = \frac{4\eta_2 - \eta_1}{3} \tag{4-261}$$

This estimate of the effectiveness factor is more accurate than either η_1 or η_2. The same extrapolation can be applied to the solution, except that a more accurate solution is obtained only at the grid points of the coarsest grid.

This extrapolation is just the first step in a Richardson extrapolation. In the Richardson method we apply the extrapolation to a series of calculations. If we make n calculations then we obtain $n - 1$ new, improved values. If the first series of n results is accurate to $0(\Delta x^2)$ the next series of $n - 1$ results is correct to $0(\Delta x^4)$. These are extrapolated further until only one answer remains, which is the best result. For most problems in engineering a single extrapolation is acceptable.

The finite element methods, and finite difference methods with more than one

Table 4-7 Effectiveness factors from finite difference method

N	Δx	ϕ	η Eqs. (4-49) and (4-144)	Error
		$R = \phi^2 c$		
2	0.5	1.0	0.71429	0.04730
4	0.25	1.0	0.74638	0.01521
8	0.125	1.0	0.75734	0.00426
16	0.0625	1.0	0.76047	0.00112
32	0.03125	1.0	0.76131	0.00029
		Extrapolated results, Eq. (4-261)		
2, 4		1.0	0.75708	0.00452
4, 8		1.0	0.76099	0.00060
8, 16		1.0	0.76151	0.00008
16, 32		1.0	0.76159	0.00001
		$R = ce^{\gamma - \gamma/T},\ T = 1 + \beta - \beta c,\ \beta = 0.4,\ \gamma = 30$		
100		0.1	1.007	
100		0.2	1.031	
100		0.3	1.077	
100		0.4	1.160	
100		0.5	1.336	
100		0.55	1.57279	
200		0.55	1.57263	
400		0.55	1.57259	
100		0.22	57.93	
100		0.23	71.70	
100		0.25	80.97	
100		0.3	85.1451	
200		0.3	85.1493	
100		0.4	77.43	
100		0.5	67.83	
100		0.7	53.00	
100		1.0	39.41	
100		2.0	21.19	
100		10.0	4.375	

dependent variable, result in matrices that have other structures. Two such matrices are illustrated in Fig. 4-11a. The banded matrix has a bandwidth BW, so that a complete row has $(2BW + 1)$ nonzero entries, centered around the diagonal. The operation count to perform a decomposition and solve for the right-hand sides is

$$\text{Operation count} = BW(BW+1)n - \tfrac{2}{3}BW^3 - BW^2 - \tfrac{1}{3}BW$$
$$+ m[(2BW+1)n - BW^2 - BW)] \quad (4\text{-}262a)$$

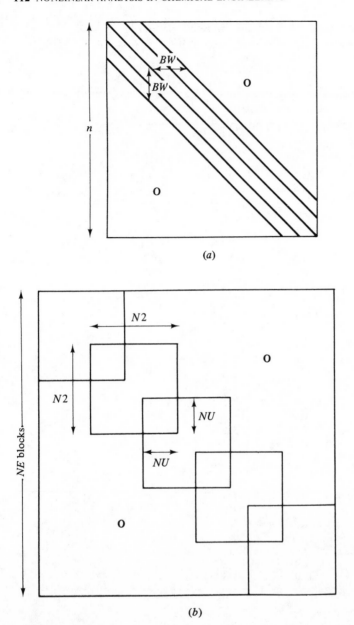

(a)

(b)

Figure 4-11 Matrix structures. (a) Banded matrix. (b) Block diagonal matrix.

The other type of matrix is a block diagonal (see Fig. 4-11b). This matrix is characterized by NE blocks, each of size $N2 \times N2$, but with a $NU \times NU$ block overlapped at the corner. The work estimate for such a matrix is

$$\text{Operation count} = \frac{NE}{3}[N2(N2^2-1)-NU(NU^2-1)]$$

$$+\tfrac{1}{3}NU(NU^2-1)+m[NE(N2^2-NU^2)+NU^2] \quad (4\text{-}262b)$$

Computer programs to perform the decomposition and/or the fore-and-aft sweep are provided in the appendix for dense matrices (DECOMP and SOLVE), the block diagonal matrix (LUD and FAS), and a tridiagonal matrix (INVTRI and SWEEP). Using $NE = 1$ in the program for block diagonal matrix also corresponds to a dense matrix. Thus we can regard as solved the problem of linear equations with a matrix that is dense or sparse with the tridiagonal or block diagonal structure.

4-9 ORTHOGONAL COLLOCATION ON FINITE ELEMENTS

Previous applications of the orthogonal collocation method have utilized a trial function that is a series of orthogonal polynomials, each of which is defined over the entire domain $0 \leqslant x \leqslant 1$. When the solution has steep gradients it is more advantageous to use trial functions that are defined only over part of the region and piece together adjacent functions to provide an approximation over the whole domain. Then small regions can be utilized near the steep gradients and the approximation improved. We are thus led to the method of orthogonal collocation on finite elements. We present two forms of the method that differ only in the trial functions. One uses lagrangian functions and adds conditions that make the first derivatives or fluxes continuous between elements, and the other form uses Hermite polynomials, which automatically have continuous first derivatives between elements.

We apply the method to the catalyst pellet problem of Eqs. (4-136), (4-137a), and (4-137b). The domain is divided into elements as shown in Fig. 4-12. Within each element we apply orthogonal collocation as we have before; the residual is evaluated at the internal collocation points. If we have NE elements and $NCOL$ internal collocation points then $NE \times NCOL$ is the total number of residual conditions. With the two boundary conditions the number of conditions falls short of that needed to define the polynomial, which is $(NCOL+1)NE+1$. We thus append $NE-1$ conditions at the element boundaries by making the first derivatives continuous there. Then the resulting solution has continuous derivatives throughout the domain. Alternatively, when there are material inhomogeneities present we can make the flux continuous across element boundaries, resulting in a solution that is continuous, with continuous flux, as in the exact solution.

The equations are written for the kth element, in which we define

$$u = \frac{x-x_{(k)}}{h_k} \qquad h_k = x_{(k+1)}-x_{(k)} \quad (4\text{-}263)$$

so that the variable u goes from zero to one in the element. Then we can use the formalism of Sec. 4-4 to provide the collocation points and matrices to represent the derivatives. We transform the equation into the u variable to obtain

$$\frac{1}{h_k^2}\frac{d^2c}{du^2} + \frac{a-1}{x_{(k)}+uh_k}\frac{1}{h_k}\frac{dc}{du} = \phi^2 f(c) \tag{4-264}$$

and at $u = 1$

$$\frac{1}{h_1}\frac{dc}{du}\bigg|_{u=0} = 0 \qquad -\frac{1}{h_{NE}}\frac{dc}{du} = \text{Bi}_m(c-1) \tag{4-265}$$

Figure 4-12 illustrates the global numbering system i and the local numbering system I in an element. We then refer to c_i as the solution at the ith point and understand it to be according to the global numbering system. We also refer to c_I and understand it to refer to the local numbering system on an element. Usually the element in question is obvious, so we do not note the element, but that is understood. It is necessary to know both I and k, the element number, to obtain the global number. These numbers are related by

$$i = (k-1)(NCOL+1)+I \tag{4-266}$$
$$x_i = x_{(k)} + u_I h_k$$

On an element we then apply orthogonal collocation to Eq. (4-264) giving

$$\frac{1}{h_k^2}\sum_{J=1}^{NP} B_{IJ}c_J + \frac{a-1}{x_i}\frac{1}{h_k}\sum_{J=1}^{NP} A_{IJ}c_J = \phi^2 f(c_I) \qquad I = 2,\ldots,NP-1 \tag{4-267}$$

The local points $I = 2,\ldots,NP-1$ designate the interior collocation points. The continuity of flux between elements requires

$$D\frac{dc}{dx}\bigg|_{x_k^-} = D\frac{dc}{dx}\bigg|_{x_k^+} \tag{4-268}$$

and this is obtained by requiring

$$\left(\frac{D_{N+2}}{h_{k-1}}\sum_{J=1}^{NP} A_{NP,J}c_J\right)_{\text{element } k-1} = \left(\frac{D_1}{h_k}\sum_{J=1}^{NP} A_{1J}c_J\right)_{\text{element } k} \tag{4-269}$$

Here we must carefully specify the element. When the diffusivity is constant across the element (i.e. no material inhomogeneities), Eq. (4-269) makes the first derivative of the solution $c(x)$ continuous in the entire region $0 \leqslant x \leqslant 1$. The two boundary conditions are applied in their respective elements. For the first element we have

$$\frac{1}{h_1}\sum_{J=1}^{NP} A_{1J}c_J = 0 \tag{4-270}$$

while the boundary condition at $x = 1$ affects the last element

$$-\frac{1}{h_{NE}}\sum_{J=1}^{NP} A_{NP,J}c_J = \text{Bi}_m(c_{NP}-1) \tag{4-271}$$

The equations are then assembled in a global way, so that the terms for c_I are put into the appropriate place corresponding to c_i. The final structure of the

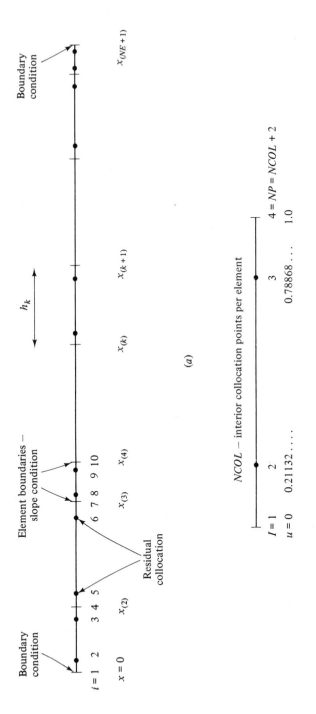

Figure 4-12 Collocation points on finite elements, lagrangian cubic polynomials. (*a*) Global numbering system *i*. (*b*) Local numbering system *I*.

115

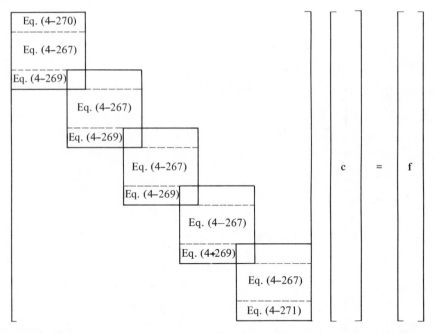

Figure 4-13 Matrix structure for orthogonal collocation of finite elements with lagrangian cubic polynomials.

matrix is shown in Fig. 4-13. The successive substitution method of solving the nonlinear equations is illustrated since it displays the matrix structure, while the Newton–Raphson method has the same structure but more complicated equations. We write the final, assembled equations in the form

$$\mathbf{AAc} = \mathbf{f} \qquad (4\text{-}272)$$

We note in passing that all the methods can be represented in the form of Eq. (4-272). Orthogonal collocation on finite elements gives the matrix **AA** with the structure illustrated in Fig. 4-13. Finite difference gives the matrix **AA** with the structure defined in Eq. (4-252), whereas orthogonal collocation gives a dense, square matrix with every element filled. The decomposition of equations of the form shown in Fig. 4-13 is done using subroutines LUD and FAS in the appendix. The operation count for such a matrix is

$$\text{Operation count} = \frac{NE}{3} NP(NP^2 - 1) + mNE(NP^2 - 1) + m \qquad NP = NCOL + 2$$

$$(4\text{-}273)$$

As a detailed illustration let us solve the problem

$$\frac{d^2 c}{dx^2} = \phi^2 c \qquad (4\text{-}274)$$

$$\frac{dc}{dx}(0) = 0 \qquad c(1) = 1 \tag{4-275}$$

using two elements $(NE = 2)$ and two internal collocation points $(NCOL = 2)$, which corresponds to a cubic polynomial on each element. The collocation points and elements are

$$h_1 = h_2 = \tfrac{1}{2} \tag{4-276}$$

$$x_1 = 0 \qquad x_2 = \tfrac{1}{2} \times 0.21132 = 0.10566 \qquad x_3 = \tfrac{1}{2} \times 0.78868 = 0.39434$$

$$x_4 = 0.5 \qquad x_5 = 0.60566 \qquad x_6 = 0.89434 \qquad x_7 = 1.0$$

and the equations are

$$\begin{bmatrix} A_{11}A_{12}A_{13} & A_{14} & & & & \\ B_{21}B_{22}B_{23} & B_{24} & & \mathbf{0} & & \\ B_{31}B_{32}B_{33} & B_{34} & & & & \\ A_{41}A_{42}A_{43} & A_{44}-A_{11} & -A_{12} & -A_{13} & -A_{14} \\ & B_{21} & B_{22} & B_{23} & B_{24} \\ \mathbf{0} & B_{31} & B_{32} & B_{33} & B_{34} \\ & 0 & 0 & 0 & 1 \end{bmatrix} \begin{bmatrix} c_1 \\ c_2 \\ c_3 \\ c_4 \\ c_5 \\ c_6 \\ c_7 \end{bmatrix} = \begin{bmatrix} 0 \\ h_1^2\phi^2 c_2 \\ h_1^2\phi^2 c_3 \\ 0 \\ h_2^2\phi^2 c_5 \\ h_2^2\phi^2 c_6 \\ 1.0 \end{bmatrix} \tag{4-277}$$

The Newton–Raphson method merely requires expanding the reaction rate term as

$$f(c_j^{s+1}) = f(c_j^s) + \left.\frac{df}{dc}\right|_{c_j^s}(c_j^{s+1} - c_j^s) \tag{4-278}$$

and putting the $(df/dc)c_j^{s+1}$ term on the left-hand side, thus affecting only the diagonal element of the matrix. Here $df/dc = 1$ and the problem is linear. Solving this for $\phi = 6$ gives $c_1 = 0.00597$, $c_2 = 0.00608$, $c_3 = 0.02863$, $c_4 = 0.05478$, $c_5 = 0.08308$, $c_6 = 0.51965$, and $c_7 = 1$. The effectiveness factor is obtained by integrating the reaction rate over the domain

$$\eta = \tfrac{1}{2}(W_2 c_2 + W_3 c_3) + \tfrac{1}{2}(W_2 c_5 + W_3 c_6) = 0.1594 \tag{4-279}$$

The exact solution gives $\eta = 0.1667$, so the answer is 4 percent off.

The same problem can be solved with orthogonal collocation. Using symmetric polynomials $N = 1$, we get the same equations as Eqs. (4-216) but for a first-order reaction rate. The result is $c_1 = 0.0649$ and $\eta = 0.221$, or a 33 percent error. If two terms are used the equations are, with matrices evaluated from Table 4-6,

$$-4.740 c_1 + 5.677 c_2 - 0.9373 = 36 c_1$$
$$8.323 c_1 - 23.26 c_2 + 14.94 = 36 c_2 \tag{4-280}$$

The solution is $c_1 = 0.0128$, $c_2 = 0.2539$, and $\eta = 0.1699$, or a 2 percent error. This solution requires much less effort than orthogonal collocation on finite elements, as it involves solving only two equations rather than six, and is more accurate.

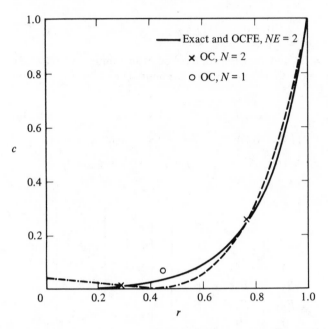

Figure 4-14 Concentration in pellet, for a first-order, irreversible reaction in slab, with $\phi = 6$.

This is because the solution does not have very steep gradients, and the necessity of finite elements is not apparent. Furthermore, the orthogonal collocation uses only even polynomials in x whereas orthogonal collocation on finite elements uses both even and odd polynomials, thus automatically doubling the number of equations without improving the accuracy.

The solution is plotted in Fig. 4-14 for these approximations. We note that the gradient of concentration is greatest near the boundary, indeed from $x_5 = 0.60566$ to $x_7 = 1.0$. Let us solve the problem again using orthogonal collocation on finite elements but using a smaller element near the boundary. We take $x_{(2)} = 0.7$, $x_1 = 0$, $x_2 = 0.14792$, $x_3 = 0.55208$, $x_4 = 0.7$, $x_5 = 0.76340$, $x_6 = 0.93660$, and $x_7 = 1.0$. Equation (4-277) is used with different h_k and a revised fourth equation, and the solution is $c_1 = 0.0093$, $c_2 = 0.0052$, $c_3 = 0.0692$, $c_4 = 0.1744$, $c_5 = 0.2437$, $c_6 = 0.6810$, and $c_7 = 1.0$. Now the effectiveness factor is

$$\eta = 0.7(W_2 c_2 + W_3 c_3) + 0.3(W_2 c_5 + W_3 c_6) = 0.1648 \qquad (4\text{-}281)$$

which is within 1 percent of the exact solution, compared with the 4 percent error for two elements of equal size. This result illustrates the advantage of finite elements: smaller elements can be concentrated in the region with steep gradients to improve the accuracy.

The use of orthogonal collocation on finite elements is particularly valuable when the gradients are even greater. Let us consider the nonisothermal problem of Eq. (4-136) with $\gamma = 20$, $\beta = 0.02$, $Bi_m = 250$, and $Bi = 5$, giving $\delta = 50$. By Eq.

(4-135) the temperature is related to the concentration by $T(r) = K_2 - \beta c(r)$. For $\phi = 14.44$ the solution is extremely steep, having the value $c(1) = 0.16$ and $c(0) = 0$, which gives $T(r) = 1.843 - 0.02c(r)$. The temperature in the pellet is thus between 1.840 and 1.843, and the reaction rate is very high. The result is that all the mass that diffuses into the pellet is reacted near the boundary, and the concentration drops to zero at $r = 0.997$. With uniform elements, in this case it takes 330 elements to have 1 element in the region in which the solution is important. Thus we need a way to locate elements efficiently. For this problem a simple two-element solution is satisfactory.

We separate the domain into two zones or elements. In the innermost zone $0 \leqslant x \leqslant b$, the components are in equilibrium and $c = c_{eq}$. In the outer reaction zone $b \leqslant x \leqslant 1$, the reaction takes place and there is a concentration gradient. We transform the domain $x = [0, 1]$ using

$$u = \frac{x - b}{1 - b}$$

so that u goes from zero to one in the reaction zone. Equation (4-136) then becomes

$$\frac{1}{(1-b)^2} \frac{d^2c}{du^2} + \frac{a-1}{b+u(1-b)} \frac{1}{(1-b)} \frac{dc}{du} = \phi^2 R(c, T) \qquad (4\text{-}282)$$

and the boundary condition of Eqs. (4-137a) and (4-137b) at $u = 1$ is

$$-\frac{1}{(1-b)} \frac{dc}{du} = \text{Bi}_m(c - 1) \qquad (4\text{-}283)$$

For continuity between zones we impose the condition at $u = 0$ that

$$c = c_{eq}, \qquad \frac{dc}{du} = 0 \qquad (4\text{-}284)$$

Orthogonal collocation is next applied to the reaction zone. The polynomials are not symmetric functions of u anymore (even though the solution is a symmetric function of x) so we use the matrices from Table 4-4. The collocation equation is applied at $u = \frac{1}{2}$ for a three-term series (quadratic polynomial)

$$\frac{1}{(1-b)^2}(B_{21}c_{eq} + B_{22}c_2 + B_{23}c_3) + \frac{a-1}{b+\frac{1}{2}(1-b)} \frac{1}{(1-b)}$$
$$\times (A_{21}c_{eq} + A_{22}c_2 + A_{23}c_3) = \phi^2 R(c_2, c_3) \qquad (4\text{-}285)$$
$$T = 1 + \beta\delta + \beta c_3(1 - \delta) - \beta c \qquad (4\text{-}286)$$

The boundary condition is

$$-\frac{1}{(1-b)}(A_{31}c_{eq} + A_{32}c_2 + A_{33}c_3) = \text{Bi}_m(c_3 - 1) \qquad (4\text{-}287)$$

Applying the conditions of Eqs. (4-284) gives the solution in the form

$$c = c_{eq} + (c_3 - c_{eq})u^2 \qquad (4\text{-}288)$$

The equations can be rearranged with $c_1 = c_{eq}$ to give

$$c_2 = \frac{c_3 + 3c_{eq}}{4} \tag{4-289}$$

$$c_3 = \frac{(1-b)\text{Bi}_m + 2c_{eq}}{2 + (1-b)\text{Bi}_m} \tag{4-290}$$

$$2\left[\frac{1}{(1-b)^2} + \frac{a-1}{1-b^2}\right](c_3 - c_{eq}) = \phi^2 R(c_2, c_3) \tag{4-291}$$

The effectiveness factor is

$$\eta = 3(1-b)\left[b^2 \int_0^1 R\,du + 2b(1-b)\int_0^1 Ru\,du + (1-b)^2 \int_0^1 Ru^2\,du\right] \quad \text{for a sphere}$$

$$\eta = 2(1-b)\left[b\int_0^1 R\,du + (1-b)\int_0^1 Ru\,du\right] \quad \text{for a cylinder} \tag{4-292}$$

$$\eta = (1-b)\int_0^1 R\,du \quad \text{for a slab}$$

and the quadrature is given by

$$\int_0^1 R\,du = \tfrac{1}{6}(R_1 + R_3) + \tfrac{2}{3}R_2 \tag{4-293}$$

We can apply these equations as follows. Given ϕ we can solve the three equations for b. Conversely, given b we can calculate directly the corresponding value of ϕ. By choosing various values of b we can obtain the solution for various ϕ values without iteration, and obtain the η versus ϕ curve. The value $b = 0$ gives a nonzero ϕ and this is the smallest b for which this method works. This case corresponds to having a zero concentration at the center. For smaller ϕ it is necessary (and both possible and efficient) to use orthogonal collocation, perhaps with $N = 1$. When $b > 0$ we are applying orthogonal collocation on finite elements using $NE = 2$ and allowing the element location to be determined.

We now solve the problem using OCFERXN for the given parameters to obtain the solution shown in Fig. 4-15. For $\phi = 14.44$ for example, the value of b is 0.99725. To apply orthogonal collocation on finite elements to this case we put elements at

$$x_{(k)} = 0, 0.5, 0.997, 0.998, 0.999, 1.0 \tag{4-294}$$

and obtain a solution with an accuracy of 0.1 percent in η. Such a value of ϕ is in the region approaching the asymptotic solution and represents a considerable extension of our capability. Additional elements can give improved accuracy.

We can also apply orthogonal collocation using trial functions that are Hermite polynomials. We restrict attention here to cubic polynomials. The trial functions on one element are shown in Fig. 4-16. Each trial function needs four parameters to define the cubic polynomial. There are also four quantities of

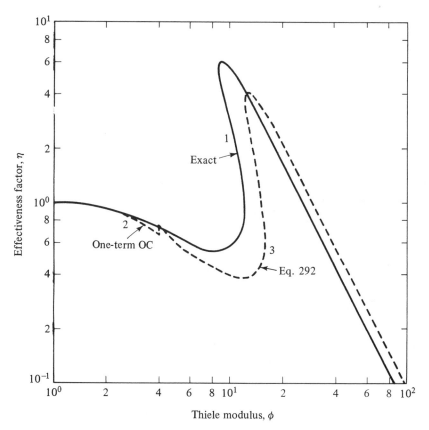

Figure 4-15 Comparison of one-term collocation with exact answer, using Eq. (4-136), with $a = 3$, $\gamma = 20$, $\beta = 0.02$, $Bi_m = 250$, $Bi = 5$, and $\delta = 50$.

interest: the value of the function and the first derivative at each end. The trial functions are defined such that three of these quantities are zero and the fourth is one. Thus we take in the kth element

$$
\begin{aligned}
H_1 &= (1-u)^2(1+2u) \\
H_2 &= u(1-u)^2 h_k \\
H_3 &= u^2(3-2u) \\
H_4 &= u^2(u-1)h_k
\end{aligned}
\tag{4-295}
$$

The representation of the function in the kth element is then

$$
c(u) = \sum_{I=1}^{4} a_I H_I(u)
\tag{4-296}
$$

and the first derivative is

$$
\frac{dc}{dx} = \frac{1}{h_k}\frac{dc}{du} = \frac{1}{h_k}\sum_{I=1}^{4} a_I \frac{dH_I}{du}
\tag{4-297}
$$

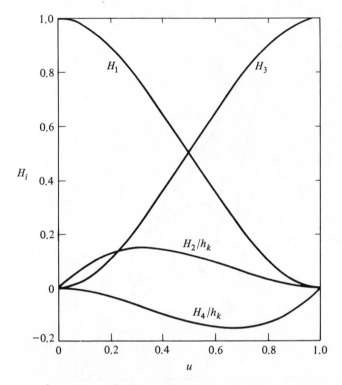

Figure 4-16 Hermite cubic polynomials.

while the second derivative is

$$\frac{d^2c}{dx^2} = \frac{1}{h_k^2}\frac{d^2c}{du^2} = \frac{1}{h_k^2}\sum_{I=1}^{4} a_I \frac{d^2H_I}{du^2} \tag{4-298}$$

At $u = 0$, $H_1 = 1$ and $H_2 = H_3 = H_4 = 0$ so that $c(0) = a_1$. Likewise, $H'_2(0) = h_k$, $H_3(1) = 1$, and $H'_4(1) = h_k$.

To apply the collocation method we need to be able to evaluate the function and its derivatives at the collocation points. These are given by

$$\left.\frac{dc}{du}\right|_{u_J} = \sum_{I=1}^{4}\left.\frac{dH_I}{du}\right|_{u_J} a_I \qquad \left.\frac{d^2c}{du^2}\right|_{u_J} = \sum_{I=1}^{4}\left.\frac{d^2H_I}{du^2}\right|_{u_J} a_I \tag{4-299}$$

which are just linear combinations of the nodal values $c(0)$, $c'(0)$, $c(1)$, and $c'(1)$. We write these equations in the form

$$c(u_J) = \sum_{I=1}^{4} H_{JI} a_I \qquad \frac{dc}{du}(u_J) = \sum_{I=1}^{4} A_{JI} a_I \qquad \frac{d^2c}{du^2}(u_J) = \sum_{I=1}^{4} B_{JI} a_I \tag{4-300}$$

where the matrices **H**, **A**, and **B** are 2×4 matrices listed in Table 4-8.

Table 4-8 Matrices for Hermite polynomials

$\mathbf{H(u_j)}$
$$\begin{pmatrix} 0.88490\ 0180 & 0.13144\ 5856h_k & 0.11509\ 9820 & -0.03522\ 0811h_k \\ 0.11509\ 9820 & 0.03522\ 0811h_k & 0.88490\ 0180 & -0.13144\ 5856h_k \end{pmatrix}$$

$\mathbf{A}$
$$\begin{pmatrix} -1 & 0.28867\ 5136h_k & 1 & -0.28867\ 5136h_k \\ -1 & -0.28867\ 5136h_k & 1 & 0.28867\ 5136h_k \end{pmatrix}$$

$\mathbf{B}$
$$\begin{pmatrix} -3.4641\ 01620 & -2.73205\ 0810h_k & 3.46410\ 1620 & -0.73205\ 0810h_k \\ 3.4641\ 01620 & 0.73205\ 0810h_k & -3.46410\ 1620 & 2.73205\ 0810h_k \end{pmatrix}$$

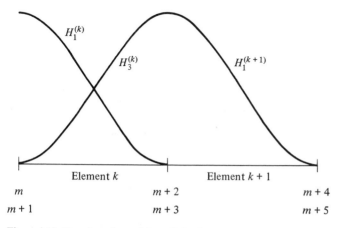

Figure 4-17 Hermite polynomials on finite elements.

$$c^{(k)}(u) = a_m H_1^{(k)}(u) + a_{m+1} H_2^{(k)} + a_{m+2} H_3^{(k)} + a_{m+3} H_4^{(k)}$$
$$c^{(k+1)}(u) = a_{m+2} H_1^{(k+1)} + a_{m+3} H_2^{(k+1)} + a_{m+4} H_3^{(k+1)} + a_{m+5} H_4^{(k+1)}$$

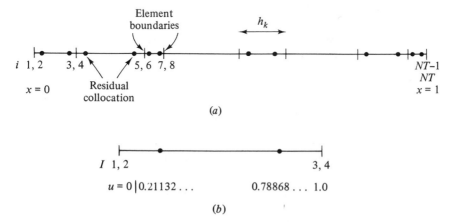

Figure 4-18 Collocation points on finite elements, Hermite cubic polynomials. (a) Global numbering system i. (b) Local numbering system I.

When we use finite elements we have as parameters the function and its derivative at each node, as illustrated in Fig. 4-17. The trial functions are combined such that the parameters at the right end of the kth element are the same as those at the left end of the $(k+1)$th element. The collocation points are still the same points, but the solution is now written in terms of the function and its derivative at the ends of each element, rather than the function at the collocation points, which are internal to the element (see Fig. 4-18). We let NE be the number of elements and then we have $NT = 2NE + 2$ parameters, with $2NE$ residual equations and two boundary conditions.

Application to the Eqs. (4-136) and (4-137) gives the residual in the kth element

$$\frac{1}{h_k^2} \sum_{I=1}^{4} B_{JI} a_I + \frac{a-1}{x_j} \frac{1}{h_k} \sum_{I=1}^{4} A_{JI} a_I = \phi^2 f \left(\sum_{I=1}^{4} H_{JI} a_I \right) \qquad (4\text{-}301)$$

We also have the boundary conditions

$$a_2 = 0 \qquad (4\text{-}302)$$

$$-a_{NT} = \text{Bi}_m (a_{NT-1} - 1) \qquad (4\text{-}303)$$

We have no need for conditions that make the first derivative continuous

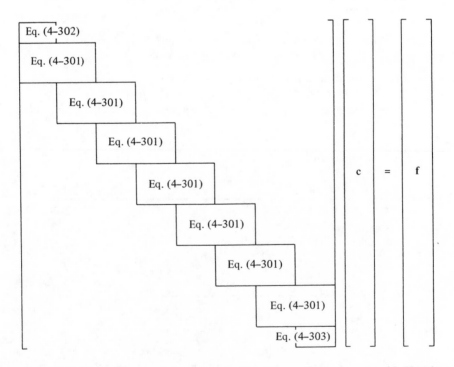

Figure 4-19 Matrix structure for orthogonal collocation on finite elements with Hermite cubic polynomials.

across the element boundaries because it is already so. Consider the point between the kth and the $(k+1)$th element. We suppose the fourth coefficient in the kth element is denoted by the global index $m+3$, as in Fig. 4-18. Then the first derivative at $u = 1$ in the kth element is

$$\frac{dc}{dx}\bigg|_{u=1} = \frac{1}{h_k} a_{m+3} \frac{dH_4}{du}\bigg|_{u=1} = a_{m+3} \tag{4-304}$$

The first derivative at $u = 0$ in the $(k+1)$th element is

$$\frac{dc}{dx}\bigg|_{u=0} = \frac{1}{h_{k+1}} a_{m+3} \frac{dH_2}{du}\bigg|_{u=0} = a_{m+3} \tag{4-305}$$

Equations (4-304) and (4-305) are the same, so that the first derivative of the solution is continuous between elements.

The equations are assembled in a global way, using the boundary condition given by Eq. (4-302). Each element then contributes two equations of the form given in Eq. (4-301). Finally we have the last boundary condition, Eq. (4-303). The equations can be written in the form given in Eq. (4-272).

The structure of the matrix **AA** is illustrated in Fig. 4-19. This can be solved using an appropriate LU decomposition. The operation count to solve an equation with m different right-hand sides using Hermite polynomials is

$$\text{Operation count} = 5NE + 4 + m(7NE + 5) \tag{4-306}$$

We next apply this method to the special case of a linear reaction rate and use two elements of equal size $h = \frac{1}{2}$. The equations are then

$$a_2 = 0$$
$$B_{11}a_1 + B_{13}a_3 + B_{14}a_4 = h_1^2\phi^2(H_{11}a_1 + H_{13}a_3 + H_{14}a_4)$$
$$B_{21}a_1 + B_{23}a_3 + B_{24}a_4 = h_1^2\phi^2(H_{21}a_1 + H_{23}a_3 + H_{24}a_4)$$
$$B_{11}a_3 + B_{12}a_4 + B_{13}a_5 + B_{14}a_6 = h_2^2\phi^2(H_{11}a_3 + H_{12}a_4 + H_{13}a_5 + H_{14}a_5) \tag{4-307}$$
$$B_{21}a_3 + B_{22}a_4 + B_{23}a_5 + B_{24}a_6 = h_2^2\phi^2(H_{21}a_3 + H_{22}a_4 + H_{23}a_5 + H_{24}a_6)$$
$$a_5 = 1.0$$

These equations can be solved most easily if the right-hand side is moved to the left. For the case $\phi = 6$, we get the solution

$$\begin{array}{lll} a_1 = 0.0060 & a_2 = 0 & a_3 = 0.0548 \\ a_4 = 0.3125 & a_5 = 1.0 & a_6 = 5.737 \end{array} \tag{4-308}$$

This solution is the same as that derived in Eq. (4-279). This can be seen by using the lagrangian solution to evaluate the derivatives at the element end points or the Hermite solution to evaluate the solution at the interior collocation points.

The advantage of using Hermite rather than lagrangian polynomials is that the former do not require a subsidiary condition to make the first derivative continuous. This reduces the number of equations by $NE - 1$, or roughly by one-third, for cubic trial functions. The Hermite polynomials may require pivoting if the diagonal element of the matrix is zero during the decomposition, but most LU decomposition routines do the pivoting automatically within a certain structure. (If the pivot can be any element in the matrix then the sparse nature of the matrix is changed.) We note that the two approaches give identical results, however, since on each element the polynomial is a cubic function of the independent variable, the two boundary conditions are satisfied by both solutions, the residuals are evaluated at the same points, and the first derivatives are continuous between the elements. These conditions are sufficient to require the polynomials to be the same, so the only preference for one formulation over another is for convenience or economy. The lagrangian formulation is definitely preferred when the flux is continuous between elements but the first derivative is not. This can arise when the diffusivity, for example, is discontinuous because two different materials are joined together between two elements. In this case the exact solution does not have continuous derivatives, and it makes no sense to use a Hermite polynomial which does.

4-10 GALERKIN FINITE ELEMENTS

The Galerkin finite element method is similar to orthogonal collocation on finite elements except that the Galerkin method is used instead of collocation. It is possible to use the same trial functions but it is more common to use lower-order functions, which are either linear or quadratic in position. We consider first the linear basis functions.

First we break the domain $0 \leqslant x \leqslant 1$ into elements, as shown in Fig. 4-20. The ith element has size h_i. We define the basic functions as N_i. As shown in Fig. 4-20a the value of N_i is one at the ith point x_i, is zero at the $(i-1)$th and $(i+1)$th points, and is a linear function in between. In the rest of the domain the function is identically zero. In two-dimensional applications, it is important to develop the equations on an element-by-element basis, and we do that here in one dimension, too. We thus define a local coordinate system in each element using the transformed coordinates. Thus in ith element

$$u = \frac{x - x_i}{h_i} \tag{4-309}$$

Now as u goes from zero to one the corresponding x goes from x_i to x_{i+1}. Within each element we define the linear basis functions

$$N_I = \begin{cases} N_1(u) = 1 - u \\ N_2(u) = u \end{cases} \tag{4-310}$$

which are shown in Fig. 4-20. We must always be able to relate the local numbering system I and the global numbering system i. For example N_i in the global numbering system is N_2 in the $(i-1)$th element but N_1 in the ith element. It is necessary to keep in mind that the function N_i is defined with a global index i,

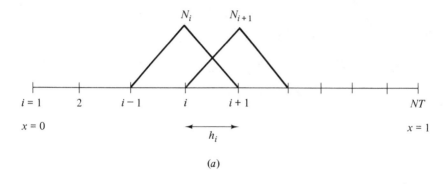

(a)

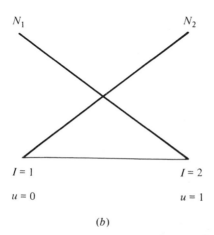

(b)

Figure 4-20 Galerkin finite element—linear trial functions. (*a*) Global numbering system *i*. (*b*) Local numbering system *I*.

while the function N_I refers to the local coordinate I. The numbering for three elements is illustrated in Fig. 4-21.

With this understanding we write the full trial functions as

$$c(x) = \sum_{i=1}^{NT} c_i N_i(x) \tag{4-311}$$

where each $N_i(x)$ is defined only on the appropriate elements, in particular the $(i-1)$th and ith elements, by Eq. (4-310). In the Galerkin method we form the residual by substituting the trial function into the differential equations. For illustration we use Eqs. (4-136), (4-137a), and (4-137b), replacing r by x. The residual is then multiplied by each trial function, in this case N_j, to obtain the Galerkin equations. We do this for the two boundary conditions as well, giving

$$\int_0^1 N_j \frac{d}{dx}\left(x^{a-1} \frac{d}{dx}(\Sigma c_i N_i) \right) dx = \phi^2 \int_0^1 R(\Sigma c_i N_i) N_j x^{a-1} dx \tag{4-312}$$

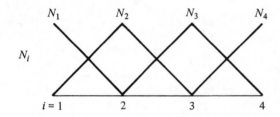

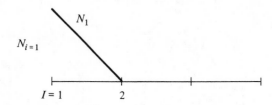

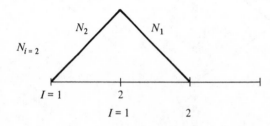

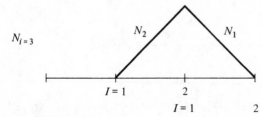

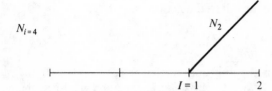

Figure 4-21 Local and global trial functions.

$$\left[N_j \sum_{i=1}^{NT} c_i \frac{dN_i}{dx} \right]_{x=0} = 0 \qquad (4\text{-}313)$$

$$\left[N_j \sum_{i=1}^{NT} c_i \frac{dN_i}{dx} \right]_{x=1} + N_j \mathrm{Bi}_m \left[\sum_{i=1}^{NT} c_i N_i(x=1) - 1 \right] = 0 \qquad (4\text{-}314)$$

These equations are combined by integrating Eq. (4-312) by parts to give

$$\int_0^1 N_j \frac{d}{dx}\left[x^{a-1} \frac{d}{dx}(\Sigma c_i N_i) \right] dx$$

$$= \left[N_j x^{a-1} \frac{d}{dx}(\Sigma c_i N_i) \right]_0^1 - \int_0^1 \frac{dN_j}{dx} \frac{d}{dx}(\Sigma c_i N_i) x^{a-1} dx \quad (4\text{-}315)$$

The weighted boundary residuals are inserted to obtain the final Galerkin equation

$$-\sum_{i=1}^{NT} \int_0^1 \frac{dN_j}{dx} \frac{dN_i}{dx} x^{a-1} dx\, c_i - [N_j \mathrm{Bi}_m(\Sigma c_i N_i - 1)]_{x=1}$$

$$= \phi^2 \int_0^1 N_j R(\Sigma c_i N_i) x^{a-1} dx \quad (4\text{-}316)$$

We now examine the evaluation of the first integral in Eq. (4-316). The weighting function N_j is nonzero only in the elements $j-1$ and j, and is identically zero elsewhere. Thus we only need to integrate over these elements. In addition the function N_i is only nonzero in the $(i-1)$th and ith elements, and we only need to integrate over them. Consequently, we can break the integration up into integrations over elements. We use in the kth element Eq. (4-309) and

$$\frac{dN_j}{dx} = \frac{1}{h_k} \frac{dN_j}{du} \qquad dx = h_k\, du \quad (4\text{-}317)$$

Then we have

$$\sum_{i=1}^{NT} \int_0^1 \frac{dN_j}{dx} \frac{dN_i}{dx} x^{a-1} dx\, c_i = \sum_{k=1}^{NE} \sum_{I=1}^{2} \frac{1}{h_k} \int_0^1 \frac{dN_J}{du} \frac{dN_I}{du} x^{a-1} du\, c_I^e \quad (4\text{-}318)$$

In each element there are at the most two trial functions that are nonzero. In the local coordinate system these are $I = 1$ and $I = 2$, hence the summation over $I = 1, 2$. The integral must be evaluated over every element, however, hence the summation over k. The local integral is zero except in two elements. Henceforth we note this summation over all elements by Σ_e. We then can write the Galerkin equation, Eq. (4-316), in terms of local coordinates

$$-\sum_e \frac{1}{h_e} \sum_{I=1}^{2} \int_0^1 \frac{dN_J}{du} \frac{dN_I}{du} x^{a-1} du\, c_I^e - \sum_e \left\{ N_J \mathrm{Bi}_m \left[\sum_{I=1}^{2} c_I^e N_I(u) \right] - 1 \right\}$$

$$= \phi^2 \sum_e h_e \int_0^1 N_J R(\Sigma c_I^e N_I) x^{a-1} du \quad (4\text{-}319)$$

The element integrals are defined as

$$B_{JI}^e = -\frac{1}{h_e} \int_0^1 \frac{dN_J}{du} \frac{dN_I}{du} x^{a-1} du \quad (4\text{-}320)$$

$$F_J^e = \phi^2 h_e \int_0^1 N_J(u) R(\Sigma c_I^e N_I) x^{a-1} du \quad (4\text{-}321)$$

and

$$x = x_e + uh_e \tag{4-322}$$

Boundary terms may be added if needed (they are only nonzero for elements on the boundary)

$$B_{JI}^e = -\mathrm{Bi}_m N_J(x = 1)N_I(x = 1) \tag{4-323}$$

and

$$F_J^e = -\mathrm{Bi}_m N_J(x = 1) \tag{4-324}$$

We can then write Eq. (4-319) as

$$\sum_e B_{JI}^e c_I^e = \sum_e F_J^e \tag{4-325}$$

where we sum over all elements. The equations can be constructed within each element using the local numbering system, and then added together to obtain the final matrix. Note in particular that for planar geometry, in which $a = 1$, the local matrix of Eq. (4-320) is the same in each element except for the scale factor h_e. Thus we do not have to calculate many integrals. We can use 100 elements but still only calculate 4 integrals ($I, J = 1$, and 2). This is the reason for using the local numbering system and local coordinates. The local matrices for certain terms are listed in Table 4-9.

To calculate the term F_J^e we need the concentration as a function of position. Within the eth element the concentration is given by

$$c^e(u) = c_I^e(1-u) + c_{I+1}^e u \tag{4-326}$$

Usually Eq. (4-321) cannot be integrated analytically so we use numerical quadrature. The same formula, quadrature points, and weights are used as in Eq. (4-104) to give

$$\int_0^1 N_J(u)R[\Sigma c_I^e N_I(u)]x^{a-1}\,du = \sum_{k=1}^{NG} W_k N_J(u_k)R[\Sigma c_I^e N_I(u_k)]x_k^{a-1} \tag{4-327}$$

The quadrature points and weights are given in Table 4-3; usually $NG = 2$ or 3 points are sufficient. Two quadrature points will integrate exactly a term that is a cubic polynomial in u whereas three points give exact integration for a quintic polynomial in u. If the reaction rate expression is not a polynomial in c, the integration may not be exact for any number of quadrature points. This introduces another source of error in the approximation, and some experimentation may be necessary to make the error resulting from inaccurate quadratures less than that resulting from an inaccurate approximation due to too few elements.

The equations resulting from Eq. (4-316) or (4-325) are tridiagonal. The LU decomposition for tridiagonal matrices can thus be used to solve the equations. The operation count to solve m systems is $2NE + m(3NE + 1)$. If the reaction rate is nonlinear then Newton–Raphson or successive substitution can be used to solve the nonlinear algebraic equations. Newton–Raphson uses one LU decomposition and one fore-and-aft sweep for each iteration, while the successive substitution

Table 4-9 Galerkin element matrices

Linear shape functions

$$N_1 = 1 - u \qquad \frac{dN_1}{du} = -1$$

$$N_2 = u \qquad \frac{dN_2}{du} = 1$$

$$\int_0^1 \frac{dN_J}{du}\frac{dN_I}{du}\,du = \begin{pmatrix} 1 & -1 \\ -1 & 1 \end{pmatrix}$$

$$\int_0^1 N_J \frac{dN_I}{du}\,du = \begin{pmatrix} -\frac{1}{2} & \frac{1}{2} \\ -\frac{1}{2} & \frac{1}{2} \end{pmatrix} \qquad \int_0^1 N_J N_I\,du = \begin{pmatrix} \frac{1}{3} & \frac{1}{6} \\ \frac{1}{6} & \frac{1}{3} \end{pmatrix}$$

$$\int_0^1 N_J\,du = \begin{pmatrix} \frac{1}{2} \\ \frac{1}{2} \end{pmatrix} \qquad \int_0^1 N_J u\,du = \begin{pmatrix} \frac{1}{6} \\ \frac{1}{3} \end{pmatrix}$$

Quadratic shape functions

$$N_1 = 2(u-1)\left(u-\tfrac{1}{2}\right) \qquad \frac{dN_1}{du} = 4u - 3$$

$$N_2 = 4u(1-u) \qquad \frac{dN_2}{du} = 4 - 8u$$

$$N_3 = 2u\left(u-\tfrac{1}{2}\right) \qquad \frac{dN_3}{du} = 4u - 1$$

$$\int_0^1 \frac{dN_J}{du}\frac{dN_I}{du}\,du = \begin{pmatrix} \frac{7}{3} & -\frac{8}{3} & \frac{1}{3} \\ -\frac{8}{3} & \frac{16}{3} & -\frac{8}{3} \\ \frac{1}{3} & -\frac{8}{3} & \frac{7}{3} \end{pmatrix}$$

$$\int_0^1 N_J \frac{dN_I}{du}\,du = \begin{pmatrix} -\frac{1}{2} & \frac{2}{3} & -\frac{1}{6} \\ -\frac{2}{3} & 0 & \frac{2}{3} \\ \frac{1}{6} & -\frac{2}{3} & \frac{1}{2} \end{pmatrix} \qquad \int_0^1 N_J N_I\,du = \begin{pmatrix} \frac{2}{15} & \frac{1}{15} & -\frac{1}{30} \\ \frac{1}{15} & \frac{8}{15} & \frac{1}{15} \\ -\frac{1}{30} & \frac{1}{15} & \frac{2}{15} \end{pmatrix}$$

$$\int_0^1 N_J\,du = \begin{pmatrix} \frac{1}{6} \\ \frac{2}{3} \\ \frac{1}{6} \end{pmatrix} \qquad \int_0^1 N_J u\,du = \begin{pmatrix} 0 \\ \frac{2}{3} \\ \frac{1}{3} \end{pmatrix}$$

method uses one LU decomposition per problem and one fore-and-aft sweep per iteration. We can see that the matrix values for the first and second derivatives are identical to those obtained for the finite difference method, but the Galerkin method gives different results for the reaction rate term. In the finite difference method this term is just evaluated at a grid point whereas in the Galerkin method it is integrated over the element.

We next solve a simple problem, Eq. (4-145), using four elements $h_i = \frac{1}{4}$ with the linear shape functions. For simplicity we do only the linear reaction ($n = 1$) in plane geometry ($a = 1$). The overall problem can be represented in the form of Eq. (4-272). Let us begin with that matrix, filled with zeros, for the case with five values c_i

$$
\begin{bmatrix}
0 & 0 & 0 & 0 & 0 \\
0 & 0 & 0 & 0 & 0 \\
0 & 0 & 0 & 0 & 0 \\
0 & 0 & 0 & 0 & 0 \\
0 & 0 & 0 & 0 & 0
\end{bmatrix}
\begin{bmatrix}
c_1 \\ c_2 \\ c_3 \\ c_4 \\ c_5
\end{bmatrix}
=
\begin{bmatrix}
0 \\ 0 \\ 0 \\ 0 \\ 0
\end{bmatrix}
\tag{4-328}
$$

The element matrices are Eqs. (4-320) and (4-321) with the appropriate terms from Table 4-9

$$
B^e_{JI} = -4 \begin{pmatrix} 1 & -1 \\ -1 & 1 \end{pmatrix} = \begin{pmatrix} -4 & 4 \\ 4 & -4 \end{pmatrix}
\tag{4-329}
$$

$$
F^e_J = \phi^2 \frac{1}{4} \begin{pmatrix} \dfrac{c^e_1}{3} + \dfrac{c^e_2}{6} \\[2mm] \dfrac{c^e_1}{6} + \dfrac{c^e_2}{3} \end{pmatrix} = \frac{\phi^2}{24} \begin{pmatrix} 2c^e_1 + c^e_2 \\ c^e_1 + 2c^e_2 \end{pmatrix}
$$

We assemble the first element. The local number system is one and two, as is the global numbering system. Thus the $\mathbf{B}$ and $\mathbf{F}$ terms are placed in the locations corresponding to the global indices one and two

$$
\begin{bmatrix}
-4 & 4 & 0 & 0 & 0 \\
4 & -4 & 0 & 0 & 0 \\
0 & 0 & 0 & 0 & 0 \\
0 & 0 & 0 & 0 & 0 \\
0 & 0 & 0 & 0 & 0
\end{bmatrix}
\begin{bmatrix}
c_1 \\ c_2 \\ c_3 \\ c_4 \\ c_5
\end{bmatrix}
= \frac{\phi^2}{24}
\begin{bmatrix}
2c_1 + c_2 \\ c_1 + 2c_2 \\ 0 \\ 0 \\ 0
\end{bmatrix}
\tag{4-330}
$$

Next we assemble the second element. The local matrices are the same, see Eq. (4-329), but the global numbering is two and three. Thus the local matrices are placed into the appropriate locations of the global matrix at the two–three position

$$
\begin{bmatrix}
-4 & 4 & 0 & 0 & 0 \\
4 & -8 & 4 & 0 & 0 \\
0 & 4 & -4 & 0 & 0 \\
0 & 0 & 0 & 0 & 0 \\
0 & 0 & 0 & 0 & 0
\end{bmatrix}
\begin{bmatrix}
c_1 \\ c_2 \\ c_3 \\ c_4 \\ c_5
\end{bmatrix}
= \frac{\phi^2}{24}
\begin{bmatrix}
2c_1 + c_2 \\ c_1 + 4c_2 + c_3 \\ c_2 + 2c_3 \\ 0 \\ 0
\end{bmatrix}
\tag{4-331}
$$

We continue in this fashion for all the elements to get the result

$$
\begin{bmatrix}
-4 & 4 & 0 & 0 & 0 \\
4 & -8 & 4 & 0 & 0 \\
0 & 4 & -8 & 4 & 0 \\
0 & 0 & 4 & -8 & 4 \\
0 & 0 & 0 & 4 & -4
\end{bmatrix}
\begin{bmatrix}
c_1 \\ c_2 \\ c_3 \\ c_4 \\ c_5
\end{bmatrix}
= \frac{\phi^2}{24}
\begin{bmatrix}
2c_1 + c_2 \\ c_1 + 4c_2 + c_3 \\ c_2 + 4c_3 + c_4 \\ c_3 + 4c_4 + c_5 \\ c_4 + 2c_5
\end{bmatrix}
\tag{4-332}
$$

The last equation for c_5 is not going to be used because c_5 is determined by a boundary condition rather than the residual. We can replace the last equation by the boundary condition to obtain the final matrix problem

$$\begin{bmatrix} -4 & 4 & 0 & 0 & 0 \\ 4 & -8 & 4 & 0 & 0 \\ 0 & 4 & -8 & 4 & 0 \\ 0 & 0 & 4 & -8 & 4 \\ 0 & 0 & 0 & 0 & 1 \end{bmatrix} \begin{bmatrix} c_1 \\ c_2 \\ c_3 \\ c_4 \\ c_5 \end{bmatrix} = \frac{\phi^2}{24} \begin{bmatrix} 2c_1 + c_2 \\ c_1 + 4c_2 + c_3 \\ c_2 + 4c_3 + c_4 \\ c_3 + 4c_4 + c_5 \\ \dfrac{24}{\phi^2} \end{bmatrix} \qquad (4\text{-}333)$$

Since this is a linear problem we can solve the system in one iteration by moving the reaction rate terms to the left-hand side. For $\phi = 6$ we get the solution $c_1 = 0.00233$, $c_2 = 0.00651$, $c_3 = 0.03414$, $c_4 = 0.18467$, and $c_5 = 1.0$. The

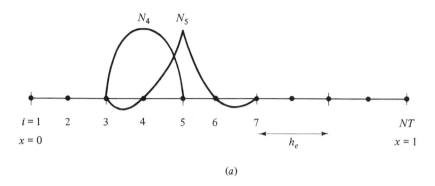

(a)

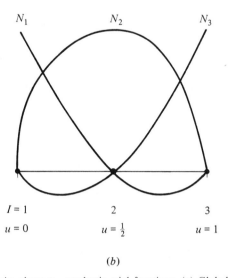

(b)

Figure 4-22 Galerkin finite element—quadratic trial functions. (a) Global numbering system i. (b) Local numbering system I.

effectiveness factor is calculated using the integrated rate expression

$$\int_0^1 c\,dx = \sum_e h_e \int_0^1 c^e(u)\,du = \sum_e h_e \sum_{I=1}^2 c_I^e \int_0^1 N_I(u)\,du \qquad (4\text{-}334)$$

The value for η is 0.1816 compared to the exact answer of 0.1667.

We can also use quadratic polynomials on each element. The trial functions are shown in Fig. 4-22. Within each element the trial functions are

$$N_I = \begin{cases} N_1 = 2(u-1)(u-\tfrac{1}{2}) \\ N_2 = 4u(1-u) \\ N_3 = 2u(u-\tfrac{1}{2}) \end{cases} \qquad (4\text{-}335)$$

The concentration is represented by the series

$$c^e(u) = c_{I-1}^e N_{I-1}(u) + c_I^e N_I(u) + c_{I+1}^e N_{I+1}(u) \qquad (4\text{-}336)$$

within the eth element. Equations (4-316) and (4-319) apply so that they are not repeated. The element matrices, Eqs. (4-320) to (4-324), are now different, coming from Table 4-9.

The equations resulting from quadratic polynomials have the structure shown in Fig. 4-23, and the LU decomposition is the same as for orthogonal collocation on finite elements with $NCOL = 1$. When there are NE elements with quadratic polynomials, the number of operations to solve m such systems is

$$\text{Operation count} = 8NE + m(8NE + 1) \qquad (4\text{-}337)$$

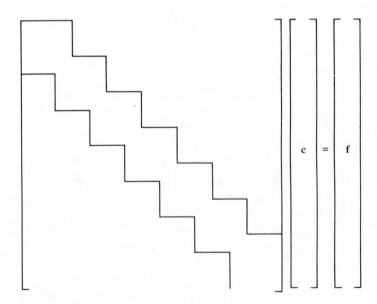

Figure 4-23 Matrix structure for Galerkin finite element with quadratic trial functions.

We next solve Eq. (4-145), for a first-order reaction ($n = 1$) and planar geometry ($a = 1$), with the quadratic shape functions. We use only two elements, which gives five unknowns, as in the case with four elements and linear shape functions. The element matrices are

$$
B^e_{JI} = -2 \begin{bmatrix} \frac{7}{3} & -\frac{8}{3} & \frac{1}{3} \\ -\frac{8}{3} & \frac{16}{3} & -\frac{8}{3} \\ \frac{1}{3} & -\frac{8}{3} & \frac{7}{3} \end{bmatrix}
$$

$$
F^e_J = \frac{\phi^2}{30} \begin{bmatrix} 2c^e_1 + c^e_2 - \frac{1}{2}c^e \\ c^e_1 + 8c^e_2 + c^e_3 \\ -\frac{1}{2}c^e + c^e_2 + 2c^e_3 \end{bmatrix}
$$

(4-338)

The first element is assembled into Eq. (4-328) to give

$$
\begin{bmatrix} -\frac{14}{3} + \frac{16}{3} & -\frac{2}{3} & 0 & 0 \\ +\frac{16}{3} & -\frac{32}{3} & +\frac{16}{3} & 0 & 0 \\ -\frac{2}{3} & +\frac{16}{3} & -\frac{14}{3} & 0 & 0 \\ 0 & 0 & 0 & 0 & 0 \\ 0 & 0 & 0 & 0 & 0 \end{bmatrix} \begin{bmatrix} c_1 \\ c_2 \\ c_3 \\ c_4 \\ c_5 \end{bmatrix} = \frac{\phi^2}{30} \begin{bmatrix} 2c_1 + c_2 - \frac{1}{2}c_3 \\ c_1 + 8c_2 + c_3 \\ -\frac{1}{2}c_1 + c_2 + 2c_3 \\ 0 \\ 0 \end{bmatrix}
$$

(4-339)

Assembling the second element gives

$$
\begin{bmatrix} -\frac{14}{3} + \frac{16}{3} & -\frac{2}{3} & 0 & 0 \\ +\frac{16}{3} & -\frac{32}{3} & +\frac{16}{3} & 0 & 0 \\ -\frac{2}{3} & +\frac{16}{3} & -\frac{28}{3} & \frac{16}{3} & -\frac{2}{3} \\ 0 & 0 & \frac{16}{3} & -\frac{32}{3} & \frac{16}{3} \\ 0 & 0 & -\frac{2}{3} & \frac{16}{3} & -\frac{14}{3} \end{bmatrix} \begin{bmatrix} c_1 \\ c_2 \\ c_3 \\ c_4 \\ c_5 \end{bmatrix} = \frac{\phi^2}{30} \begin{bmatrix} 2c_1 + c_2 - \frac{1}{2}c_3 \\ c_1 + 8c_2 + c_3 \\ -\frac{1}{2}c_1 + c_2 + 4c_3 + c_4 - \frac{1}{2}c_5 \\ c_3 + 8c_4 + c_5 \\ -\frac{1}{2}c_3 + c_4 + 2c_5 \end{bmatrix}
$$

(4-340)

Again the last equation is replaced by $c_5 = 1$. The resulting system of equations is solved for $\phi = 6$ to give the solution $c_1 = 0.00784$, $c_2 = 0.01442$, $c_3 = 0.06286$, $c_4 = 0.21677$, and $c_5 = 1.0$. The effectiveness factor is given by Eq. (4-334) with $I = 1$–3; $\eta = 0.1715$. We note the improved accuracy obtained from quadratic as opposed to linear shape functions.

The Galerkin method has one advantage over the collocation method in certain cases: the boundary conditions that involve derivatives need not be satisfied by the trial function, and boundary terms can be included in the formulation. Of course this means that the approximate solution does not satisfy the boundary conditions exactly, whereas the collocation method results in the satisfaction of the boundary conditions. If the highest derivative in the differential equation is second-order, then any boundary conditions involving first-order derivatives can be treated in this way and are called natural boundary conditions. Boundary conditions involving only the function must be satisfied explicitly, which is why we have always used $c_5 = 1$ but have not needed to apply any special condition for $dc/dx = 0$ at $x = 0$.

In addition to the ease of incorporating the natural boundary conditions the Galerkin method is advantageous for cases in which a variational principle exists.[3]

For most nonlinear problems this is not true, and then the Galerkin method has no particular advantage over the collocation method. Both methods use polynomials on an element, but the Galerkin finite element methods generally use low-order basis functions. The collocation on finite element methods generally use polynomials that are cubic functions, or higher, of the independent variable. These differences can have important implications (see Sec. 4-13). If both Galerkin and collocation on finite element methods use the same trial functions, the collocation method is preferred because of its simplicity and because fewer calculations are needed to derive the equations. These considerations are discussed more quantitatively in Sec. 4-13.

There is an alternative for the treatment of nonlinear terms, such as the reaction rate term, or terms involving a diffusivity or thermal conductivity that depends on the solution. Instead of evaluating the reaction rate term at the gaussian quadrature points and integrating, it is possible to evaluate the reaction rate term at the nodes, and then interpolate using the shape functions. For example, we have used

$$\text{Rate} = R(c^e(u)) = R\left[\sum_I c_I^e N_I(u)\right] \tag{4-341}$$

where $c^e(u)$ is given by Eq. (4-326) or (4-335).

Alternatively, we could calculate the rate at the nodes

$$R_I^e = \begin{cases} R_1^e = R(c^e(0)) \\ R_2^e = R(c^e(\tfrac{1}{2})) \\ R_3^e = R(c^e(1)) \end{cases} \tag{4-342}$$

The reaction rate term is then expressed as

$$\text{Rate} = \sum_I R_I^e N_I(u) \tag{4-343}$$

Then the integrals are easily evaluated explicitly, or exactly, since they are polynomial expressions in the shape functions. If the reaction rate is highly nonlinear and cannot be well represented by linear or quadrative functions on an element, this approach gives additional errors. It is useful, however, in one problem treated in Sec. 5-9.

4-11 INITIAL-VALUE TECHNIQUES

Another approach for solving boundary-value problems is to convert them to parabolic partial differential equations that are integrated to steady state. A complete treatment of the methods for solving the parabolic partial differential equation is given in Chapter 5, but here the ideas are introduced.

Consider Eq. (4-136). We write this equation with a time derivative on left-hand side

$$\frac{\partial c}{\partial t} = \frac{1}{r^2}\frac{\partial}{\partial r}\left(r^2\frac{\partial c}{\partial r}\right) - \phi^2 R(c) \tag{4-344}$$

We begin the calculation with an initial guess of the solution

$$c(r, 0) = c_0(r) \tag{4-345}$$

and integrate Eq. (4-344) until steady state is reached.
    Let us write at the ith point

$$c(r_i, t) = c_i(t) \tag{4-346}$$

and replace Eq. (4-344) by

$$\frac{dc_i}{dt} = AA_{ij}C_j - \phi^2 R(c_i) \tag{4-347}$$

where the matrix **AA** depends on the spatial discretization used (finite difference, collocation, etc.). We now have a set of ordinary differential equations that can be integrated by the methods discussed in Chapter 3. We integrate to steady state (i.e. $t \to \infty$) thus we want to use an implicit method so that large time steps can be introduced. The method should also have a variable time-step feature so that small time steps can be used initially when the calculations might be unstable for too large a Δt. Large time steps can be used as the steady state is approached and the solution becomes stationary in time.
    This method of integration has a parallel to the successive substitution method of solving algebraic equations if an Euler explicit scheme is used. Let us take the representative equation

$$f(x) = 0 \tag{4-348}$$

Successive substitution would use the iterative scheme

$$x^{s+1} = x^s + \beta f(x^s) \tag{4-349}$$

and for β small enough the method would converge. If we change Eq. (4-348) to an initial-value problem

$$\frac{\partial x}{\partial t} = f(x) \tag{4-350}$$

and apply the Euler method with $x^n = x(t_n)$ we get

$$\frac{x^{n+1} - x^n}{\Delta t} = f(x^n) \tag{4-351}$$

This is clearly the same as Eq. (4-349). We know that the Euler method is not a suitable method of integration if Eq. (4-350) is stiff, and this suggests that the successive substitution method is not suitable either. The backward Euler method, an implicit method, is satisfactory for stiff problems, and would give in place of Eq. (4-351)

$$\frac{x^{n+1} - x^n}{\Delta t} = f(x^{n+1}) \tag{4-352}$$

The right-hand side is expanded in a Taylor series

$$\frac{x^{n+1} - x^n}{\Delta t} = f(x^n) + \frac{\partial f}{\partial x}(x^{n+1} - x^n) \tag{4-353}$$

and rearrangement gives

$$\left(1 - \Delta t \frac{\partial f}{\partial x}\right)(x^{n+1} - x^n) = \Delta t f(x^n) \tag{4-354}$$

For large step size Δt the method is Newton–Raphson, but for small Δt it is a successive substitution method. This suggests a combined strategy for solving sets of nonlinear algebraic equations. Of course any method of integration can be used to solve Eq. (4-347).

There is an alternative method of applying initial-value techniques to solve boundary-value problems. The only reason we cannot apply initial-value techniques to integrate in x is that two or more boundary conditions are applied at different positions. If we knew all the conditions at one position x we could integrate with x as a time-like variable. The next method uses this approach.

We suppose the two boundary conditions are that the function takes specified values at $x = 0$ and $x = 1$. We do not know a priori the value of the first derivative at $x = 0$, although once we have the exact solution that value is known. Let us guess the value of $y'(0)$ and use the known value of $y(0)$. Then we have two conditions at the same point, and these are sufficient to solve a second-order equation by integrating forward from $x = 0$. We integrate until $x = 1$ and check the value of $y(1)$. If it is correct we made a good guess of $y'(0)$; if not we must make another guess and try again.

For linear problems we proceed as follows. Suppose the problem is

$$Ly = g(x) \tag{4-355}$$

$$y(0) = a \qquad y(1) = b \tag{4-356}$$

where Ly is an arbitrary second-order differential operator. The forcing function $g(x)$ and the boundary values a and b are all specified. Consider then three problems:

problem I—solution $y_1(x)$

$$Ly = g(x) \qquad y(0) = a \qquad y'(0) = 0 \tag{4-357}$$

problem II—solution $y_2(x)$

$$Ly = 0 \qquad y(0) = 0 \qquad y'(0) = 1 \tag{4-358}$$

problem III—solution $y_3(x)$

$$Ly = 0 \qquad y(0) = 1 \qquad y'(0) = 0 \tag{4-359}$$

Each of these problems is an initial-value one, and we can apply the methods of Chapter 3 to solve them numerically. We construct the full solution as

$$y(x) = y_1(x) + c_1 y_2(x) + c_2 y_3(x) \tag{4-360}$$

This function satisfies the differential equation for all choices of c_1 and c_2. It satisfies the boundary conditions if we require

$$a = a + c_1 y_2(0) + c_2 y_3(0)$$
$$b = y_1(1) + c_1 y_2(1) + c_2 y_3(1)$$

(4-361)

or

$$c_2 = 0 \qquad c_1 = \frac{b - y_1(1)}{y_2(1)}$$

Thus the solution to the two-point boundary-value problem is to solve two initial-value problems (three in the general case) to find $y_1(x)$ and $y_2(x)$. This proceduré works unless c_1 becomes very large, in which case round-off errors are important, or unless y_1 and $c_1 y_2$ are both large but with opposite signs, in which case the solution is poorly determined due to round-off errors. Unfortunately, we cannot predict either occurrence while we are choosing a method. This approach has the advantage that if the solution has a steep gradient, in the initial-value method, a variable step size is used with a small step size at that region and a large step size elsewhere. If a variable step is used, however, the various solutions y_1, y_2, and y_3 may not be known at the same points, so that construction of the complete solution must use interpolated values.

The same type of initial-value method can be used for nonlinear problems in an iterative fashion. Keller[5] gave a good treatment and called this approach the shooting method. Consider the second-order problem

$$y'' = f(x, y, y')$$
$$a_0 y(0) - a_1 y'(0) = \alpha \qquad \text{for } a_i \geqslant 0$$
$$b_0 y(1) + b_1 y'(1) = \beta \qquad \text{for } b_i \geqslant 0$$

(4-362)

We convert this to

$$u'' = f(x, u, u')$$
$$u(0) = a_1 s - c_1 \alpha$$
$$u'(0) = a_0 s - c_0 \alpha$$

(4-363)

where we choose the c_0 and c_1 such that

$$a_1 c_0 - a_0 c_1 = 1$$

(4-364)

We next convert the second-order initial-value problem of Eq. (4-363) to two first-order problems

$$u' = v$$
$$v' = f(x, u, v)$$
$$u(0) = a_1 s - c_1 \alpha$$
$$v(0) = a_0 s - c_0 \alpha$$

(4-365)

and define the quantity

$$\chi(s) = b_0 u(1, s) + b_1 u'(1, s) - \beta = b_0 u(1, s) + b_1 v(1, s) - \beta$$

(4-366)

that we would like to make zero.

$$\chi(s) = 0 \tag{4-367}$$

We then employ both a successive substitution and a Newton method to do this. In the successive substitution iteration we replace Eq. (4-367) by

$$s = s - m\chi(s) \qquad m \neq 0 \tag{4-368}$$

Keller[5] showed that if

$$\frac{\partial f}{\partial y} \leqslant N \tag{4-369}$$

for some N and $0 < m < 2\Gamma$, where Γ increases as N increases, then the iteration scheme

$$s^{k+1} = s^k - m\chi(s^k) \tag{4-370}$$

converges as $k \to \infty$. The procedure is then to choose an s, solve the initial-value problems of Eqs. (4-365), check the function given by Eq. (4-367), and iterate with Eq. (4-370).

For Newton's method of iteration we replace Eq. (4-367) by the Newton formula

$$s^{k+1} = s^k - \frac{\chi(s^k)}{\dot{\chi}(s^k)} \tag{4-371}$$

The function $\dot{\chi} = d\chi/ds$ is determined as the solution to a subsidiary problem. We let

$$\zeta \equiv \frac{\partial u(x,s)}{\partial s} \qquad \eta \equiv \frac{\partial v(x,s)}{\partial s} \tag{4-372}$$

$$\zeta' = \eta$$

$$\eta' = \frac{\partial f}{\partial v}\eta + \frac{\partial f}{\partial u}\zeta$$

$$\zeta(0) = a_1 \qquad \eta(0) = a_0 \tag{4-373}$$

$$\dot{\chi}(s) = b_0\zeta(1,s) + b_1\eta(1,s)$$

With shooting methods we can "shoot" in either direction, and we can use any of the methods for solving initial-value problems. Applied to Eq. (4-136) by Weisz and Hicks,[9] the method proved very powerful because it could be used even when ϕ was large and the concentration was small (say 10^{-30}) at the center. The integration uses the boundary condition $c'(0) = 0$ and guesses the value of $c(0)$. If we want to find the entire curve of η versus ϕ, rather than solve just for one specified ϕ, the method has even more advantages because the nonlinear problem can be solved without iteration.

Next we change the problem

$$\frac{1}{r^{a-1}}\frac{d}{dr}\left(r^{a-1}\frac{dc}{dr}\right) = \phi^2 R(c) \tag{4-374}$$

$$-\frac{dc}{dr}(0) = 0 \qquad c(1) = 1$$

into an initial-value one

$$\frac{1}{z^{a-1}} \frac{d}{dz}\left(z^{a-1} \frac{dc}{dz}\right) = d^2 R(c) \tag{4-375}$$

$$-\frac{dc}{dz}(0) = 0 \qquad c(bz = 1) = 1$$

by the choice $r = bz$ and $d = b\phi$. For any d we choose an arbitrary $c(0)$ and integrate this last problem until the concentration reaches one. Suppose this happens at $z = z_1$. Now we let

$$b = \frac{1}{z_1} \qquad \phi = \frac{d}{b} = dz_1 \tag{4-376}$$

and we have the exact solution without iteration to Eq. (4-374) for the case $\phi = dz_1$. We don't know the value of ϕ ahead of time, of course, so that this method is suitable when we want to traverse the entire curve of η versus ϕ, which corresponds to making successive choices of $c(0)$. For each choice we get a solution without iteration. We also note that the method is not possible, without iteration, when the boundary condition is of the third kind as in Eqs. (4-137a) and (4-137b), because then the reaction rate expression depends on $c(1)$, which is not known in the initial-value technique. The shooting method would be applicable for the special case of $\delta = 1$.

For illustration we apply the shooting method to Eq. (4-374). The ordinary differential equations are

$$u' = v$$

$$v' = \phi^2 R(u) + \frac{a-1}{x} v$$

$$\zeta' = \eta \tag{4-377}$$

$$\eta' = \phi^2 \frac{dR}{du} \zeta + \frac{a-1}{x} \eta$$

where the concentration is u and its first derivative with respect to position is v. The variables ζ and η are defined in Eq. (4-372). We must solve these equations with the boundary conditions

$$u(0) = s \qquad v(0) = 0 \qquad \zeta(0) = 1 \qquad \eta(0) = 0 \tag{4-378}$$

and the functions χ and $\dot{\chi}$ are given by

$$\chi(s) = u(1) - 1 \qquad \dot{\chi}(s) = \zeta(1, s) \tag{4-379}$$

These equations can be easily integrated using the initial-value techniques. (See program IVRXN in the appendix.) The iteration scheme of Eq. (4-371) works well for simple reactions, such as $R = c$ or $R = c^2$, but is not robust for large ϕ. The same is true for the first-order, irreversible reaction with $\beta = 0.4$ and $\phi = 30$. Intermediate steady states in Fig. 4-5 are easily determined with an appropriate guess of s, but the iteration scheme may oscillate for large ϕ. When the scheme

Table 4-10 Results from shooting technique

s	ϕ	η	Exact
		$R = c$	
0.9950	0.1	0.9967	0.9967
0.9566	0.3	0.9710	0.9710
0.6481	1.0	0.7616	0.7616
0.0993	3.0	0.3317	0.3317
		$R = c^2$	
0.9950	0.1	0.9934	
0.9581	0.3	0.9449	
0.7123	1.0	0.6525	
	3.0	Failed	
	$R = ce^{\gamma - \gamma/T}, T = 1 + \beta - \beta c, \beta = 0.4, \gamma = 30$		
0.9983	0.10	1.007	
0.9930	0.2	1.032	
0.9829	0.3	1.077	
0.9654	0.4	1.160	
0.9307	0.5	1.336	
0.7661	0.55	2.368	
0.6498	0.5	3.265	
0.4135	0.4	5.445	
0.1442	0.3	10.84	
0.0499	0.25	20.46	
0.02177	0.23	30.84	
0.6583–4	0.25	80.98	
0.1097–5	0.3	85.167	
Guess 10^{-7}	0.7	Oscillated	

works the answers are quite accurate and the calculation is fast. Typical results are given in Table 4-10.

4-12 QUASILINEARIZATION

The quasilinearization method can be illustrated simply as a way of solving two-point boundary-value problems. Suppose the equation is

$$\frac{d^2 y}{dx^2} = f(y, x) \tag{4-380}$$

We expand the nonlinear function as

$$f(y, x) = f(y^s, x) + \left.\frac{\partial f}{\partial y}\right|_{y^s} (y - y^s) + \dots \tag{4-381}$$

and rewrite Eq. (4-380) as

$$\frac{d^2 y^{s+1}}{dx^2} = f(y^s, x) + \left.\frac{\partial f}{\partial y}\right|_{y^s} (y^{s+1} - y^s) \tag{4-382}$$

We use the original boundary conditions. This is linear in y^{s+1} and we can solve this using several methods. If we use shooting methods, we call the method quasilinearization. If we use finite differences, we get the same results as if finite difference is applied to Eq. (4-380) and the Newton–Raphson method is used to solve the resulting nonlinear algebraic equations. If we use orthogonal collocation, the method is the same as if orthogonal collocation is applied to Eq. (4-380) and Newton–Raphson is used to solve the nonlinear algebraic equations. One new result appears from quasilinearization: the iterations converge for all $y(x)$ if

$$\frac{\partial^2 f}{\partial y^2} \geq 0 \tag{4-383}$$

or if the reverse inequality holds everywhere. Further details are available in the book by Lee (1966). This result does provide a convergence theorem for finite difference, collocation, or finite element methods; the Newton–Raphson method converges provided Eq. (4-383) is satisfied.

4-13 COMPARISON

The previous sections introduce a variety of methods for solving two-point boundary-value problems. While the advantages of each method are mentioned as it is introduced it is instructive to discuss all the methods together now that the details of solution are understood. We do that in the context of three problems: heat conduction in a slab, diffusion and reaction in a catalyst pellet, and viscoelastic fluid flow in a pipe.

Many of the numerical methods—orthogonal collocation OC, finite difference FD, orthogonal collocation on finite elements OCFE, and Galerkin finite elements method GFEM—differ in the method of approximating the solution and the principle generating the governing equations. To compare these methods we need to summarize the known information about errors, storage requirements, and work required to set up and solve the problems.

An important consideration is how the error decreases as more points or unknowns are added to the approximation. Information for the various methods is given in Table 4-11. The error estimates give the principle term in the expression

$$\|c - c_{ex}\| = K \Delta x^m \tag{4-384}$$

The mean square error is defined as

$$\|c - c_{ex}\| = a \int_0^1 (c - c_{ex})^2 x^{a-1} \, dx \tag{4-385}$$

Table 4-11 Comparison of numerical methods

Method and basis	Error	Storage of matrix	LU decomposition	Right-hand side	Both
					Operation count
OC, NCOL interior points	$\left(\dfrac{1}{NCOL}\right)^{dNCOL}$*	$(NCOL+1)^2$	$\frac{1}{3}(NCOL+1)^3$	$(NCOL+1)^2$	$\frac{1}{3}(NCOL+1)^3+(NCOL+1)^2$
OCFE–L, NE elements, order NP	Δx^{NP}†	$NE\cdot NP^2$	$\frac{1}{3}NP^3NE$	$NE\cdot NP^2$	$NE\cdot NP^2\left(1+\dfrac{NP}{3}\right)$
OCFE–H, degree $NP-1$	Δx^{NP}†	$8NE$	$5NE$	$7NE$	$12NE$
GFEM–2, NE elements	Δx^3	$8NE$	$8NE$	$8NE$	$16NE$
GFEM–1, NE elements	Δx^2	$3NE$	$2NE$	$3NE$	$5NE$
FD, $n=\dfrac{1}{\Delta x}$	Δx^2	$3n$	$2n$	$3n$	$5n$

* The term d in the error formula depends on the solution being approximated.
† The error at the knots goes as $\Delta x^{2(NP-2)}$.

The constants are different for each method. For each method the error estimate is actually

$$\text{Error} < K\Delta x^{\min(k,m)} \tag{4-386}$$

where the power k depends on properties of the exact solution. If the exact solution is highly continuous (it has many derivatives which are bounded), k is large and the error bounds in Table 4-11 apply. If the exact solution is not highly continuous (perhaps first derivatives are continuous, but second derivatives are not), the power k overrules the error bounds listed in Table 4-11. We speak of the rate of convergence of the errors as the number of terms or elements increases. The rate of convergence is fixed by the power of Δx in Eq. (4-386). For a case with low k (i.e. the solution does not have many continuous derivatives) the rate of convergence of all methods is essentially the same. Then the preferred method is determined by the work requirements, as discussed below. For a case with large k the rate of convergence of each method is different: finite difference methods converge as Δx^2, finite element methods with cubic polynomials converge as Δx^4, and so forth.

To compare the methods we assume that an orthogonal collocation solution with three internal collocation points gives the same accuracy as a ten-term finite difference solution. The finite element methods are then scaled to have the number of elements that give the same error using Table 4-11. Results are listed in Table 4-12 under low accuracy. The number of elements needed for each method decreases as the degree of the polynomial increases. This is because each element of a high-order method has more parameters than each element of a low-order method. Most of the finite element methods need about ten terms under this assumption, with high-order orthogonal collocation on finite elements being the exception needing only seven.

Now suppose we wish to improve the accuracy. If we increase the number of interior collocation points in orthogonal collocation from 3 to 5, the error decreases by a factor of $5^5/3^3 = 116$. The number of elements for each of the other methods is then increased to improve the error 116 times, giving the number of

Table 4-12 Number of terms required for similar accuracy

Method		Low accuracy			High accuracy	
		NT	Operation count*		NT	Operation count*
OC	$NCOL = 3$	4	37	$NCOL = 4$	5	67
OCFE–H, NP = 4	$NE = 3$	7	36	$NE = 10$	21	120
OCFE–L, NP = 4	$NE = 3$	10	112	$NE = 10$	31	373
GFEM–2	$NE = 5$	11	80	$NE = 22$	45	160
GFEM–1	$NE = 10$	11	55	$NE = 108$	109	545
FD	$NE = 10$	11	55	$NE = 108$	109	545

* Number of multiplications to perform one LU decomposition and one fore-and-aft sweep.

elements listed in Table 4-12. Over 100 grid points are needed for finite difference or Galerkin finite element with linear polynomials. Twenty-two elements suffice for Galerkin finite element with quadratic polynomials, and only ten elements are needed for a cubic collocation method on finite elements. The fewest number of terms is needed for the high-order method, using cubic polynomials on finite elements. Of course, even fewer terms are required for global orthogonal collocation, using one high-degree polynomial over the entire domain. This example reinforces the point that high-order methods converge rapidly and require fewer terms for a given error than do low-order methods, such as finite difference or finite element methods with linear polynomials.

The next question to ask is how much work is required to solve the problem? Except for very difficult problems, or those solved thousands of times, the work is not a significant factor, given present-day computer speeds and costs. It is worthwhile to discuss the work requirements, though, because they are significant for parabolic differential equations, which are treated in Chapter 5, and those work requirements are directly related to ordinary differential equations. Likewise the storage requirements are modest for one-dimensional problems.

First we consider the work needed to solve one iteration once the matrix is available. The work is computed by counting the number of multiplications and divisions, since these operations are usually the slower ones on a computer compared with addition and subtraction. The operation counts are given in Table 4-11, as obtained from Eqs. (4-257), (4-273), and (4-306). We notice that the work requirements increase with element size in the reverse order as the error. For a given number of elements or grid points the low-order methods (finite difference, Galerkin with linear polynomials) require fewer operations, but many more elements may be required by the low-order methods. For equivalent accuracy the operation counts are listed in Table 4-12. We see that high-order methods require fewer multiplications and divisions, because the number of elements needed is very much less than the number needed by low-order methods. The rapid rate of convergence of the high-order methods overshadows the slow increase in work as the number of elements is increased.

Finally the work needed to set up the matrix can be evaluated by the work required to determine the terms in the matrix. Here we assume that the information such as listed in Tables 4-4, 4-6, and 4-8 is available, since the calculation of those results is only a small part of the total program and is only done once. We count then only the operations needed to obtain the remaining terms as listed in Table 4-13. For different problems we can combine these results to obtain an estimate of the total work.

Consider the typical diffusion–reaction problem with the following choices. A collocation solution with three internal collocation points is often equivalent in accuracy to a finite difference solution with 10 intervals or 11 nodes. Take $NP = 4$ in orthogonal collocation and $n = 10$ in the finite difference method. Then for the other methods we choose the parameters such that the number of unknowns is about ten. For orthogonal collocation on finite elements we use cubic polynomials and three elements. For Galerkin finite elements with linear polynomials we use

Table 4-13 Operation counts to formulate problem

Method	Evaluate at all needed points					Evaluate equations‡‡			
	NT	$c*$	$R(c)$*†	(c')‡	(c'')‡	$\dfrac{d}{dx}$	$\dfrac{d^2}{dx^2}$	$\dfrac{d}{dx}\left(D\dfrac{d}{dx}\right)$**	$R(c)$††
OC	NP	0	$m_1(NP-2)$	NP^2	NP^2	0	0	$NP(NP-2+m_2)$	$m_1(NP-2)$
OCFE-H	$2NE+2$	$2NP\cdot NE$	$2m_1NE$	$2NP\cdot NE$	$NP(3NE+1)$	0	0	$2NE(NP+4+m_3+m_2)$	$2m_1NE$
OCFE-L	$(NP-1)NE+1$	0	$m_1NE(NP-2)$	$NP\cdot NT$	$NP\cdot NT$	0	0	$2NP\cdot NE(NP-2)+m_2NT$	$m_1NE(NP-2)$
GFEM-2	$2NE+1$	$3NQ\cdot NE$ and $R(c)$	$m_1NQ\cdot NE$	$3NT$	$3NT$	0	0	$NE\cdot NQ(3+m_2)$	$NE\cdot NQ(m_1+3)$
GFEM-2	Interpolate $D(c)$ and $R(c)$							$NE(2m_2+9)+m_2$	$NE(2m_1+9)+m_1$
GFEM-1	$NE+1$	$2NQ\cdot NE$ and $R(c)$	$m_1NQ\cdot NE$	NT	—	0	0	$NE\cdot NQ(2+m_2)$	$NE\cdot NQ(m_1+2)$
GFEM-1	Interpolate $D(c)$ and $R(c)$							$NE(m_2+4)+m_2$	$NE(m_1+4)+m_1$
FD	$n+1$	0	$m_1(n-1)$	$n+1$	$2(n+1)$	0	0	$n(m_2+3)+m_2-3$	$m_1(n-1)$

* At collocation points or Gauss' points.
† Takes m_1 operations to find R given c.
‡ At grid points or collocation points.
** Takes m_2 operations to find D given c, m_3 to find dD/dc given c. Evaluate at collocation points, grid points, or with nodal weighting function.
†† To find the term involving $R(c)$ in Eq. (4-321), assuming c is known but including the evaluation of R.
‡‡ All operations independent of the solution c are performed once and the results are stored. These operations are not included.

Table 4-14 Simplification of Table 4-13. Operation counts for $m_1 = 15, m_2 = 4, m_3 = 4$

Method	Evaluate at needed points					Evaluate equations			Solve one iteration*	
	NT	c	R(c)	c'	c''	$\dfrac{d}{dx}\left(D\dfrac{d}{dx}\right)$	R(c)	R(c) & c	Set up problem	Solve one iteration
OC, NP = 4	4	0	30	16	16	24	30	30	36	90
OCFE-H, cubic, NE = 3	8	24	90	24	40	96	90	114	45	255
OCFE-L, cubic, NE = 3	10	0	90	40	40	88	90	90	106	284
GFEM-2, NE = 5, NQ = 3	11	45	225	33	33	105	270	315	81	501
GFEM-2, NE = 5, NQ = 3		Interpolate D(c) and R(c)				89	210	210	81	380
GFEM-1, NE = 10	11	60	450	11	—	180	510	570	51	801
GFEM-1, NE = 10		Interpolate D(c) and R(c)		11		84	205	205	51	340
FD, n = 10	11	0	135	11	22	71	135	135	51	256

* Set up problem; solve one iteration.

ten elements while for quadratic polynomials we use five elements. In the Galerkin method three quadrature points per element are used to evaluate the integrals. It is assumed that the reaction rate can be evaluated with 15 multiplications once c is known, where the number 15 includes several multiplications for evaluating the exponential. We also assume that the diffusivity, which depends on c, is evaluated in four multiplications, given c, and dD/dc requires four more multiplications. Table 4-13 then simplifies to Table 4-14.

We see using the above example that the orthogonal collocation method is by far and away the best method. The finite difference method and high-order orthogonal collocation on finite element method are equivalent, and the Galerkin methods are not competitive due to the expensive quadratures for nonlinear problems. High-order orthogonal collocation on finite elements is better than the equivalent low-order method except for solving the equations. These general conclusions, based on many assumptions, seem to be confirmed by detailed computations for specific problems.

The first example problem is heat conduction in a slab with a temperature-dependent thermal conductivity

$$\frac{d}{dx}\left[(1+a\theta)\frac{d\theta}{dx}\right] = 0 \qquad (4\text{-}387)$$

$$\theta(0) = 0 \qquad \theta(1) = 1$$

This problem has been solved by hand calculation for several methods. The author found that the Method of Weighted Residuals was easy to apply in the first approximation and gave reasonable results. The accuracy was unknown, however, so that higher approximations were needed to deduce the accuracy. These were complicated to set up and laborious to solve by hand. The finite difference method was easier to set up but also laborious to solve for several terms, which unfortunately were needed. The perturbation method was easy to apply in the first approximation, but the results were limited to low values of a. Even the second approximation gave inaccurate results for $a = 1$. Higher approximations were laborious to find. It was found that the orthogonal collocation method gave very accurate results, that a few terms sufficed, it was easy to set up, and it was easy to solve, since only a few terms were needed. Indeed, using $N = 2$ gave only a 2×2 matrix to invert at each iteration and the fluxes agreed to within 0.3 percent of each other and were within 0.7 percent of the exact solution. By contrast, straight collocation with uniform distribution of collocation points and two terms (i.e. the same degree polynomial) only gave an accuracy of 5 percent. The finite difference method with two interior nodes (i.e. the same degrees of freedom and the same work to solve) gave the fluxes at the two sides within 5 percent of each other and the average flux was within 5 percent of the exact solution. A summary of results given in Table 4-15 suggests that the orthogonal collocation method is a very powerful one.

There are problems, however, for which the orthogonal collocation method is not suitable, and this prompts introduction of the finite element methods. This is

Table 4-15 Approximate solution to heat conduction Eq. (4-387)

x	Moments		Finite difference		Orthogonal collocation		Exact
	$N = 1$	$N = 2$	$n = 2$	$n = 3$	$N = 1$	$N = 2$	
0.1	0.130	0.143	0.116	0.124	0.129	0.142	0.140
0.25	0.313	0.332	0.291	0.311	0.309	0.331	0.323
0.5	0.583	0.594	0.581	0.573	0.579	0.593	0.581
0.75	0.813	0.809	0.791	0.799	0.809	0.809	0.803
0.9	0.930	0.925	0.916	0.920	0.929	0.925	0.924
$\theta'(0)$	1.333	1.500	1.325	1.387	1.317	1.488	1.500
$2\theta'(1)$	1.333	1.500	1.351	1.459	1.367	1.493	1.500
Average flux	1.333	1.500	1.338	1.423	1.339	1.490	1.500

illustrated by reference to the problem of diffusion and reaction in a pellet with a first-order, irreversible reaction

$$\frac{d^2c}{dx^2} - \phi^2 c = 0 \tag{4-388}$$

$$\frac{dc}{dx}(0) = 0 \qquad c(1) = 1$$

Results for the effectiveness factor as a function of the number of terms are given in Fig. 4-8 for various values of ϕ. When ϕ is large (say 30), the solution is very steep (see Fig. 4-4c) and many terms ($N = 6$) are necessary to achieve a 1 percent accuracy. When ϕ is small (say 3), the solution is well approximated by a quadratic polynomial and only two terms are needed for 1 percent accuracy, whereas six terms give accuracy greater than 10^{-11} percent. Thus we are led to the conclusion that the orthogonal collocation method is most suitable for problems with smooth solutions and becomes less suitable if the solution has steep gradients. We then need finite elements or finite differences.

One of the advantages of the orthogonal collocation method is that if the solution is symmetric (a function of x^2 and not of x alone) this fact can be built into the trial function thereby reducing by a factor of two the number of unknowns. In the other methods this cannot be done, except in the first element at the center. An alternative is to transform the differential equation using

$$u = x^2 \tag{4-389}$$

Doing this gives

$$\frac{1}{x^{a-1}} \frac{d}{dx}\left(x^{a-1}\frac{dc}{dx}\right) = 4u\frac{d^2c}{du^2} + 2a\frac{dc}{du} \tag{4-390}$$

If the problem is solved on $0 \leqslant u \leqslant 1$, rather than $0 \leqslant x \leqslant 1$, the solution is automatically a symmetric function of x^2. The differential equation is now singular, since the coefficient of the highest derivative is zero at $u = 0$. This type of transformation has not been performed for the solutions given below.

The next problem is for a nonisothermal reaction and provides a useful comparison of the different numerical methods. Let us consider

$$\frac{1}{r^2}\frac{d}{dr}\left(r^2\frac{dc}{dr}\right) = \phi^2 R(c) = \phi^2 c\, e^{\gamma - \gamma/T}$$

$$T = 1 + \beta - \beta c \qquad (4\text{-}391)$$

$$\frac{dc}{dr}(0) = 0 \qquad c(1) = 1$$

$$\beta = 0.3 \qquad \gamma = 18 \qquad \phi = 0.5$$

We first examine the nonlinearity to see how severe it is. The needed quantity is

$$\frac{dR}{dc} = \phi^2 e^{\gamma - \gamma/T}\left(1 - \frac{\beta\gamma c}{T^2}\right) \qquad (4\text{-}392)$$

and this ranges from -2.8 to 16 as c ranges from zero to one. Clearly this is not a large variation. We know that the solution also depends on ϕ. For small ϕ the solution is a smoothly varying function, such as the one represented for $\phi = 0.1$ in

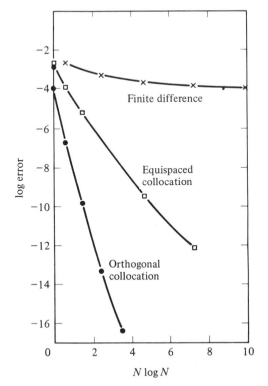

Figure 4-24 Error in boundary flux.

Fig. 4-4c. If we limit consideration to $0.9 \leqslant c \leqslant 1.0$ then the rate derivative obeys $-2.2 < dR/dc < -1.1$. In that case the problem is even easier. Results obtained with finite difference (equally spaced grid points), collocation (equally spaced collocation points), and orthogonal collocation are presented in Fig. 4-24. Clearly the best results are for orthogonal collocation since with only one term the accuracy in the effectiveness factor is 10^{-4} percent, which is only achieved with the finite difference method for $n = 10$. In addition, the orthogonal collocation results can be derived on a hand calculator, while the finite difference method requires a computer. We also see in Fig. 4-24 the importance of using the collocation points from the orthogonal polynomials, since equispaced collocation points degrade the accuracy from one to six orders of magnitude.

The finite difference and orthogonal collocation on finite elements using lagrangian polynomials give results plotted in Fig. 4-25 to display the decrease in error as the number of elements is increased. By the error estimates listed in Table 4-11 the slope of the curve should have the values -2, -4, and -6; as they do. The most accurate solutions with $N = 5$ are reaching the machine accuracy for single precision arithmetic on a CDC computer which keeps 15 digits.

The final choice of method, of course, depends on the accuracy achieved for a given computation time. Results of this type are shown in Fig. 4-26. The successive substitution method was used for the finite difference and orthogonal collocation on finite element methods, while the Newton–Raphson method was used for

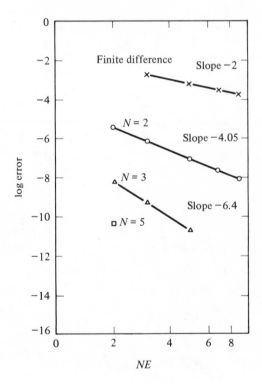

Figure 4-25 Error in boundary heat flux as a function of the number of elements. Orthogonal collocation on finite elements.

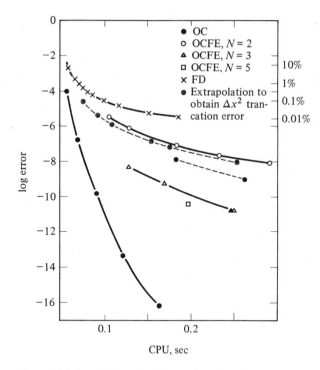

Figure 4-26 Error in boundary flux as a function of computation time.

orthogonal collocation. For a given computation time the error is largest for finite difference, smaller for orthogonal collocation on finite elements, and very much smaller for orthogonal collocation. Clearly for problems of this type the orthogonal collocation method is the best.

We next turn to a problem whose solution has a steep gradient. The problem is for a nonisothermal reaction in a catalyst pellet with external resistances

$$\frac{1}{r^2}\frac{d}{dr}\left(r^2\frac{dc}{dr}\right) = R(c) = \phi^2 c\, e^{\gamma - \gamma/T}$$

$$\frac{dc}{dr}(0) = 0 \qquad -\frac{dc}{dr}(1) = \text{Bi}_m[c(1)-1] \qquad (4\text{-}393)$$

$$T(r) = 1 + \beta c(1) + \beta\delta[1 - c(1)] - \beta c(r)$$

We examine the case with parameters $\beta = 0.02$, $\gamma = 20$, $\text{Bi}_m = 250$, $\text{Bi} = 5$, $\delta = 50$, and $\phi = 14.44$. We first examine the strength of the nonlinearity. Evaluation for $c = c(1) = 1$ and $T = 1$ gives $dR/dc = -1.10$, while for $c = 0$, $c(1) = 1$, and $T = 1.3$ gives 15.92. Results below show that $c(1) = 0.16$ so that $T(1) = 1.84$. With these results we get $dR/dc = 1.89 \times 10^6$ at the boundary $r = 1$, and $dR/dc = 1.96 \times 10^6$ at the center where $c(0) = 0$ and $T(0) = 1.8432$. Clearly this problem is a difficult one. We have already discussed in Sec. 4-7 that orthogonal collocation needs at least 40 terms just to have one collocation point in the region where $c(x) \gg 0$, and so for this problem we need finite elements. We apply two elements with the solution $c = 0$ in one element and a parabola in the other one and find [see Eqs. (4-282) to (4-291)] that the separation between the elements is at $r = 1 - b$ and $b = 0.00275$. Thus we know a priori that the solution is contained in a small region near the boundary. A finite difference method or finite element method with uniform grid points or elements is going to require a large number of them (700). We can use the information contained in the simple solution, though, to locate the elements. The solution is presented in Sec. 4-9 when the elements are placed at $x_{(k)} = 0, 0.5, 0.997, 0.998, 0.999$, and 1.0. Carey and Finlayson[2] continued the solution by examining the residual after obtaining the solution. The residual was, of course, zero at the collocation points, but it was nonzero in between. Some regions of space had a larger residual. The elements that had the five largest

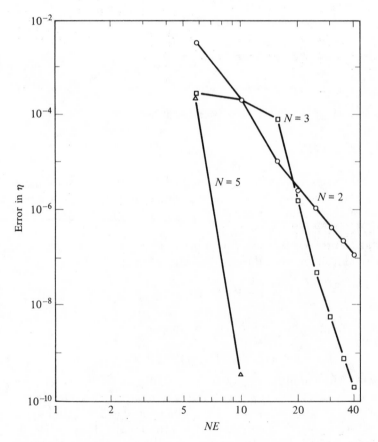

Figure 4-27 Error in effectiveness factor for orthogonal collocation on finite elements applied to Eq. (4-136) with $\gamma = 20$, $\beta = 0.02$, $Bi_m = 250$, $Bi = 5$, and $\delta = 50$.

residuals were subdivided, and the calculations repeated with five more elements. This process was continued, and the residual gradually decreased. The errors decreased, too, as shown in Fig. 4-27. In this fashion the elements can be located to optimum advantage. Let us compare the rates of convergence with the predicted ones for uniform elements. For $N = 2$ we get -4 as expected. For $N = 3$ and 5 we get -12 and -20, which are much higher than the expected values of -6 and -10. Thus for this problem finite elements or finite differences are required, but even so the location of grid points and elements must be specified. Use of the residual provides an effective means for doing that in this case. Orthogonal collocation (global polynomial) is completely ineffective in solving this problem, but orthogonal collocation on finite elements is quite useful.

For the final example we turn to a problem not yet treated: flow of a non-newtonian fluid through a pipe or between two plates. Under the assumption of fully developed flow the equations are

$$\frac{1}{r^{a-1}} \frac{d}{dr}\left(\eta r^{a-1} \frac{du}{dr}\right) + b = 0$$

$$b = \frac{\Delta p R^2}{\mu L} \tag{4-394}$$

$$\frac{du}{dr}(0) = 0 \qquad u(1) = 0$$

where R is the radius of the pipe or half the distance between the two plates, and $\Delta p/L$ is the pressure drop per unit length, which is a constant. The equation is written so that b is positive, and μ is some characteristic viscosity. The η is the viscosity, which can depend on the velocity gradient. The goal is to compute the average velocity as a function of the pressure drop or b. We integrate Eq. (3-394) once to obtain

$$\eta r^{a-1} \frac{du}{dr} = -\frac{b}{a} r^a \tag{4-395}$$

where the constant of integration is set to zero to satisfy the boundary condition at $r = 0$. We thus need to solve

$$\eta \frac{du}{dr} = -\frac{b}{a} r \tag{4-396}$$

when $u(1) = 0$ and calculate the average velocity

$$\langle u \rangle = a \int_0^1 u(r) r^{a-1}\, dr \tag{4-397}$$

The viscosity function can take several forms and here we use

$$\eta = \frac{1}{\left(1 + \lambda \left|\dfrac{du}{dr}\right|^2\right)^{(1-n)/2}} \qquad \lambda = \lambda_0 R^{-2} \tag{4-398}$$

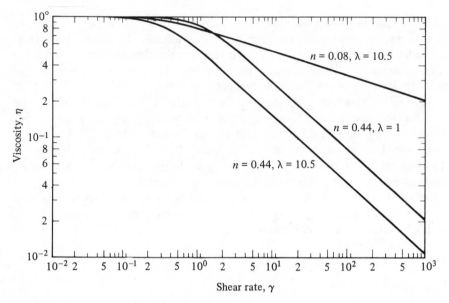

Figure 4-28 Apparent viscosity based on Eq. (4-398).

Both λ_0 and n are constants characteristic of a material; different polymers have different values of the constants. The viscosity function is displayed in Fig. 4-28 for several choices. For small $du/dr \equiv \dot{\gamma}$ the viscosity is constant and we have a newtonian fluid. For large du/dr the viscosity approaches

$$\eta = \left|\frac{du}{dr}\right|^{n-1} \frac{1}{\lambda^{(1-n)/2}} \qquad (4\text{-}399)$$

and we have a power-law fluid. The same is true for small and large λ, which is a parameter characterizing the elasticity of the material. The case $\lambda = 0$ gives an inelastic fluid and large λ represents an elastic fluid.

First we examine Eq. (4-396) to see if the solution is a symmetric function of r. If we insert the power series and attempt to equate like powers of r, we find the method breaks down except for integral values of n. For certain of these values $(n = 1, -1)$ all the odd powers of r drop out. For other values of n we assume that only symmetric solutions are needed, which is reasonable for flow in a pipe because the velocity at one radius r is the same as at the same radius $-r$ on the other side of the center line. We try the transformation $z = r^2$ to obtain the problem

$$\eta 2\sqrt{z}\frac{du}{dz} = -\frac{b}{a}\sqrt{z} \qquad (4\text{-}400)$$

Next we write Eq. (4-396) as

$$\eta(\dot{\gamma})\dot{\gamma} = -\frac{b}{a}r \qquad (4\text{-}401)$$

This is just an algebraic equation in the shear rate $\dot{\gamma}$. For any r we can solve the equation to obtain $\dot{\gamma}$ at that r. We then have $\dot{\gamma} = du/dr$ as a function of r, calling it $\dot{\gamma}(r)$. Next we solve the equation

$$\frac{du}{dr} = \dot{\gamma}(r) \qquad u(1) = 0 \tag{4-402}$$

This solution method means that the nonlinear part of the problem has been isolated into a single algebraic equation, Eq. (4-401), that is to be solved several times for different r. The differential equation part of the problem is now linear— Eq. (4-402)—and can be solved without iteration. Thus the method takes the large system of nonlinear algebraic equations resulting from solving Eq. (4-394) directly by any numerical method and resolves them into the same number of nonlinear algebraic equations, each of which has only one unknown plus a set of linear equations. This greatly reduces the computation time necessary to solve the system because the matrix to be inverted need be inverted only once rather than once each iteration.* Unfortunately, the transformed Eq. (4-400) does not admit this easy solution method (η depends on $\sqrt{z}(du/dz)$), so that transformation is not used here.

The limiting behavior is useful to obtain before deriving numerical solutions. When $\lambda = 0$ we have a newtonian fluid. Then the problem is linear and an analytic solution exists. When $\lambda = 0$ the problem reduces to

$$\frac{du}{dr} = -\frac{b}{a}r$$

$$u = \frac{b}{2a}(1 - r^2) \tag{4-403}$$

$$\langle u \rangle = \frac{b}{a(a+2)}$$

The velocity profile is a parabola and so orthogonal collocation with $N = 1$ would give the exact solution. When $\lambda \to \infty$ or $\gamma \to \infty$ the problem reduces to

$$\frac{1}{\lambda}|\dot{\gamma}|^{n-1}\dot{\gamma} = -\frac{b}{a}r \tag{4-404}$$

which has the solution

$$u = \left(\frac{b}{a\lambda}\right)^{1/n} \frac{n}{n+1}(1 - r^{(n+1/n)}) \tag{4-405}$$

$$\langle u \rangle = \left(\frac{b}{a\lambda}\right)^{1/n} \frac{1}{a(a+1+1/n)} \tag{4-406}$$

This is the solution for a power-law fluid. If n is small ($n \to 0$) the velocity profile is a high-order polynomial in r, and is nearly constant in the entire region except for

---

* This solution procedure was first suggested to the author by Thomas Patten.

a small boundary layer near $r = 1$. The profile is quite flat, and many terms would be needed to solve for it using orthogonal collocation. Finite elements or finite differences would then be useful. We now apply each of the numerical methods.

Orthogonal collocation applied to Eq. (4-402) gives the equations

$$\sum_{i=1}^{N=1} A_{ji} u_i = \dot{\gamma}(r_j) \qquad j = 1, \ldots, N$$

$$u_{N+1} = 0 \qquad\qquad (4\text{-}407)$$

or simply

$$\sum_{i=1}^{N} A_{ji} u_i = \dot{\gamma}(r_j) \qquad j = 1, \ldots, N \qquad (4\text{-}408)$$

The average velocity is given by

$$\langle u \rangle = a \sum_{i=1}^{N} W_i u_i \qquad (4\text{-}409)$$

This method is easy to apply for small N. For example $N = 2$ is easily done on a hand calculator. It is expected to be a good method for small and intermediate values of b or λ.

The finite difference and finite element methods are all applied with a uniform grid or element spacing. All methods give equations of the form

$$\mathbf{AAu} = \mathbf{f}$$

where the matrix $\mathbf{AA}$ has a different structure for different methods. For the finite difference method the equations are

$$AA_{ij} = \begin{cases} 0 & j < i-1 \\[2mm] -\dfrac{1}{2\Delta r} & j = i-1 \\[2mm] 0 & j = i \\[2mm] \dfrac{1}{2\Delta r} & j = i+1 \\[2mm] 0 & j > i+1 \end{cases} \qquad (4\text{-}410)$$

$$F_i = \dot{\gamma}(r_i) \qquad (4\text{-}411)$$

For the first node the equation must be modified to retain the Δx^2 accuracy. We cannot introduce a false boundary because we do not have an additional equation to use at $r = 0$. There is no boundary condition to be applied there, only a differential equation, so that we cannot introduce a new unknown. Instead we use a one-sided derivative at $r = 0$, Eq. (4-48). This destroys the tridiagonal nature of the equations. To prevent having to modify the matrix solution technique we combine the first two equations in such a way to make them both fit the tridiagonal structure. This is like doing an LU decomposition in reverse, to eliminate the A_{13} term using the A_{23} term. Finally, the equation at the last node is the boundary condition $u_{NT} = 0$. To calculate the average velocity we use Simpson's rule when the points are equally spaced, and the trapezoid rule if they are not.

The method of orthogonal collocation on finite elements gives equations of the form shown in Fig. 4-13, but with Eq. (4-267) replaced by

$$\frac{1}{h_k} \sum_{J=1}^{NP} A_{IJ} u_J = \hat{\gamma}(r_i) \tag{4-412}$$

The average velocity is

$$\langle u \rangle = a \sum_{l=1}^{NE} h_l \sum_{J=1}^{NP} W_J u_J r_J^{a-1} \tag{4-413}$$

With Hermite polynomials we get equations in the form of Fig. 4-19 with Eq. (4-301) replaced by

$$\frac{1}{h_k} \sum_{J=1}^{NP} A_{IJ} u_J^e = \hat{\gamma}(r_i) \tag{4-414}$$

The same formula applies for the average velocity.

The Galerkin methods give equations of tridiagonal (linear trial functions) or block diagonal (quadratic trial functions) form. The terms replacing Eq. (4-325) are

$$\sum_e A_{JI}^e u_I^e = \sum_e F_J^e \tag{4-415}$$

$$A_{JI}^e = \int_0^1 N_J(u) \frac{dN_I}{du} r^{a-1} \, du \tag{4-416}$$

$$F_J^e = \Delta x_e \int_0^1 \hat{\gamma}(r) N_J(u) \, du \tag{4-417}$$

The average velocity is given by

$$\langle u \rangle = a \sum_e u_I^e \int_0^1 N_I(u) r^{a-1} \, du \tag{4-417}$$

We apply each of these methods to the problem with b changing from a small number (where the fluid is essentially newtonian) to a large number (where it is essentially a power-law fluid). We do this for parameters typical of low-density polyethylene ($\lambda = 10.5$, $n = 0.44$, $\mu = 1.1 \times 10^5$ g/cm/sec, $R = 1$ cm, and $L = 46.9$ cm). The flow curve is shown in Fig. 4-29. The corresponding velocity profiles are illustrated in Fig. 4-30.

For this problem the major portion of the computation time is associated with the iterative solution of Eq. (4-401) for each node. The computation time and work are mostly proportional to the number of nodes, so the results are presented in that fashion. The number of elements is proportional to the number of nodes, so we know the error curves should have slopes of -2, -3, or -4 depending on the method chosen. First we consider a low flow rate, for $b = 0.1$ in planar geometry. This corresponds to a pressure drop of 7.5 lb/in² and a flow rate of 0.034 cm/sec. The

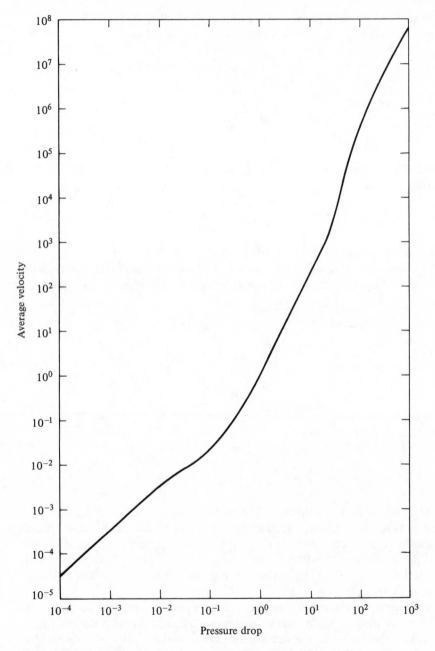

Figure 4-29 Pressure drop versus flow rate curve with $n = 0.44$ *and* $\lambda = 10.5$.

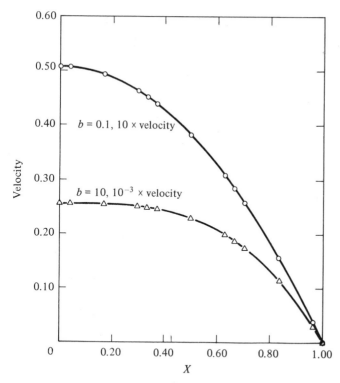

Figure 4-30 Velocity profile.

apparent shear rate at the wall for a newtonian fluid is $\dot{\gamma}_w = 3\langle v\rangle/R$, and here $\dot{\gamma}_w = 0.6\,\text{sec}^{-1}$. The error in average velocity is plotted in Fig. 4-31; the exact value is 0.03393013. Orthogonal collocation gives the best results since it provides the best accuracy with the fewest unknowns. Using symmetric polynomials for a three-term solution gives seven-digit accuracy. Such accuracy is achieved with the other methods only for much larger numbers of terms. Orthogonal collocation on finite elements with quartic polynomials needs two elements and nine terms. Orthogonal collocation on finite elements with cubic polynomials needs eight elements and 25 terms. The Galerkin method using quadratic polynomials needs 80 elements and 161 nodes while using linear polynomials needs 250 terms, and the finite difference method needs 310 terms.

Next we consider a more severe case with $b = 10$. This case corresponds to a pressure drop of $748\,\text{lb/in}^2$ and an apparent shear rate of $3,660\,\text{sec}^{-1}$, which is at the high end of possible shear rates. The errors in average flow rate are shown in Fig. 4-32; the exact result is 195.8262. Here we get 0.08 percent accuracy using orthogonal collocation with $N = 6$, cubic orthogonal collocation of finite elements with 2 elements and 7 unknowns, the Galerkin method using quadratic polynomials with 21 elements and 43 unknowns, the Galerkin method using linear polynomials with 31 elements and 32 unknowns, and finite difference with 50 unknowns. Even for this case the orthogonal collocation method proved very successful.

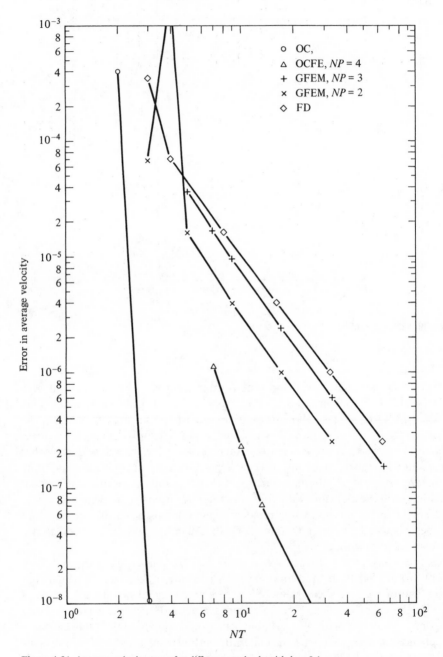

Figure 4-31 Average velocity error for different methods with $b = 0.1$.

This problem has results that are similar to those for the chemical reaction problem, even though the source of the nonlinearity is completely different. We conclude on the basis of the examples given in this chapter that if the solution is

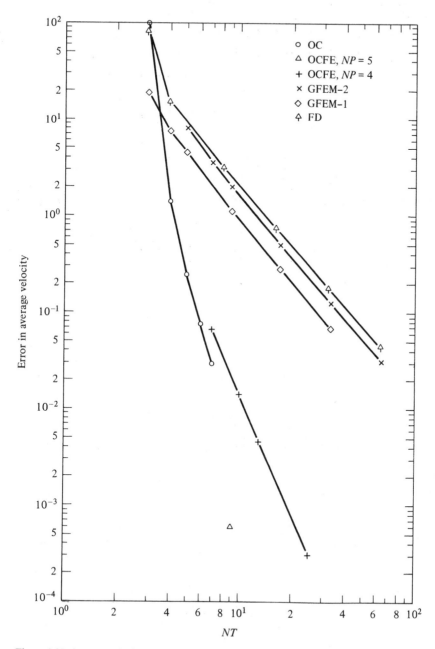

Figure 4-32 Average velocity error for different methods with $b = 10$.

smooth the orthogonal collocation method is best, but if steep gradients occur then one of the finite difference or finite element methods is better. The high-order finite element methods are best if high accuracy is desired. Any of the finite element or finite difference methods are suitable for low accuracy.

4-14 ADAPTIVE MESHES

Successful solution of certain boundary-value problems depends on using variable grid spacings and element sizes, and on the proper location of small elements. The reaction problem displayed in Fig. 4-15 is one such problem. There we use physical information about the solution to justify putting small elements near the boundary $x = 1$. In Sec. 4-13 the use of the residual to locate elements is mentioned. Here three mathematical strategies for locating grids and elements are discussed.

The first strategy, which is due to Pearson,[6] was applied to the finite difference methods. Suppose we have a solution at n points $\{c_i\}$. We then calculate the maximum allowed difference in solution between grid points

$$\delta^* = 0.01 \left(\max_i \{c_i\} - \min_i \{c_i\} \right) \tag{4-418}$$

The solution at successive grid points is compared to δ^*. When

$$|c_i - c_{i+1}| > \delta^* \tag{4-419}$$

grid points are inserted between x_i and x_{i+1}. The number of grid points is taken as

$$\text{Number} = \frac{|c_i - c_{i+1}|}{\delta^*} \tag{4-420}$$

The new set of grid points is called $\{x_i\}$. The locations are smoothed to avoid abrupt changes in x_i by using the algorithm

$$x_i' = \tfrac{1}{2}(x_{i-1}' + x_{i+1}) \tag{4-421}$$

in turn, beginning with $i = 1$. The final location of grid points is $\{x_i'\}$. The problem is then solved again on this better grid.

It is clear that this method can be used for finite element methods also, by just substituting the element nodes for the grid points. Thus we have a method of automatically locating the grid points or elements. The only difficulty is that a lower-order solution may not be very good, or it may be difficult to obtain a low-order solution, so that it is difficult to start the process. The technique has been applied to Eq. (4-393) with good results very similar to those shown in Fig. 4-27.[2]

The second method of locating the elements is based on the residual. We know that the residual is zero for the exact solution. For some problems (see Finlayson,[3] p. 388) it is possible to show that the error in the solution is bounded by the residual: small residuals mean small errors in the solution. We can use this principle for all problems; locate the elements in order to make the residuals approach zero. This method has been applied to Eq. (4-393) by Carey and Finlayson.[2] The values of c_i at the ends of elements are compared to δ^*. The number of elements inserted in place of the one with $|c_i - c_{i+1}| > \delta^*$ is taken as

$$\text{Number of elements} = \frac{|c_i - c_{i+1}| NP}{\delta^*} \tag{4-422}$$

rounded to an integral value. This is the approach used to obtain the results in Fig. 4-27. As can be seen, using a little insight into the physical problem gives a hint to use small elements near $x = 1$. The calculation begins with element nodes at 0, 0.5, 0.997, 0.998, 0.999, and 1.0. The problem is solved and the residual evaluated. Additional elements are added in elements with a large residual. The element locations are then smoothed using Eq. (4-421), and the calculations are repeated with this new, larger problem. The procedure is repeated until the residual is small enough. The analyst may note that as the process proceeds the overall residual decreases, lending confidence in the results.

The final method is based on work by Ascher, et al.[1] and Russell.[7] For a second-order differential equation and orthogonal collocation on finite elements with cubic trial functions, the error in the solution $u(x)$ is given in the ith element by

$$\|\text{Error}\|_i = C h_i^4 \|u^{(4)}\|_i \tag{4-423}$$

If we know the fourth derivative of the exact solution we can choose Δx so that the right-hand side is small in precisely the right place to make the solution meet our accuracy requirements. Unfortunately, we cannot even calculate the fourth derivative of the approximate solution since, with cubic polynomials, the fourth

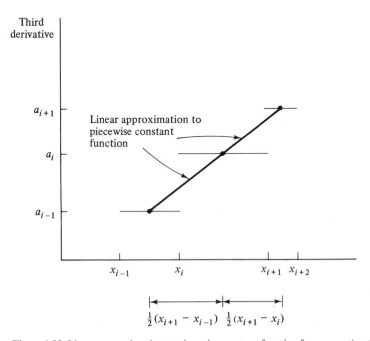

Figure 4-33 Linear approximation to piecewise constant function for error estimation.

derivative is zero. So let us evaluate the third derivative in successive elements

$$a_i = \frac{1}{\Delta x_i^3} \frac{d^3 c^i}{du^3} \qquad a_{i+1} = \frac{1}{\Delta x_{i+1}^3} \frac{d^3 c^{i+1}}{du^3} \tag{4-424}$$

Let the ith element extend from x_i to x_{i+1} and call the (constant) third derivative in the element a_i. The third derivative is a piecewise constant function, as shown in Fig. 4-33. We approximate this function by the piecewise linear function as shown. One derivative can be taken of this piecewise linear function, and this is an approximation to the fourth derivative. One value exists at the node x_i and a different one at x_{i+1}. The average of these values is taken as the estimate of the fourth derivative

$$\|u^{(4)}\|_i = \frac{1}{2} \left[\frac{a_i - a_{i-1}}{\frac{1}{2}(x_{i+1} - x_{i-1})} + \frac{a_{i+1} - a_i}{\frac{1}{2}(x_{i+2} - x_i)} \right] \tag{4-425}$$

Now we choose the element sizes so that the following error bounds are satisfied for the user-specified ε:

$$C h_i^4 \|u^{(4)}\|_i < \varepsilon \qquad \text{for each } i \tag{4-426}$$

Russell has built a computer program COLSYS around this idea. In this program the user needs to define the type of differential equation and provide a subroutine to evaluate the terms in the equation. The user specifies the desired tolerance ε and the program finds a solution that is correct to that tolerance. After one solution the element sizes are readjusted to meet Eq. (4-426) and the problem is resolved. When Eq. (4-426) holds in every element the solution is complete. Such programs are the analog for boundary-value problems of the packages, such as GEAR and Runge–Kutta, for initial-value problems. When they become more widely available and used, such packages will be employed routinely to solve two-point boundary-value problems of the type discussed in this chapter.

STUDY QUESTIONS

1. Technical details
   a. Method of Weighted Residuals
   b. Finite difference method
   c. Orthogonal collocation
   d. High-order orthogonal collocation on finite elements
   e. Low-order orthogonal collocation on finite elements
   f. Galerkin finite elements method using linear polynomials
   g. Galerkin finite elements method using quadratic polynomials
2. Perturbation method
   a. Regular
   b. Asymptotic
3. Solution of nonlinear algebraic equations using
   a. Successive substitution
   b. Newton–Raphson method

4. Convergence of each method
   a. As the element size decrease to zero
   b. As the solution changes from a smooth one to one with steep gradients
5. Variation in work required for each method as a function of
   a. Number of elements
   b. Number of trial functions
6. Initial-value techniques—methods of integration to integrate to steady state
7. Shooting methods
   a. How to apply them
   b. When to expect problems
8. Interpolation between grid points, collocation points, or nodes using
   a. Finite difference method
   b. Collocation methods
   c. Galerkin finite element methods
9. Extrapolation techniques
   a. Finite difference method
   b. Finite element methods

PROBLEMS

Methods of Weighted Residuals

4-1 Apply the Method of Weighted Residuals to the problem

$$\theta'' = -1$$

$$\theta(0) = 0 \qquad \theta(1) = 0$$

Deduce a trial function by assuming a polynomial, Eq. (4-4), and applying the boundary condition. Compare to the exact solution.

4-2 (a) Apply a one-term collocation method at $x = \frac{1}{2}$ to solve $u'' = e^u$, $u(0) = u(1) = 0$. Compare to the exact answer

x	0.1	0.2	0.3	0.4	0.5
$u(x)$	-0.0414	-0.0733	-0.0958	-0.1092	-0.1137

(b) Write the equations for a one-term Galerkin method. What difficulties do you see?

4-3 Apply the Method of Weighted Residuals to integrate $dy/dt = y$, using the linear trial function $y = y_0(\Delta t - t)/\Delta t + y_1 t/\Delta t$. Show that the Galerkin criterion leads to

$$\frac{y_1 - y_0}{\Delta t} = \tfrac{1}{3}y_0 + \tfrac{2}{3}y_1$$

and the collocation method leads to

$$\frac{y_1 - y_0}{\Delta t} = \tfrac{1}{2}(y_0 + y_1)$$

These values of weightings, when applied to $y' = f(y)$ correspond to implicit methods; the collocation method is just the trapezoid rule.

Orthogonal collocation

4-4 Evaluate the integral $I = \int_0^1 f(x^2)dx$ for $f(x^2) = 1$, x^2, x^4, and x^6 using the quadrature formula for $N = 1$ and 2. Note that the results should be exact for $f(x^2)$ a polynomial of degree $2N$ in x^2. Evaluate $\int_0^1 x^2 dx$ using the weights and quadrature points in Table 4-3.

4-5 Which polynomials must be used to apply orthogonal collocation to the following problems? Identify the polynomials by listing W, x_j, and the Tables giving the matrices and roots.

  (a) $y'' + 2y = x^2$ in $0 < x < 1$ subject to the boundary conditions
    (i) $y'(0) = 0$ $y(1) = 1$
    (ii) $y'(0) = 0$ $y'(1) + 3y(1) = 1$
    (iii) $y(0) = 0$ $y(1) = 1$
    (iv) $-y'(0) + y(0) = 0$ $y(1) = 1$
  (b) The same boundary conditions as in (a) but with the differential equation $y'' + 2y = x$.

4-6 Reevaluate Eq. (4-213), $N = 1$, except for spherical geometry. Compare the solutions of η versus ϕ for the planar and spherical geometry.

4-7 Apply a one-term orthogonal collocation method to

$$\frac{1}{r^2}\frac{d}{dr}\left(r^2\frac{dc}{dr}\right) = \phi^2 f(c)$$

$$-\frac{dc}{dr}\bigg|_{r=0} = 0 \qquad -\frac{dc}{dr}(1) = \mathrm{Bi}_m[c(1) - 1]$$

for $\mathrm{Bi}_m = 100$ and
  (a) $f = c$, $\phi^2 = 1$,
  (b) $f = c^2$, $\phi^2 = 1$,
  (c) $f = c/(1 + \alpha c)^2$, $\alpha = 20$, $\phi = 32$.
  Calculate the effectiveness factor. Compare these results to the results for $N = 2$ that were solved in problem 2-4. The equations there are for $W = 1$, and the quadrature weights are $W_1 = 0.1387779991$, $W_2 = 0.1945553342$, and $W_3 = 0$.

4-8 Determine if the following equation can have multiple solutions

$$\frac{1}{r^2}\frac{d}{dr}\left(r^2\frac{dc}{dr}\right) = \phi^2 R(c)$$

$$\frac{dc}{dr}(0) = 0 \qquad c(1) = 1 \qquad R(c) = \frac{c(E + c)}{(1 + Kc)^2}$$

The reaction rate occurs with Langmuir–Hinshelwood kinetics and one of the reactants is in stoichiometric excess (the fraction excess is represented by E). Solve for

	a	b	c	d	e
K	1	10	100	100	100
E	0	0	0	1	10

4-9 Solve problem 4-7 using the program OCRXN for $N = 1, 2, \dots, 6$.

4-10 Solve problems 4-8a to 4-8e using the program OCRXN for $N = 6$. Based on your estimates of where multiple solutions occur, do calculations for small ϕ, values of ϕ giving multiple solutions, and large ϕ.

Finite Difference Method

4-11 Solve problem 4-1 using $\Delta x = 0.25$.

4-12 Solve Eq. (4-145) for a second-order reaction. Transform the coordinate system to $u = x^2$ and apply the finite difference method with $\Delta x = \frac{1}{2}$. Prepare a curve of η versus ϕ and compare to the orthogonal collocation result in Fig. 4-7.

4-13 Solve problem 4-7 using the finite difference method and program FDRXN. Apply extrapolation techniques to η and c.

4-14 Write the equations replacing Eqs. (4-247) and (4-250) with equations correct only to first order in Δr. Modify the program FDRXN and redo problem 4-11. Compare.

4-15 Write the finite difference equations for solving problem 4-8 with a variable grid spacing under conditions of high ϕ. Modify the program FDRXN and apply to $f = c^2$, $\phi = 6$.

4-16 Solve problems 4-8a to 4-8e using the program FDRXN. Compare to results from problem 4-10.

4-17 Substitute Eq. (4-42) into Eq. (4-39), and express k_{i+1} and k_{i-1} in Taylor series in θ. Show the result is the second-order Δx^2 approximation to $k\theta'' + (dk/d\theta)/(\theta')^2$.

Perturbation Method

4-18 Derive a perturbation solution to problem 4-7b. Obtain the η versus ϕ curve up to the ϕ^2 term and for $\phi \to \infty$. Derive the solution $c(r, \phi)$. Plot η versus ϕ, and the concentration versus position for various ϕ. Identify points on the η versus ϕ curve with the concentration profiles.

4-19 Problem 4-18 but for reaction rate in problem 4-7c.

4-20 Derive Eq. (4-66).

Orthogonal Collocation on Finite Elements

4-21 Apply the method to problem 4-1 using two elements and lagrangian as well as Hermite cubic polynomials.

4-22 Apply Eqs. (4-289) to (4-291) to problem 4-7 for planar geometry. Construct the η versus ϕ curve for large ϕ. Write the equations for a four-element solution with lagrangian cubic polynomials.

4-23 Write the equations for problem 4-7, planar geometry and four elements using Hermite polynomials.

4-24 Use the program OCFERXN to solve problem 4-7. Apply extrapolation techniques to the effectiveness factor, with the original calculations correct to $0(\Delta x^4)$ when cubic polynomials are used.

4-25 Apply orthogonal collocation on finite elements with lagrangian polynomials to initial-value problems of the form $dy/dt = f(y)$. At the beginning of each element the initial value y_1 is known. Collocation is then applied at the remaining collocation points of the element and the set of equations is solved for y_{N+2}. This value is used as the initial value for the next element.

4-26 Solve problems 4-8a to 4-8e using the program OCFERXN.

Galerkin Finite Elements Method

4-27 Apply the Galerkin method to problem 4-1 using linear elements.

4-28 Apply the Galerkin method to problem 4-1 using two quadratic elements.

4-29 Apply the Galerkin finite element method to the problem in Eq. (4-145) with $n = 2$ using:
 (a) Two linear elements, derive an analytic solution.
 (b) Write the equations for many linear elements.
 (c) Write the equations for many quadratic elements.

4-30 Solve the heat transfer problem in layered slabs with heat generation in some of them

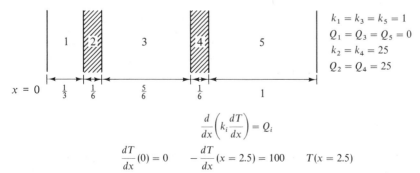

$$k_1 = k_3 = k_5 = 1$$
$$Q_1 = Q_3 = Q_5 = 0$$
$$k_2 = k_4 = 25$$
$$Q_2 = Q_4 = 25$$

$$\frac{d}{dx}\left(k_i \frac{dT}{dx}\right) = Q_i$$

$$\frac{dT}{dx}(0) = 0 \qquad -\frac{dT}{dx}(x = 2.5) = 100 \qquad T(x = 2.5)$$

Use one linear element in each domain without heat generation and two linear elements in the domains with heat generation.

Hint: derive the element equations and compare to Eqs. (4-320) and (4-321).

Shooting Method

4-31 Apply a shooting method to problem 4-7 using the program IVRXN.

4-32 Solve the problems 4-8a to 4-8e using the program IVRXN.

4-33 Apply a shooting method to problems 4-35a to c using (if possible):
(a) Forward integration from $x = 0$.
(b) Backward integration from $x = 1$.
Draw conclusions about the effectiveness of (a) and (b).

General

4-34 Consider the problem for axial conduction and diffusion in a tubular reactor

$$\frac{1}{Pe_M} \frac{d^2c}{dx^2} - \frac{dc}{dx} - R(c, T) = 0$$

$$\frac{1}{Pe_H} \frac{d^2T}{dx^2} - \frac{dT}{dx} - \beta R(c, T) = 0$$

$$\frac{1}{Pe_M} \frac{dc}{dz}(0) = c(0) - 1 \qquad \frac{1}{Pe_H} \frac{dT}{dz}(0) = T(0) - 1$$

$$\frac{dc}{dz}(1) = \frac{dT}{dz}(1) = 0$$

(a) Write the orthogonal collocation equations for $NCOL = 3$.

(b) Write the equations for the finite difference method, ten interior nodes.

(c) Write the equations for orthogonal collocation on finite elements with three elements and lagrangian cubic polynomials.

(d) Write the equations for orthogonal collocation on finite elements with Hermite polynomials, three elements.

(e) Write the equations for the Galerkin method with linear trial functions, ten elements.

(f) Write the equations for the Galerkin method with quadratic trial functions, five elements.

4-35 Program and solve one of the methods listed in problem 4-34 for the cases
(a) $\beta = 0$, $Pe_M = 1$, $R = 2c^2$
(b) $\beta = -0.056$, $Pe_H = Pe_M = 2$, $\gamma = 17.6$, $R = 3.36c\,e^{\gamma - \gamma/T}$
(c) $\beta = -0.056$, $Pe_H = Pe_M = 96$, $\gamma = 17.6$, $R = 3.817037c\,e^{\gamma - \gamma/T}$

4-36 Solve Eq. (4-394) for a fluid with viscosity function given by Eq. (4-398) and $\lambda = 1$, $n = 0.5$. Apply one of the methods listed in problem 4-9 and write down the equations to be solved.

4-37 Solve the equations derived in problem 4-36 for the cases $b = 0.1, 1, 10$, and 100.

Initial-Value Problems

4-38 Consider the successive substitution method of solving Eq. (4-349). Using your knowledge of stability of systems for integrating differential equations using the Euler method, Eq. (4-351), what can you say about the size and sign of the constant β in Eq. (4-349)? What difficulties do you see in applying successive substitution to a system of equations $f_i(\mathbf{x}) = 0$ when the signs of the equations f_i are dependent on how the analyst writes down the equations?

BIBLIOGRAPHY

The Method of Weighted Residuals is covered in depth by Finlayson.[3] Least squares methods have been reviewed recently by

Eason, E. D., and C. D. Mote, Jr.: "Solution of Nonlinear Boundary-Value Problems by Discrete Least Squares," *Int. J. Num. Methods Eng.*, vol. 11, pp. 641–652, 1977.

Finite difference methods are treated by Keller[5] as well as by

Carnahan, B., H. A. Luther, and J. O. Wilkes: *Applied Numerical Methods*, John Wiley & Sons, Inc., New York, 1969.

The latter book has computer programs for many engineering problems, but often they are linear problems.

Perturbation methods are treated by

Bellman, R.: *Perturbation Techniques in Mathematics, Physics, and Engineering*, Holt, Rinehart and Winston, New York, 1964.

Nayfeh, A. H.: *Perturbation Methods*, John Wiley & Sons, Inc., New York, 1972,

and the orthogonal collocation method is covered in Chapter 5 by Finlayson[3] as well as in

Villadsen, J., and M. L. Michelsen: *Solution of Differential Equation Models by Polynomial Approximation*, Prentice-Hall, Englewood Cliffs, 1978.

There is no book treating orthogonal collocation on finite elements, and books on Galerkin finite element methods are listed in Chapter 6. The following review article describes the methods that are available for solving boundary-value problems, and nearly all the methods are included in this chapter. The review does provide a convenient list of references giving applications and original papers.

Aktas, A., and H. J. Stetter: "A Classification and Survey of Numerical Methods for Boundary-Value Problems in Ordinary Differential Equations," *Int. J. Num. Methods Eng.*, vol. 11, pp. 771–796, 1977.

Extensions of the idea introduced in problem 4-3 are given by

Hulme, B. L.: "One-Step Piecewise Polynomial Galerkin Methods for Initial-Value Problems," *Math. Comp.*, vol. 26, pp. 415–426, 1972.

The A(0) stability of both Galerkin and orthogonal collocation methods are deduced. Quasilinearization is treated in depth by

Lee, E. S.: *Quasilinearization and Invariant Imbedding*, Academic Press, New York, 1968.

REFERENCES

1. Ascher, U., J. Christiansen, and R. D. Rossell: "A Collocation Solver for Mixed-Order Systems of Boundary-Value Problems," *Math. Comp.*, vol. 33, pp. 659–679, 1979.
2. Carey, G. F., and B. A. Finlayson: "Orthogonal Collocation on Finite Elements," *Chem. Eng. Sci.*, vol. 30, pp. 587–596, 1975.
3. Finlayson, B. A.: *The Method of Weighted Residuals and Variational Principles*, Academic Press. New York, 1972.
4. Forsythe, G., and C. B. Moler: *Computer Solution of Linear Algebraic Systems*, Prentice-Hall, Englewood Cliffs, 1967.
5. Keller, H. B.: *Numerical Methods for Two-Point Boundary-Value Problems*, Blaisdell, New York, 1972.
6. Pearson, C. E.: "On a Differential Equation of Boundary Layer-Type," *J. Math. Phys.*, vol. 47, pp. 351–358, 1968.
7. Russell, R. D., and J. Christiansen: "Adaptive Mesh Selection Strategies for Solving Boundary-Value Problems," *SIAM J. Num. Anal.*, vol. 15, pp. 59–80, 1978.
8. Shanks, D.: "Nonlinear Transformations of Divergent and Slowly Convergent Sequences," *J. Math. Phys.*, vol. 34, pp. 1–42, 1955.
9. Weisz, P. B., and J. S. Hicks: "The Behavior of Porous Catalyst Particles in View of Internal Mass and Heat Diffusion Effects," *Chem. Eng. Sci.*, vol. 17, pp. 265–275, 1962.

FIVE

PARABOLIC PARTIAL DIFFERENTIAL EQUATIONS— TIME AND ONE SPATIAL DIMENSION

Chapter 3 treats evolution problems beginning at some point in time and continuing indefinitely. Chapter 4 treats two-point boundary-value problems in which the conditions at the far end influence the solution everywhere. Generally, evolution problems have time as the independent variable, and the two-point boundary-value problems have space as the independent variable. Here we combine the two problems and treat parabolic partial differential equations. Now the independent variables are both time and space, and the problem is evolutionary in time and similar to two-point boundary-value problems in space. As a consequence the techniques of solution are combinations of those found in Chapters 3 and 4. We first present similarity methods, however, because if a similarity transformation exists the analyst should always employ it. It reduces a problem having two independent variables to a problem having only one such variable, with resultant savings in solution time and effort.

5-1 SIMILARITY TRANSFORMATION

We now study diffusion and reaction in a medium under transient conditions. The prototype problem is

$$\frac{\partial c}{\partial t} = \frac{\partial}{\partial x}\left[D(c)\frac{\partial c}{\partial x}\right] + R(c) \tag{5-1}$$

$$c(x, 0) = h(x) \quad \text{one initial condition} \tag{5-2}$$

$$c(0, t) = g(t)$$
$$ \quad \text{two boundary conditions}$$
$$c(\infty, t) = c_\infty \tag{5-4}$$

$$\text{(5-3)}$$

We ask ourselves if there is some way the t and x coordinates can be combined so that the solution is only a function of their combination, not of time and space individually. The answer to this question requires group theory (see Ames[1]), and here we give a simplified method of answering the question following Ames.[2]

The simplified approach is to transform both the independent variables t and x and the dependent variable c to see if the equation may be simplified. The boundary conditions must also simplify, and indeed the three conditions must collapse into two, since the resulting equation will be second-order with only one independent variable. Only two boundary conditions are necessary to solve such a problem. The transformation we try is

$$\bar{t} = a^\alpha t \qquad \bar{x} = a^\beta x \qquad \bar{c} = a^\gamma c \tag{5-5}$$

where for the present the values of a, α, β, and γ are parameters to be freely chosen. It is at this step that we have restricted the problem from the general realm of group theory; we will only answer the question of the similarity transformation for transformations of the type given in Eqs. (5-5). Putting these into Eq. (5-1) gives

$$a^{\alpha-\gamma} \frac{\partial \bar{c}}{\partial \bar{t}} = a^{2\beta-\gamma} \frac{\partial}{\partial \bar{x}} \left[D(a^{-\gamma}\bar{c}) \frac{\partial \bar{c}}{\partial \bar{x}} \right] + R(a^{-\gamma}\bar{c}) \tag{5-6}$$

Group theory says a system is conformally invariant if it has the same form in the new variables. For this system to be conformally invariant the transformation must be independent of a. We then need

$$\gamma = 0 \tag{5-7}$$

$$\alpha - \gamma = 2\beta - \gamma \quad \text{or} \quad \alpha = 2\beta \tag{5-8}$$

When we have a general reaction rate term we see that the result still depends on a unless $\alpha = \beta = 0$, in which case we have no transformation. Thus we conclude that a similarity transformation of the type given in Eq. (5-5) does not exist for all expressions $R(c)$. For $R(c) = c^n$, with $D(c) = $ constant, similarity transformation is possible (see problem 5-1). Next we leave out the reaction rate term and consider further the case with $D(c)$. The invariants are

$$\eta = \frac{x}{t^\delta} \qquad \delta = \frac{\beta}{\alpha} \tag{5-9}$$

and the solution is

$$f(\eta) = \frac{c(x, t)}{t^{\gamma/\alpha}} \tag{5-10}$$

Here

$$\delta = \frac{\beta}{\alpha} = \tfrac{1}{2} \qquad \eta = \frac{x}{t^{1/2}} \qquad f(\eta) = c(x, t) \tag{5-11}$$

We check the boundary conditions

$$c(\bar{x}a^{-\beta}, 0) = h(\bar{x}a^{-\beta}) \tag{5-12}$$

$$c(\infty, a^{-\alpha}\bar{t}) = c_\infty \tag{5-13}$$

These must combine and be conformally invariant. The first one cannot depend on x and the second one cannot depend on t. They must also have the same value for the right-hand side. Thus the following boundary conditions are allowed:

$$c(\bar{x}a^{-\beta}, 0) = c_\infty \tag{5-14}$$

$$c(\infty, a^{-\alpha}\bar{t}) = c_\infty \tag{5-15}$$

The other boundary condition is

$$c(0, \bar{t}a^{-\alpha}) = g(\bar{t}a^{-\alpha}) \tag{5-16}$$

and this cannot depend on t, so that the allowed boundary condition is

$$g(t) = c_1 \tag{5-17}$$

We note that

$$\eta = \frac{x}{t^{1/2}} \tag{5-18}$$

is infinite at either $x \to \infty$ or $t = 0$, and this allows the initial condition and the far boundary condition to be combined. Likewise, $x = 0$ gives $\eta = 0$ so that we have $c(\eta = 0) = c_1$.

We next rephrase the problem with

$$C = \frac{c - c_\infty}{c_1 - c_\infty} \qquad \eta = \frac{x}{\sqrt{4D_0 t}} \tag{5-19}$$

Equation (5-1) then becomes, with $R = 0$,

$$\frac{d}{d\eta}\left[K(C)\frac{dC}{d\eta}\right] + 2\eta\frac{dC}{d\eta} = 0 \tag{5-20}$$

$$K(C) = \frac{D(C)}{D_0} \qquad D_0 = D(c_1) \tag{5-21}$$

and the boundary conditions are

$$C(\infty) = 0 \tag{5-22}$$

$$C(0) = 1 \tag{5-23}$$

This is obtained using the chain rule of differentiation $C = C(\eta(x, t))$

$$\frac{\partial C}{\partial t} = \frac{dC}{d\eta}\frac{\partial \eta}{\partial t} \qquad \frac{\partial C}{\partial x} = \frac{dC}{d\eta}\frac{\partial \eta}{\partial x} \tag{5-24}$$

$$\frac{\partial^2 C}{\partial x^2} = \frac{d^2 C}{d\eta^2}\left(\frac{\partial \eta}{\partial x}\right)^2 \tag{5-25}$$

$$\frac{\partial \eta}{\partial t} = -\frac{x/2}{\sqrt{4D_0 t^3}} \qquad \frac{\partial \eta}{\partial x} = \frac{1}{\sqrt{4D_0 t}} \tag{5-26}$$

Several solutions to this problem have been tabulated by Crank.[3] To reach this far we have to assume uniform initial concentration, a steady boundary value, a semi-infinite planar geometry, and that the diffusivity depends on concentration. We leave it as an exercise (see problem 5-2) to determine if the same equation for cylindrical geometry has a similarity transformation. We have simplified the problem greatly but have not yet solved it.

Another transformation is useful for simplifying the problem even further. Let us apply the Kirchhoff transformation

$$\psi = \int_0^C K(z)dz \tag{5-27}$$

Then the first and second derivatives are

$$\frac{d\psi}{d\eta} = K(C)\frac{dC}{d\eta} \qquad \frac{d^2\psi}{d\eta^2} = \frac{d}{d\eta}\left[K(C)\frac{dC}{d\eta}\right] \tag{5-28}$$

We must also invert Eq. (5-27) so that we can write

$$K(\psi) = K(C) \tag{5-29}$$

Then Eq. (5-20) becomes

$$K(\psi)\frac{d^2\psi}{d\eta^2} + 2\eta\frac{d\psi}{d\eta} = 0 \tag{5-30}$$

This may or may not be simpler depending on the difficulty of inverting the transformation Eq. (5-27).

We have transformed the problem into a two-point boundary-value problem, but what do we do when one boundary is at infinity? There are several alternatives. The first approach is to transform the domain by letting

$$x = e^{-\eta} \tag{5-31}$$

The resulting equation is

$$x\frac{d}{dx}\left[K(C)x\frac{dC}{dx}\right] + 2x\ln x\frac{dC}{dx} = 0 \tag{5-32}$$

which is now on a finite domain, although it has singular coefficients (which are zero at $x = 0$). All the methods of Chapter 4 apply to Eq. (5-32).

Another approach is to transform the equation to an integral equation. We write Eq. (5-20) as

$$\frac{d}{d\eta}\left(K\frac{dC}{d\eta}\right) = -\frac{2\eta}{K}K\frac{dC}{d\eta} \tag{5-33}$$

and form

$$\frac{dy}{d\eta} = -\frac{2\eta}{K(C(\eta))}y \tag{5-34}$$

Then

$$y = A\exp\left[-\int_0^\eta \frac{2z}{K(C(z))}dz\right] \tag{5-35}$$

and from Eq. (5-33)

$$y = K \frac{dC}{d\eta} \tag{5-36}$$

Thus

$$\frac{dC}{d\eta} = \frac{A}{K} \exp\left[-\int_0^\eta \frac{2z}{K(C(z))} \, dz \right] \tag{5-37}$$

We can integrate this once to obtain

$$C(\eta) = B + A \int_0^\eta \frac{d\xi}{K(C(\xi))} \exp\left[-\int_0^\xi \frac{2z}{K(C(z))} \, dz \right] = B + A f(\eta) \tag{5-38}$$

For the boundary condition $c(0) = 1$ we need $B = 1$, and for $c(\infty) = 0$ we need

$$1 + A f(\infty) = 0 \tag{5-39}$$

Thus the solution is written as

$$C(\eta) = 1 - \frac{f(\eta)}{f(\infty)} \tag{5-40}$$

This is an exact solution to Eq. (5-20), even though it is not directly soluble except when K is a constant. We can, however, solve it iteratively by choosing a $c(z)$, integrating to get $c(\eta)$, and repeating. When $K = 1$ we get the complementary error function

$$f(\eta) = \int_0^\eta e^{-\xi^2} \, d\xi \tag{5-41}$$

$$C(\eta) = 1 - \frac{f(\eta)}{f(\infty)} = 1 - \operatorname{erf} \eta = \operatorname{erfc} \eta \tag{5-42}$$

which is a tabulated function.

Another alternative is to place the condition $\eta \to \infty$ at some arbitrary location $\eta = \eta_\infty$, and then apply the methods for two-point boundary-value problems. The location of η_∞ is chosen by experience with similar problems and the eventual solution. The location can be varied to ensure that the particular choice does not influence the results.

Still another alternative is to recognize that the solution approaches an asymptote as $\eta \to \infty$. A shooting method is started at $\eta = 0$, and eventually the solution quits changing for large η. By using an implicit method with a variable step size, in the region with a constant solution the step size can be quite large. Then we can integrate to a large η, which although still finite is much larger than in any other option. We must then iterate to fit the boundary condition, but we expect to iterate in this nonlinear problem anyway. A method such as Gear's for stiff problems or just Runge–Kutta for easier problems should suffice.

What happens if we actually have a finite domain? Suppose we have the same differential equation, Eq. (5-1), but the boundary conditions are

$$c(0, t) = 1 \qquad c(1, t) = 0 \tag{5-43}$$

and initial condition is

$$c(x, 0) = 0 \tag{5-44}$$

The transformation Eq. (5-18) gives

$$x = 0 \rightarrow \eta = 0 \qquad x = 1 \rightarrow \eta = \frac{1}{t^{1/2}} \tag{5-45}$$

The analogs of Eqs. (5-12) and (5-13) are

$$c(\bar{x}a^{-\beta}, 0) = 0 \tag{5-46}$$

$$c(a^{\alpha/2}\bar{t}^{-1/2}, a^{-\alpha}\bar{t}) = 0 \tag{5-47}$$

and these must combine into one condition and be conformally invariant. Despite the fact that the right-hand sides are the same it is not possible to combine the conditions; Eq. (5-46) applies at $\eta = \infty$, while Eq. (5-47) applies at $\eta = 1/t^{1/2}$. The similarity transformation fails for this problem with a finite domain.

One clue to the existence of a similarity transformation for the original problem expressed in Eqs. (5-1) to (5-4) is the infinite domain. Examination of the equations reveals that there is no natural length scale, such as provided by the domain thickness if the problem had a finite domain. Yet the equations contain the dimensions of length, and if they are made nondimensional a length scale must be introduced. What length scale should be used when the problem has no natural one? There is none. In such situations a similarity transformation is always suggested.

Similarity transformations can often be applied to problems with finite domains as an approximation for small times. Consider Fig. 5-1 for diffusion in a slab. At small times the solution is nonzero only near $x = 0$, and the other boundary $x = 1$ may as well be at infinity. In fact, of course, the exact mathematical solution is not exactly zero near $x = 1$ but it may be so small, such as 10^{-38}, that we can regard it as being zero. Only when time proceeds and the solution begins to change significantly near the point $x = 1$ does the finite boundary have

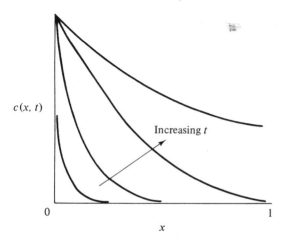

$c(x, t)$

Increasing t

x

Figure 5-1 Diffusion in a slab.

to be introduced. For small times, then, we can regard the domain as semi-infinite and apply a similarity transformation.

Let us apply this technique to the following linear problem for diffusion of heat or mass in a slab where

$$\frac{\partial c}{\partial t} = \frac{\partial^2 c}{\partial x^2} \tag{5-48}$$

$$c(x, 0) = 0 \qquad c(0, t) = 1 \tag{5-49}$$

$$c(1, t) = 0 \tag{5-50}$$

For small times let us replace the condition of Eq. (5-50) by the approximation

$$c(\infty, t) = 0 \tag{5-51}$$

The solution to Eqs. (5-48), (5-49) and (5-51) is Eq. (5-42). We are interested in when the solution at $x = 1$ is noticeable, and we choose 10^{-3} as a noticeable value. The erfc takes the value 10^{-3} for $\eta = 2.327$. The point x at which η takes the value 2.327 moves with $t^{1/2}$ since

$$x = 2.327t^{1/2} \tag{5-52}$$

When the time is large enough that this point reaches $x = 1$ the approximation of the similarity solution is no longer valid. For larger times another technique must be employed, as illustrated in Sec. 5-2.

The problem to be considered is the transient version of Eq. (4-3) for heat conduction in a slab with a temperature-dependent thermal conductivity. We take only the case $a = 1$, thus

$$\frac{\partial \theta}{\partial t} = \frac{\partial}{\partial x}\left[(1+\theta)\frac{\partial \theta}{\partial x}\right] \tag{5-53}$$

$$\theta(x, 0) = 0 \qquad \theta(0, t) = 1 \tag{5-54}$$

$$\theta(1, t) = 0 \tag{5-55}$$

Because of the finite domain the problem has no similarity solution. We replace the last boundary condition by the condition at infinity

$$\theta(\infty, t) = 0 \tag{5-56}$$

Equations (5-53), (5-54) and (5-56) have a similarity solution, and this solution is a good approximation to the original problem with Eq. (5-55) for small time.

We introduce the similarity variable

$$\eta = \frac{x}{\sqrt{4t}} \tag{5-57}$$

and transform [see Eq. (5-20)] the equation to

$$\frac{d}{d\eta}\left[(1+\theta)\frac{d\theta}{d\eta}\right] + 2\eta\frac{d\theta}{d\eta} = 0 \tag{5-58}$$

The transformation Eq. (5-27) gives

$$\psi = \theta + \tfrac{1}{2}\theta^2 \tag{5-59}$$

but Eq. (5-30) is no easier to solve than Eq. (5-58). The integral solution given by Eq. (5-38) cannot be evaluated analytically. The alternatives are to resort to numerical solution of Eq. (5-32), iterative solutions of Eq. (5-38), or numerical solutions obtained with shooting methods. All of these require the use of a computer. Instead, we solve Eq. (5-58) using the Method of Weighted Residuals.

We assume that Eq. (5-58) is written for the trial function so that it is the residual. Multiplying by the weighting function, denoted here by $\delta\theta$,

$$\int_0^\infty \delta\theta \frac{d}{d\eta}\left[(1+\theta)\frac{d\theta}{d\eta}\right]d\eta + 2\int_0^\infty \delta\theta\eta\frac{d\theta}{d\eta}\,d\eta = 0 \tag{5-60}$$

and integrating by parts we obtain

$$\int_0^\infty \frac{d}{d\eta}\left[\delta\theta(1+\theta)\frac{d\theta}{d\eta}\right]d\eta - \int_0^\infty (1+\theta)\frac{d\delta\theta}{d\eta}\frac{d\theta}{d\eta}\,d\eta + 2\int_0^\infty \delta\theta\eta\frac{d\theta}{d\eta}\,d\eta = 0 \tag{5-61}$$

The first term can be evaluated at the boundaries, but we are going to require that $\delta\theta = 0$ at both boundaries. Thus the Galerkin equation is

$$-\int_0^\infty (1+\theta)\frac{d\delta\theta}{d\eta}\frac{d\theta}{d\eta}\,d\eta + 2\int_0^\infty \delta\theta\eta\frac{d\theta}{d\eta}\,d\eta = 0 \tag{5-62}$$

We need to choose a trial function that takes the value one at $\eta = 0$ and approaches zero as $\eta \to \infty$. It would also be advantageous if the first derivative approached zero as $\eta \to \infty$. A polynomial function meeting these criteria is

$$\theta = \begin{cases} (1-a\eta)^2 & \eta < \dfrac{1}{a} \\[2mm] 0 & \eta > \dfrac{1}{a} \end{cases} \tag{5-63}$$

The parameter a is to be chosen by the Galerkin criterion. For this function we have

$$\frac{d\theta}{d\eta} = -2a(1-a\eta) \tag{5-64}$$

$$K = 1 + \theta = 1 + (1-a\eta)^2 \tag{5-65}$$

The variation, or weighting function, is

$$\delta\theta = \frac{\partial\theta}{\partial a} = -2\eta(1-a\eta) \tag{5-66}$$

$$\frac{d\delta\theta}{d\eta} = -2(1-2a\eta) \tag{5-67}$$

The Galerkin criterion then requires that

$$-\int_0^{1/a} [1+(1-a\eta)^2][-2(1-2a\eta)][-2a(1-a\eta)]d\eta$$

$$+2\int_0^{1/a} [-2\eta(1-a\eta)]\eta[-2a(1-a\eta)]d\eta = 0 \qquad (5\text{-}68)$$

The upper limit of integration is taken as $1/a$, since the function and residual are zero for $\eta \geqslant 1/a$. By taking $u = a\eta$ we transform the integrals into

$$-I_2 + \frac{2}{a^2}I_1 = 0 \qquad (5\text{-}69)$$

$$I_2 = \int_0^1 (2-2u+u^2)(1-2u)(1-u)du$$

$$I_1 = \int_0^1 u^2(1-u)^2 du \qquad (5\text{-}70)$$

For the linear problem with a constant thermal conductivity the first term in the integral I_2 is 1.0. The values are

$$\begin{array}{ll} I_2 = \frac{19}{60} & I_1 = \frac{1}{30} \\ I_2 = \frac{1}{6} & \text{for constant } k \end{array} \qquad (5\text{-}71)$$

$$\begin{array}{ll} a^2 = \frac{4}{19} & \text{for nonlinear} \\ a^2 = \frac{2}{5} & \text{for linear} \end{array} \qquad (5\text{-}72)$$

The final solution is Eq. (5-63) with a^2 defined by Eqs. (5-72). This solution is an approximation of the finite domain problem, too, provided that the outer boundary $\eta = 1/a$ does not reach $x = 1$. This condition is

$$\eta = \frac{1}{a} = \frac{x}{\sqrt{4t}}\bigg|_{x=1} = \frac{1}{\sqrt{4t}} \qquad (5\text{-}73)$$

so that the same solution is used for the finite domain and $t < a^2/4$. The approximate solution is compared to the exact solution of the problem on an infinite domain in Fig. 5-2 and gives an approximation that is within engineering accuracy.

5-2 SEPARATION OF VARIABLES

When the domain is finite the classical method of solving the linear problem is to apply the separation of variables. Let us consider diffusion in a slab with reaction

$$\frac{\partial c}{\partial t} = \frac{\partial}{\partial x}\left[D(c)\frac{\partial c}{\partial x}\right] + R(c) = D(c)\frac{\partial^2 c}{\partial x^2} + \frac{dD}{dc}\left(\frac{\partial c}{\partial x}\right)^2 + R(c) \qquad (5\text{-}74)$$

$$c(x,0) = 0 \qquad (5\text{-}75)$$

$$c(0,t) = 1 \qquad c(1,t) = 0 \qquad (5\text{-}76)$$

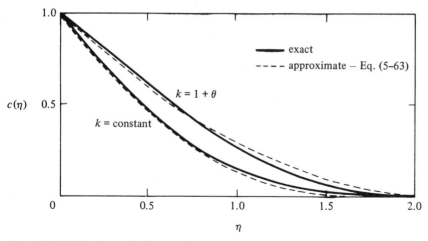

Figure 5-2 Diffusion in semi-infinite medium.

We try a solution of the form

$$c(x, t) = X(x)T(t) \tag{5-77}$$

If we can separate the problem into one ordinary differential equation for $X(x)$ and another for $T(t)$ the solution will be simpler than if we have to solve for $c(x, t)$ together. Putting Eq. (5-77) into Eq. (5-74), dividing by TX, and rearranging we get

$$\frac{1}{T}\frac{dT}{dt} = \frac{D}{X}\frac{d^2X}{dx^2} + \frac{dD}{dc}\frac{T}{X}\left(\frac{dX}{dx}\right)^2 + \frac{R(XT)}{XT} \tag{5-78}$$

If the diffusion coefficient depends on concentration $D(c)$ or if the reaction rate term $R(c)$ is nonlinear, it is not possible to separate Eq. (5-78) into one side that depends on time only and one side that depends on position only. Separation of variables then fails.

Simplifying Eq. (5-74) by taking D constant and the reaction rate linear (i.e. $R = -kc$), we get

$$\frac{\partial c}{\partial t} = D\frac{\partial^2 c}{\partial x^2} - kc \tag{5-79}$$

Equation (5-78) then gives

$$\frac{1}{DT}\frac{dT}{dt} = \frac{1}{X}\frac{d^2X}{dx^2} - \frac{k}{D} \tag{5-80}$$

One side of Eq. (5-80) is a function of t alone, and the other side is a function of x alone. Yet both sides equal each other. The only way this can be true is if the function is a constant. Otherwise, if x is fixed one side is fixed and the other side could be changed by changing t, the equality would not hold. We let the constant be λ and write the separate equations

$$\frac{1}{DT}\frac{dT}{dt} = -\lambda \tag{5-81}$$

$$\frac{1}{X}\frac{d^2X}{dx^2} - \frac{k}{D} = -\lambda \tag{5-82}$$

Equation (5-81) is easily solved in the form

$$T(t) = Ae^{-\lambda Dt} \tag{5-83}$$

and Eq. (5-82) is written as

$$\frac{d^2X}{dx^2} + \left(\lambda - \frac{k}{D}\right)X = 0 \tag{5-84}$$

We next simplify the boundary conditions

$$c(1, t) = 0 = T(t)X(1) \tag{5-85}$$

$$c(0, t) = 1 = T(t)X(0) \tag{5-86}$$

The first condition, Eq. (5-85), gives $X(1) = 0$ but the second one, Eq. (5-86), does not separate. We need to make the boundary conditions of the problem homogeneous. This is done by finding a function that satisfies the nonhomogeneous boundary conditions $(1-x)$, and then solving for the remainder. Let us then solve for

$$u(x, t) = (x-1) + c(x, t) \tag{5-87}$$

The differential equation in u is

$$\frac{\partial u}{\partial t} = D\frac{\partial^2 u}{\partial x^2} \tag{5-88}$$

which is the same, while the initial condition is

$$u(x, 0) = x - 1 + c(x, 0) = x - 1 \tag{5-89}$$

The boundary conditions on u are then

$$u(0, t) = -1 + c(0, t) = 0 \tag{5-90}$$

$$u(1, t) = c(1, t) = 0 \tag{5-91}$$

We use $u(x, t) = X(x)T(t)$ and arrive at the same equations for T and X, but with $X(0) = X(1) = 0$.

We solve Eq. (5-84) for the case without reaction ($k = 0$)

$$\frac{d^2X}{dx^2} + \lambda X = 0 \tag{5-92}$$

$$X(0) = X(1) = 0 \tag{5-93}$$

The equation is linear and easily solved. The general solution is obtained by trying e^{mx} and finding that $m^2 + \lambda = 0$, thus $m = \pm i\sqrt{\lambda}$. The exponential term

$\exp(\pm i \sqrt{\lambda} x)$ is written in terms of sines and cosines, so that the general solution is

$$X = B \cos \sqrt{\lambda} x + E \sin \sqrt{\lambda} x \tag{5-94}$$

The boundary conditions are

$$X(1) = B \cos \sqrt{\lambda} + E \sin \sqrt{\lambda} = 0 \tag{5-95}$$

$$X(0) = B = 0 \tag{5-96}$$

Now if $B = 0$ and we want $D \neq 0$ (otherwise the solution is zero), we must have

$$\sin \sqrt{\lambda} = 0 \tag{5-97}$$

This is true only for certain values of λ, which are called the eigen or characteristic values,

$$\lambda_n = n^2 \pi^2 \tag{5-98}$$

For each eigen value we have a corresponding eigen function

$$X_n(x) = E \sin n\pi x \tag{5-99}$$

The composite solution is then

$$X_n(x) T_n(t) = EA \sin n\pi x \, e^{-\lambda_n D t} \tag{5-100}$$

This function satisfies the boundary conditions and the differential equation but not the initial condition. To do that we add up several of these solutions, each with a different eigen function, and replace EA by A_n

$$u(x, t) = \sum_{n=1}^{\infty} A_n \sin n\pi x \, e^{-n^2 \pi^2 D t} \tag{5-101}$$

The constants A_n are chosen by making $u(x, t)$ satisfy the initial condition

$$u(x, 0) = \sum_{n=1}^{\infty} A_n \sin n\pi x = x - 1 \tag{5-102}$$

We define the residual as the error in the initial condition

$$R(x) = x - 1 - \sum_{n=1}^{\infty} A_n \sin n\pi x \tag{5-103}$$

We next apply the Galerkin method and make the residual orthogonal to a complete set of functions, which are the eigen functions,

$$\int_0^1 (x - 1) \sin m\pi x \, dx = \sum_{n=1}^{\infty} A_n \int_0^1 \sin m\pi x \sin n\pi x \, dx = \frac{A_m}{2} \tag{5-104}$$

It turns out that the Galerkin criterion for finding A_m also satisfies a least squares criterion: the residual squared and integrated over x is minimized by the A_m given by Eq. (5-104) (see problem 5-6). The final solution is then

$$c(x, t) = 1 - x + \sum_{n=1}^{\infty} A_n \sin n\pi x \, e^{-n^2 \pi^2 D t} \tag{5-105}$$

This solution is an exact solution to the linear problem. To evaluate the solution an infinite number of terms must be evaluated, but a finite number of terms can give a good approximation. For large times a single term is sufficient since the exponential term decreases so rapidly with n, but for small times, a large number of terms is necessary. For $t \to 0$ only an infinite number suffices. In that case it may be better to use the similarity approximation (assuming the domain at $x = 1$ is really at $x \to \infty$). Problem 5-5 illustrates this point.

For nonlinear problems the method of separation of variables fails, and we must use either similarity solutions (for small times). Alternatively, the Method of Weighted Residuals or the numerical methods described below may be employed.

5-3 METHOD OF WEIGHTED RESIDUALS

The method is first illustrated using the linear problem, Eqs. (5-48) to (5-50). This problem has been solved twice before. Then the nonlinear problem of Eq. (5-53) is solved. The latter problem has been solved before, but only with an approximate solution, and only for an infinite domain. If the domain is finite and the solution is desired for all times, none of the techniques in Secs. 5-1 and 5-2 apply. The Method of Weighted Residuals or one of the numerical methods is the only choice then.

For the linear problem we expand the trial function in the series

$$c(x, t) = c_0(x) + \sum_{i=1}^{N} A_i(t)c_i(x) \tag{5-106}$$

We choose $c_0(x)$ to satisfy the nonhomogeneous boundary conditions

$$c_0(0) = 1 \qquad c_0(1) = 0 \tag{5-107}$$

The simplest form is

$$c_0(x) = 1 - x \tag{5-108}$$

Next we choose the $c_i(x)$ to satisfy the homogeneous boundary conditions

$$c_i(0) = c_i(1) = 0 \tag{5-109}$$

Possible choices are

$$c_i(x) = \sin i\pi x \tag{5-110}$$

$$= x^i(1-x) \qquad \text{or} \qquad x(1-x)^i \tag{5-111}$$

$$= x(1-x)P_{i-1}(x) \tag{5-112}$$

First we consider Eq. (5-110) and apply the Galerkin method. The residual is

$$\frac{1}{D}\frac{\partial c}{\partial t} - \frac{\partial^2 c}{\partial x^2} = \frac{1}{D}\sum_{i=1}^{N}\frac{dA_i}{dt}\sin i\pi x + \sum_{i=1}^{N}(i\pi)^2 A_i \sin i\pi x \tag{5-113}$$

In the Galerkin method we make the residual orthogonal to $\sin j\pi x$. This gives

$$\frac{1}{D}\frac{dA_i}{dt} = -(i\pi)^2 A_i \tag{5-114}$$

which is solved to give

$$A_i = A_i(0)e^{-i^2\pi^2 Dt} \tag{5-115}$$

and the final solution is then

$$c(x, t) = 1 - x + \sum_{i=1}^{N} A_i(0)\sin i\pi x\, e^{-i^2\pi^2 Dt} \tag{5-116}$$

The constants $A_i(0)$ are obtained by applying the Galerkin method to the initial residual $c(x, 0) = 0$. As before we get

$$\tfrac{1}{2}A_i(0) = \int_0^1 (x-1)\sin i\pi x\, dx \tag{5-117}$$

and we get the same solution. Thus the Galerkin method applied to linear problems gives the first N terms of the exact solution found by separation of variables when the expansion functions are the exact eigen functions.

Next, we make the choice

$$c_i(x) = x^i(1 - x) \tag{5-118}$$

Then the derivatives of $c_i(x)$ are

$$\begin{aligned}
\frac{\partial c_i}{\partial x} &= ix^{i-1} - (i+1)x^i \\
\frac{\partial^2 c_i}{\partial x^2} &= ix^{i-2}[i - 1 - (i+1)x]
\end{aligned} \tag{5-119}$$

and the residual is

$$\text{Residual} = \frac{1}{D}\sum_{i=1}^{N}\frac{dA_i}{dt}x^i(1-x) - \sum_{i=1}^{N} A_i ix^{i-2}[i - 1 - (i+1)x] \tag{5-120}$$

We make this orthogonal to $x^j(1 - x)$

$$\begin{aligned}
\frac{1}{D}\sum_{i=1}^{N}\frac{dA_i}{dt}\int_0^1 x^{i+j}(1-x)^2 dx \\
= \sum_{i=1}^{N} A_i i\int_0^1 x^{i+j-2}(1-x)[i - 1 - (i+1)x]dx
\end{aligned} \tag{5-121}$$

These equations can be written in the format

$$\frac{1}{D}\sum_{i=1}^{N} B_{ji}\frac{dA_i}{dt} = \sum_{i=1}^{N} C_{ji}A_i \tag{5-122}$$

which can be solved with matrix methods or numerically. We note that the only difference between this solution and Eq. (5-116) is the choice of trial function—sine functions versus polynomials—making the matrices in Eq. (5-122) different. Because the sine functions are the exact eigen functions in that case, Eq. (5-122) simplifies to Eq. (5-114), but otherwise the methods are similar. Equation (5-114) is much simpler than Eq. (5-122) because the time equations decouple and are easily solved. The initial conditions are evaluated by making the initial residual orthogonal to $x^j(1-x)$. This gives a set of N equations to solve for the $NA_i(0)$. We expect that as N is increased the accuracy of the solution is improved, just as we know that if only N terms of the exact solution given by Eq. (5-101) are used we have an error that decreases as more terms are included.

For a single term we have from Eq. (5-122)

$$\frac{1}{D} \cdot \frac{1}{30} \frac{dA_1}{dt} = -\frac{1}{3} A_1 \tag{5-123}$$

$$A_1(t) = A_1(0)e^{-10Dt}$$

$$c(x, 0) = 1 - x + A_1(0)x(1-x) = 0 \tag{5-124}$$

Applying the Galerkin method to Eq. (5-124) gives $A_1(0) = -2.5$. The complete solution is then

$$c(x, 0) = 1 - x - 2.5e^{-10Dt}x(1-x) \tag{5-125}$$

We do not expect this solution to be good for small times, and it is not. It is a reasonable approximation for larger times, however, and Eq. (5-63) is valid for small times.

Whenever we need the solution for small times we know that we must use many terms of the exact solution in order to achieve accurate results. Note at this point that we must use a computer to evaluate the exact solution even if we can write it down in analytic form. Thus any Method of Weighted Residuals that expands the solution in a similar way must need many terms to achieve good results for small times. If we want the solution for small times, it is easier to replace the finite domain by an infinite domain and solve that problem. Then when the concentration begins to change at the far boundary we go back to the expansion solution. One advantage of the Method of Weighted Residuals is that we can write the approximate solution as the solution derived from the similarity transformation plus a series. Initially, the series is small, but later it becomes important. In this problem we use

$$c(x, t) = 1 - \text{erf} \frac{x}{\sqrt{4Dt}} + u(x, t) \tag{5-126}$$

The first term satisfies the differential equation and all boundary conditions, except that at $x = 1$. We make $u(x, t)$ satisfy

$$\frac{1}{D} \frac{\partial u}{\partial t} = \frac{\partial^2 u}{\partial x^2} \tag{5-127}$$

$$u(0, t) = 0$$

$$u(1, t) = -1 + \text{erf} \frac{1}{\sqrt{4Dt}} \tag{5-128}$$

$$u(x, 0) = 0$$

For small time u is small, but for larger times u increases, but the total solution given in Eq. (5-126) always satisfies all conditions: differential equation, boundary and initial conditions. If the Method of Weighted Residuals is used to satisfy Eq. (5-127), the differential equation is satisfied approximately. However, a single term often suffices for good accuracy. As an example, we use a single term

$$u(x, t) = x\left(-1 + \text{erf} \frac{1}{\sqrt{4Dt}}\right) + A(t)x(1-x) \tag{5-129}$$

The in..ial condition is satisfied if $A(0) = 0$. The Galerkin weighted residual is

$$\frac{1}{D}[f(t)] \int_0^1 x^2(1-x)dx + \frac{1}{D}\frac{dA}{dt}\int_0^1 x^2(1-x)^2 dx$$

$$+ 2A(t)\int_0^1 x(1-x)dx = 0 \tag{5-130}$$

where

$$\frac{d}{dt}\left(\text{erf}\frac{1}{\sqrt{4Dt}}\right) \equiv f(t) = \frac{-e^{-1/(4Dt)}}{\sqrt{4\pi Dt^3}} \tag{5-131}$$

This gives

$$\frac{1}{D}\frac{dA}{dt} = -10A(t) + \frac{2.5}{D}[f(t)] \tag{5-132}$$

$$A(0) = 0 \tag{5-133}$$

The complete solution, Eq. (5-126), is a good approximation for all times, as shown in Fig. 5-3. The exact and approximate solutions are indistinguishable on the graph.

Suppose we have a nonlinear diffusion problem with $D(c)$ in a finite domain. A similarity transformation is not possible on the finite domain, and separation of variables is not possible for a nonlinear problem. However, the Method of Weighted Residuals is applicable and can give good results. We apply it to Eqs. (5-53) to (5-55).

For small time we simply use Eq. (5-63) for $t < a^2/4$; $a^2 = 4/19$ for the nonlinear problem, and $a^2 = 2/5$ for the linear problem. For larger times we use the expansion

$$\theta(x, t) = a(t) + b(t)x + c(t)x^2 \tag{5-134}$$

The boundary conditions require

$$a = 1 \qquad a + b + c = 0 \tag{5-135}$$

The trial function is then

$$\theta(x, t) = (1 - x^2) + b(t)x(1 - x) \tag{5-136}$$

We apply the Galerkin method to Eq. (5-53) and integrate by parts as in Eqs. (5-60) to (5-62)

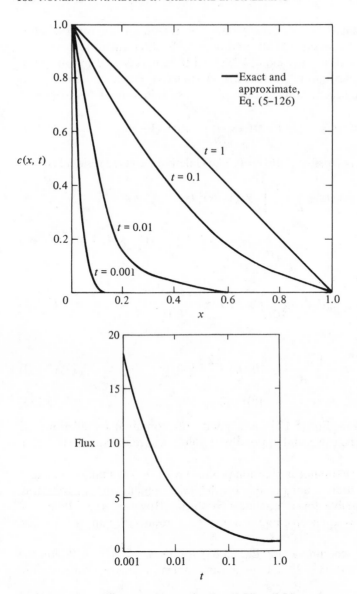

Figure 5-3 Linear diffusion problems using Eqs. (5-74) to (5-76).

$$\int_0^1 \delta\theta \, \frac{\partial\theta}{\partial t} \, dx = - \int_0^1 (1+\theta) \frac{d\delta\theta}{dx} \frac{\partial\theta}{\partial x} \, dx \qquad (5\text{-}137)$$

The weighting function is

$$\delta\theta = x(1-x) \qquad \frac{d\delta\theta}{dx} = 1-2x \qquad (5\text{-}138)$$

The other terms are

$$\frac{\partial \theta}{\partial t} = \frac{db}{dt} x(1-x) \qquad \frac{\partial \theta}{\partial x} = -2x + b(1-2x) \qquad (5\text{-}139)$$

Combining all terms gives

$$I_1 \frac{db}{dt} = I_2 + I_3 b + I_4 b^2 \qquad (5\text{-}140)$$

$$I_1 = \int_0^1 x^2(1-x)^2 dx = \tfrac{1}{30}$$

$$I_2 = 2 \int_0^1 x(2-x^2)(1-2x)dx = -\tfrac{11}{30}$$

$$I_3 = 2 \int_0^1 x^2(1-x)(1-2x)dx = -\tfrac{17}{30} \qquad (5\text{-}141)$$

$$I_4 = -\int_0^1 x(1-x)(1-2x)^2 dx = -\tfrac{1}{30}$$

Thus

$$\frac{db}{dt} = -11 - 17b - b^2 \qquad (5\text{-}142)$$

For the linear problem $I_2 = -1/30$, $I_3 = -1/3$, and $I_4 = 0$ giving

$$\frac{db}{dt} = -10 - 10b \qquad (5\text{-}143)$$

We fit the initial conditions at $t = t_0 = a^2/4$, where a^2 is given by Eqs. (5-72). At that time Eq. (5-63) gives

$$\theta = (1-x)^2 \qquad (5\text{-}144)$$

while Eq. (5-136) gives

$$\theta = 1 - x^2 + bx(1-x) \qquad (5\text{-}145)$$

These agree if we take

$$b(t_0) = -2 \qquad (5\text{-}146)$$

Therefore the nonlinear problem of Eqs. (5-53) to (5-55) is solved by Eqs. (5-63), (5-57) and (5-72) for $t < a^2/4$ and by Eq. (5-136) for $t > a^2/4$, with $b(t)$ determined by Eqs. (5-142) and (5-146). The linear problem is solved by the same function for $t < a^2/4$ with different a^2, and with $b(t)$ the solution to Eqs. (5-143) and (5-146). These approximations are shown in Fig. 5-4 and are reasonable compared with the exact solution. Interestingly, these approximate solutions are simple enough to be solved entirely on a programmable calculator with about 30 program steps and eight memory registers. Equation (5-142) is solved using Euler's method with a fixed Δt for several Δt. The results are then extrapolated to get the best estimate as $\Delta t \to 0$.

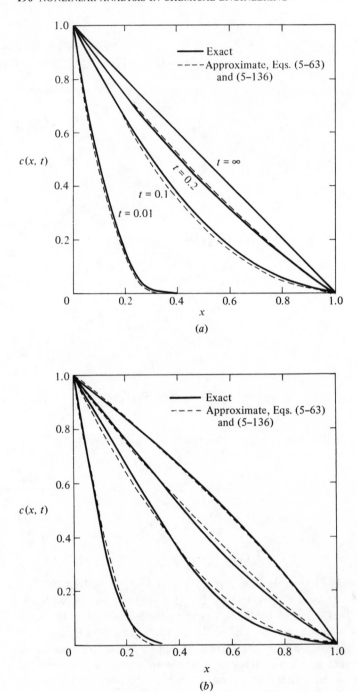

Figure 5-4 Linear and nonlinear diffusion problems. (a) $k = 1 + \theta$. (b) $k = 1 + \theta$.

5-4 ORTHOGONAL COLLOCATION

Orthogonal collocation is easy to apply to transient problems. Take the nonlinear diffusion problem, Eqs. (5-74) to (5-76). The solution exhibits no particular symmetry properties so that we expand in polynomials in x

$$c(x, t) = a(t) + b(t)x + x(1-x) \sum_{i=1}^{N} a_i(t)P_{i-1}(x) \tag{5-147}$$

We can write this as

$$c(x, t) = \sum_{i=1}^{N+2} d_i(t)x^{i-1} \tag{5-148}$$

and evaluate it at the collocation points

$$c(x_j, t) = \sum_{i=1}^{N+2} d_i(t)x_j^{i-1} \quad \text{or} \quad \mathbf{c}(t) = \mathbf{Qd}(t) \tag{5-149}$$

The first space derivative evaluated at the collocation points gives

$$\left.\frac{\partial c}{\partial x}\right|_{x_j} = \sum_{i=1}^{N+2} d_i(t)(i-1)x_j^{i-2} \quad \text{or} \quad \frac{\partial \mathbf{c}}{\partial x} = \mathbf{Cd}(t) \tag{5-150}$$

However, $Q_{ji} = x_j^{i-1}$ is independent of t so we can invert it once to obtain

$$\frac{\partial \mathbf{c}}{\partial x} = \mathbf{CQ}^{-1}\mathbf{c} \equiv \mathbf{Ac} \tag{5-151}$$

This is the same as Eq. (4-103), except that now both $\mathbf{c}$ and $\partial \mathbf{c}/\partial \mathbf{x}$ are functions of time. In addition

$$\left.\frac{\partial c}{\partial t}\right|_{x_j} = \frac{d}{dt}c(x_j, t) = \frac{dc_j}{dt} \tag{5-152}$$

Thus for the diffusion problem of Eq. (5-74) we write the collocation equations either as

$$\frac{dc_j}{dt} = \sum_{i=1}^{N+2} A_{ji}D(c_i) \sum_{l=1}^{N+2} A_{il}c_l \tag{5-153}$$

or as

$$\frac{dc_j}{dt} = D(c_j) \sum_{i=1}^{N+2} B_{ji}c_i + \frac{dD}{dc}(c_j)\left(\sum_{i=1}^{N+2} A_{ji}c_i\right)^2 \tag{5-154}$$

with the usual boundary and initial conditions. Equation (5-153) can be written as

$$\frac{dc_j}{dt} = \sum_{l=1}^{N+2} AA_{jl}(c)c_l \qquad AA_{jl} = \sum_{i=1}^{N+2} A_{ji}D(c_i)A_{il} \tag{5-155}$$

By applying orthogonal collocation we can reduce the partial differential equation Eq. (5-74) to a set of ordinary differential equations as Eq. (5-155) that are initial-value problems. We solve them using any of the methods discussed in Chapter 3. For both linear and nonlinear problems near $t = 0$ we may want to solve the subsidiary problem of Eqs. (5-126) and (5-127), and apply orthogonal collocation to get $u(x, t)$.

The most widespread application of orthogonal collocation is in reaction and diffusion problems, such as the transient counterpart to problems solved in Chapter 4. The first example is for the equations governing a packed bed reactor with radial dispersion. The basic equations in cylindrical geometry are

$$\frac{\partial T}{\partial z} = \alpha' \nabla^2 T + \beta' R(c, T)$$

$$\frac{\partial c}{\partial z} = \alpha \nabla^2 c + \beta R(c, T) \tag{5-156}$$

$$\frac{\partial T}{\partial r} = \frac{\partial c}{\partial r} = 0 \qquad \text{at} \qquad r = 0 \tag{5-157}$$

$$-\frac{\partial T}{\partial r}\bigg|_{r=1} = \text{Bi}_w[T(1, z) - T_w(z)] \qquad -\frac{\partial c}{\partial r}\bigg|_{r=1} = 0 \tag{5-158}$$

$$T(r, 0) = T_0 \qquad c(r, 0) = c_0 \tag{5-159}$$

$$\alpha = \frac{Ld_p}{R^2 \text{Pe}_m} \qquad \alpha' = \frac{Ld_p}{R^2 \text{Pe}_h}$$

$$\text{Pe}_m = \frac{Gd_p}{\rho D_e} \qquad \text{Pe}_h = \frac{C_p Gd_p}{k_e}$$

$$\beta = \frac{k_0 L}{G} \qquad \beta' = \frac{(-\Delta H_{rxn})k_0 c_0 L}{C_p G T_0} \tag{5-160}$$

$$\text{Bi}_w = \frac{h_w R}{k_e}$$

The reactor length is L, its radius is R, and the catalyst diameter is d_p. The mass flux is G (mass per total cross sectional area per unit time) and the density is ρ, so that the superficial velocity is G/ρ. Dispersion in the radial directions is modeled by the dispersion coefficients D_e for mass and k_e for heat. The fluid heat capacity is C_p, while the reaction rate constant is k_0. The heat of reaction is $-\Delta H_{rxn}$. Heat is transferred at the wall from the reactor to the surroundings, and the rate is governed by the heat transfer coefficient h_w in Eq. (5-158). In Eqs. (5-156), the first term represents the convection of heat and mass, while the next term is for radial dispersion caused by the flow around the packing. The last term is the reaction rate term.

It is possible to have a problem in which this reaction rate depends on the diffusion of heat and mass in a catalyst pellet. In such cases it is necessary to solve a two-point boundary-value problem (involving several dependent variables) at each (r, z) location in order to evaluate $R(c, T)$ in Eqs. (5-156). Orthogonal collocation, which is also useful then, is illustrated below for a stirred tank problem. Here we assume that the reaction rate is just a known function of the bulk temperature and concentration

$$R(c, T) = (1-c)e^{\gamma - \gamma/T} \tag{5-161}$$

Applying orthogonal collocation gives the following equations from the residuals:

$$\frac{dT_j}{dz} = \alpha' \sum_{i=1}^{N+1} B_{ji} T_i + \beta' R(c_j, T_j)$$

$$\frac{dc_j}{dz} = \alpha \sum_{i=1}^{N+1} B_{ji} c_i + \beta R(c_c, T_j) \tag{5-162}$$

The initial conditions are

$$T_j(0) = T_0 \qquad c_j(0) = c_0 \tag{5-163}$$

and the boundary conditions are

$$-\sum_{i=1}^{N+1} A_{N+1,i} T_i = \text{Bi}_w(T_{N+1} - T_w) \qquad -\sum_{i=1}^{N+1} A_{N+1,i} c_i = 0 \tag{5-164}$$

The boundary conditions at $r = 0$ are automatically satisfied by using polynomials that are functions of r^2. This gives $2N$ ordinary differential equations coupled with two linear algebraic equations. It is convenient to solve the boundary conditions for c_{N+1} and T_{N+1} and to introduce these values into the differential equations

$$\frac{dT_j}{dz} = \alpha' \sum_{i=1}^{N} B'_{ji} T_i + \frac{B_{j,N+1} \text{Bi}_w T_w}{\text{Bi}_w + A_{N+1,N+1}} + \beta' R_j$$

$$\frac{dc_j}{dz} = \alpha \sum_{i=1}^{N} B'_{ji} c_i + \beta R_j \tag{5-165}$$

$$B'_{ji} = B_{ji} - \frac{B_{j,N+1} A_{N+1,i}}{\text{Bi}_w + A_{N+1,N+1}} \tag{5-166}$$

and $\text{Bi}_w = 0$ for the concentration equation. These are integrated from the initial conditions, Eq. (5-163), using a variety of methods for comparison.

The orthogonal collocation method gives a first approximation that can be used to gain a physical insight into the solution. For the first approximation and cylindrical geometry the equations are

$$\frac{dT_1}{dz} = -\frac{6\alpha' \text{Bi}}{3 + \text{Bi}} (T_1 - T_w) + \beta' R_1$$

$$\frac{dc_1}{dz} = \beta R_1 \tag{5-167}$$

These equations take the same form as the lumped parameter model, in which case no gradients are allowed,

$$\frac{dT}{dz} = -\text{Nu}_w(T - T_w) + \beta' R$$

$$\frac{dc}{dz} = \beta R \tag{5-168}$$

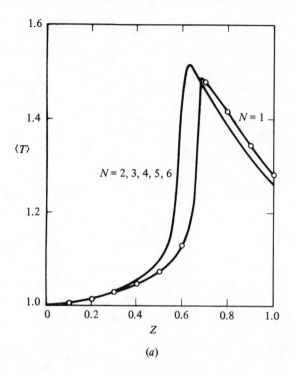

(a)

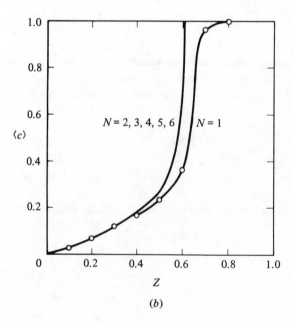

(b)

Figure 5-5 (a) Average temperature for Eq. (5-165), with $Bi_w = 1$ and $T_w = 0.92$. (b) Average conversion.

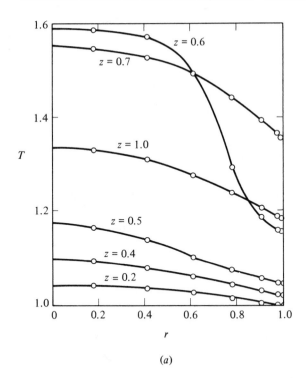

(a)

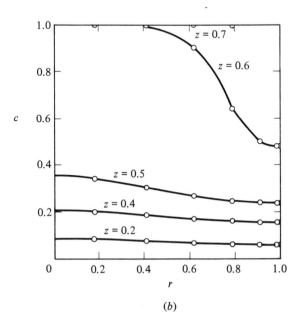

(b)

Figure 5.6 (a) Radial temperature profiles for Eq. (5-165) with $Bi_w = 1$, $T_w = 0.92$, and $N = 6$. (b) Radial conversion profiles.

$$\text{Nu}_w = \frac{2UL}{c_p GR} \tag{5-169}$$

Equations (5-167) to (5-169) are the same provided

$$\text{Nu}_w = \frac{6\alpha' \text{Bi}_w}{3 + \text{Bi}_w} \quad \text{or} \quad \frac{1}{\text{Nu}_w} = \frac{1}{2\alpha'} \left(\frac{1}{\text{Bi}_w} + \frac{1}{3} \right) \tag{5-170}$$

or when

$$\frac{1}{U} = \frac{1}{h_w} + \frac{R}{3k_e} \tag{5-171}$$

The relative importance of the wall resistance is evident by computing the two terms in Eq. (5-171). Comparing $1/h_w$ to $1/U$ gives an idea of the fraction of the total heat transfer resistance that occurs at the wall. The equivalent comparison is

$$\frac{U}{h_w} = \frac{1}{1 + \text{Bi}_w/3} \tag{5-172}$$

For $\text{Bi}_w = 1$, 75 percent of the resistance is at the wall, and the temperature profiles are expected to be relatively uniform inside the bed. A low-order approximation (say $N = 1$ or 2) is appropriate. For $\text{Bi}_w = 10$ the wall resistance is 23 percent of the total, and for $\text{Bi}_w = 20$ it is only 13 percent. In the latter case steep temperature

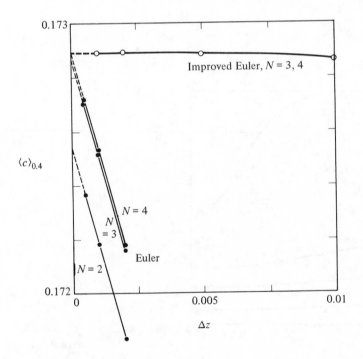

Figure 5-7 Extrapolation of orthogonal collocation solution with Δz and $\text{Bi}_w = 1.0$.

gradients are possible in the radial direction and more collocation points ($N = 6$) are needed.

For the first example we choose $\alpha = \alpha' = 1$, $\beta = 0.3$, $\beta' = 0.2$, $\gamma = 20$, $Bi_w = 1$, and $T_w = 0.92$. Since this case has a low Biot number we expect that the temperature profiles are relatively uniform in the radial direction, and a low-order collocation solution may suffice. The temperature averaged in the radial direction is shown in Fig. 5-5 as a function of length. The solution with $N = 1$ is close to the exact solution, and the solutions with $N = 2, 3, 4, 5$, and 6 are indistinguishable from each other. The temperature and concentration profiles at specific axial positions are illustrated in Fig. 5-6. The small radial gradients are evident, permitting a low-order polynomial in the radial direction.

Results are first given for methods that use a fixed step size Δz. The Euler method and the improved Euler method (second-order Runge–Kutta) are all explicit, whereas the trapezoid rule and the backward Euler method are implicit. To illustrate the accuracy of the different methods we choose the average concentration at $z = 0.4$ as a quantity of interest. The exact value is 0.172903 as

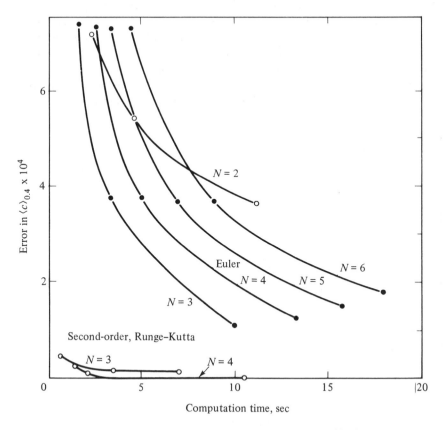

Figure 5-8 Error versus computation time for orthogonal collocation with $Bi_w = 1.0$.

determined by extrapolation of the most accurate results. At the outlet of the reaction the concentration is 1.0, since the reaction has proceeded to completion.

The errors in solutions calculated with different N and the Euler and the improved Euler methods are shown in Fig. 5-7 as a function of step size Δz. It is apparent that no improvement occurs for N larger than three and that the error in the Euler results is proportional to Δz. The error in the results using the improved Euler method is proportional to Δz^2, which is appropriate for a second-order method. The errors are plotted versus computation time in Fig. 5-8. The computation time increases with N, and for N greater than three no improvement is noted. The second-order Runge–Kutta method is very accurate, fast, and not much more difficult to program than the Euler method. Three methods are compared in Fig. 5-9 for $N = 3$. Results obtained with a backward Euler method and fixed step length are too inaccurate to be shown in Fig. 5-9. The trapezoid rule is more expensive than the explicit methods, even though large steps can be taken. For $\Delta z = 0.01$ we see that the use of four iterations rather than just one improves the accuracy very little but increases the computation time by 50 percent. It is more efficient to reduce the step size and take one iteration per step than to use a larger step and several iterations per step. The extra time spent solving the matrix problem is illustrated in Table 5-1, which shows the computation time as a function of step size. Comparing explicit and implicit methods for the same step size reveals that the explicit methods are from five to ten times faster, but require smaller step sizes for equivalent accuracy. Among the methods that have a fixed step size the second-order Runge–Kutta method is preferred. This is based on the best accuracy for the least computation time and on simple programming.

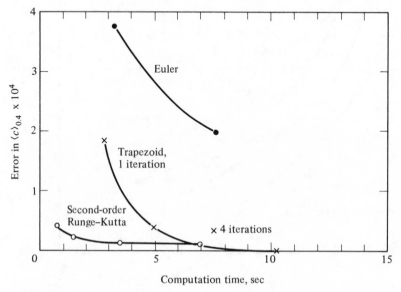

Figure 5-9 Comparison of methods with fixed step size, using orthogonal collocation with $N = 3$ and $Bi = 1.0$.

Table 5-1 Step sizes and errors for Eq. (5-165) in which
$\text{Bi}_w = 1$, $T_w = 0.92$, **and** $N = 3$

Δz	$\langle c \rangle_{0.4}$	Error in $\langle c \rangle_{0.4} \times 10^{4*}$	Computation time, sec
		Euler, $0(\Delta z)$†	
0.002	0.1721659	7.4	1.6
0.001	0.1725256	3.8	3.4
0.0005	0.1727072	2.0	6.7
		Improved Euler or second-order Runge–Kutta, $0(\Delta z^2)$†	
0.01	0.1728603	0.49	0.74
0.005	0.1728821	0.21	1.5
0.002	0.1728886	0.14	3.5
0.001	0.1728896	0.13	7.0
		Trapezoid rule, $0(\Delta z^2)$†	
0.02	1 iteration	1.8	2.8
0.01	1 iteration	0.35	4.9
	4 iterations	0.35	7.5
0.005	1 iteration	0	10.0
		Backward Euler, $0(\Delta z)$‡	
0.04	—	Unstable	0.66
0.02	—	Unstable	2.0
0.01	0.2066	336	3.9

* Using 0.172903 as the exact solution.
† Using FORTRAN compiler on CDC 6400.
‡ Using MNF compiler on CDC 6400.

Next we consider the variable-step methods: fourth-order Runge–Kutta and GEARB. For GEARB the matrix is assumed to be banded, whereas the actual matrix is dense. GEARB is used with the proviso that the bandwidth is arbitrarily taken as half the dimension of the matrix. The average concentration at $z = 0.4$ is listed in Table 5-2. For a given choice of N the Runge–Kutta and GEARB methods give about equivalent results, with GEARB being slightly faster for high N. For $N = 3$ the second-order Runge–Kutta method with fixed step length $\Delta z = 0.01$ is faster, but less accurate, than the results obtained using a variable step length. None of the computation times are expensive, however, and either Runge–Kutta or GEARB could be used because of the good accuracy and reasonable computation time.

For the second case we choose $\text{Bi}_w = 20$ and $T_w = 1$. Now more of the heat transfer resistance is interior to the bed, as given by Eq. (5-172). As a consequence, more terms are needed in the polynomial expansion (higher N). The average

Table 5-2 Errors for Eq. (5-165) and variable-step methods, in which $Bi_w = 1$ and $T_w = 0.92$

	Runge–Kutta, RKINIT			GEARB, $MF = 22, \varepsilon = 10^{-5}$				
N	$\langle c \rangle_{0.4}$	ε	Computation time, sec	$\langle c \rangle_{0.4}$	Computation time, sec	Number of steps	Number of function evaluations	Number of jacobian evaluations
1	0.16475	10^{-5}	0.46	0.16495	0.62	131	276	26
2	0.17254	10^{-5}	1.0	0.17256	1.3	156	328	25
3	0.172890	10^{-5}	1.9	0.1729864	2.1	181	396	25
4	0.1729016	10^{-6}	3.8	0.1730009	2.3	194	500	31
5	0.1729020	10^{-6}	7.7	0.1729843	3.3	178	548	29
6	0.1729020	10^{-6}	15.0	0.1729808	4.6	182	627	31

Exact solution = 0.1729028.

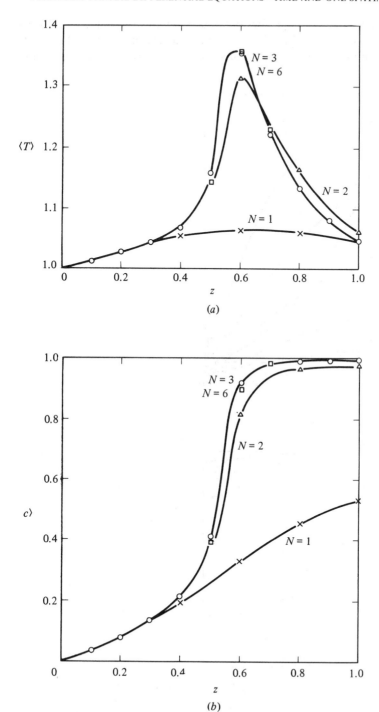

Figure 5-10 Profiles for Eq. (5-165) with $Bi_w = 20$ and $T_w = 1$. (a) Temperature. (b) Conversion.

temperature and concentration profiles in the bed are illustrated in Fig. 5-10. The radial profiles illustrated in Fig. 5-11 demonstrate the need for a high-order polynomial to approximate such a profile. This time we choose the average

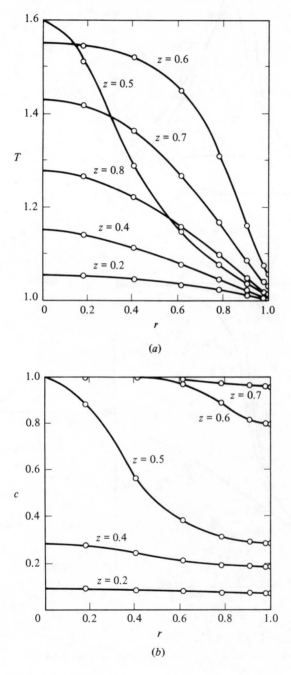

(a)

(b)

Figure 5-11 Radial profiles for Eq. (5-165) with $Bi_w = 20$, $T_w = 1$, and $N = 6$. (a) Temperature. (b) Conversion.

concentration at $z = 0.6$, which is at the peak of the temperature profiles, and hence is a sensitive indicator of accuracy. The exact value is 0.91926; the average concentration at the exit of the bed is 0.99230. For this case the temperature does not rise as much and the conversion is not complete due to the better heat transfer at the wall resulting from the higher Biot number.

Results are first presented for methods using fixed step sizes. The average concentrations are listed in Table 5-3 for several methods. Apparently $N = 5$ is necessary to achieve adequate results. The fourth-order Runge–Kutta method is

Table 5-3 Errors for Eq. (5-165), with $Bi_w = 20$ and $T_w = 1$

| | | | | | | | Number of | |
| | | | | | | | --- | --- |
N	Δz	$\langle c \rangle$	Error* $\times 10^4$	ε	Computation time, sec	Steps	Function evaluations	Jacobian evaluations
				Fourth-order Runge–Kutta, RKINIT				
1		0.3348	5840	10^{-5}	0.22			
2		0.8031	1110	10^{-5}	0.97			
3		0.8985	208	10^{-5}	1.8			
4		0.9037	156	10^{-5}	3.8			
5		0.91979	5.3	10^{-5}	7.9			
6		0.91937	1.1	10^{-6}	—			
				GEARB, MF = 22				
1		0.3347	5840	10^{-5}	0.18	44	79	8
2		0.8088	1110	10^{-5}	1.0	126	276	21
3		0.8956	237	10^{-5}	1.3	155	367	25
4		0.9045	147	10^{-5}	2.1	175	445	27
5		0.92059	13	10^{-5}	3.1	172	531	29
6		0.91969	4.3	10^{-5}	4.0	160	545	26
				Second-order Runge–Kutta, fixed Δz				
1	0.005	0.3348	5840		0.47			
	0.002	0.3348	5840		1.1			
2	0.005	0.8068	1120		0.97			
	0.002	0.8080	1110		2.3			
	0.001	0.8081	1110		4.7			
3	0.005	0.8952	240		1.5			
	0.002	0.8982	211		3.7			
	0.001	0.8984	208		7.4			
4	0.005	0.8989	204		2.1			
	0.002	0.9035	158		5.3			
	0.001	0.9037	156		10.0			
5	0.002	0.9196	3.4		7.1			
	0.001	0.9197	4.4		14.1			

* Using 0.91926 as the exact answer.

about equivalent to the GEARB method for low N, but GEARB is faster for N above three. One important point to note in the results from GEARB is that the number of steps needed does not change drastically as N is increased. With Runge–Kutta, on the other hand, the computation time increases drastically as N is increased. This is because the system becomes stiff (see below). Thus as N increases the GEARB method is preferable, taking larger step sizes so that it is faster overall. Also, the variable step methods are faster, for the same accuracy, than the second-order Runge–Kutta method (see Table 5-3).

The difficulty of integration depends on the eigen values of the jacobian on the right-hand side. The jacobian, of course, depends on the reaction rate, which depends on the solution. For "easy" problems useful information can be obtained by ignoring the reaction rate term and just looking at the diffusion term. Then an explicit scheme is stable provided

$$\Delta t \leqslant \frac{p}{|\lambda|_{\max}} \min\left(\frac{1}{\alpha}, \frac{1}{\alpha'}\right) \tag{5-173}$$

where $|\lambda|_{\max}$ is the maximum eigen value of the matrix $\mathbf{B}'$, and p is about two (see Table 3-1). We can calculate the eigen values of the matrix, but a simpler approach is to estimate the largest one. The largest eigen value of the matrix $\mathbf{B}'$ is bounded by

$$|\lambda|_{\max} < \max_{1 \leqslant j \leqslant N} \sum_{i=1}^{N} |B'_{ji}| \equiv LB \tag{5-174}$$

Table 5-4 Parameters for maximum eigen value in orthogonal collocation

			N			
	1	2	3	4	5	6
Planar						
$f(0)$	0.0	10.5	50.1	142	325	642
$f(\infty)$	3.0	42.5	185	536	1244	2491
q	0.323	0.0932	0.0423	0.0248	0.0160	0.0112
Cylindrical						
$f(0)$	0.0	16.0	65.3	175	385	743
$f(\infty)$	8.0	66.2	250	677	1503	2921
q	0.248	0.0750	0.0373	0.0222	0.0146	0.0103
Spherical						
$f(0)$	0.0	22.5	82.4	211	450	851
$f(\infty)$	15.0	95.1	326	835	1787	3386
q	0.204	0.0625	0.0327	0.0200	0.0134	0.00957

When Bi is large $\mathbf{B'} = \mathbf{B}$. Values of LB can easily be calculated once the matrix $\mathbf{B}$ is determined and the Biot number chosen. The maximum eigen value of the matrix is correlated by the following equation, with parameters given in Table 5-4,

$$f(\text{Bi}) \equiv |\lambda(\text{Bi})|_{\max} = f(0) + [f(\infty) - f(0)]\frac{q\text{Bi}}{1 + q\text{Bi}} \qquad (5\text{-}175)$$

The importance of this information is best illustrated by an example. For planar geometry and polynomials determined by $W = 1$ (see Table 4-5) the eigen value for $N = 1$ is 0.097 for Bi $= 0.1$, and 2.9 for Bi $= 100$. Clearly the stable step size is a strong function of the Biot number since the maximum eigen value varies by a factor of 30 in this example. For $N = 6$ the corresponding eigen values are 644 and 1620 and the dependence on the Biot number is less dramatic. For the same Biot number (say Bi $= 10$) and changing N the maximum eigen value changes from 5.7 at $N = 1$ to 950 at $N = 6$. Thus large N requires a smaller step size. Furthermore, the stiffness ratio (i.e. the ratio of largest to smallest eigen value) increases dramatically with N. For an infinite Biot number the stiffness ratio is 1 for $N = 1$, 17 for $N = 2$, 75 for $N = 3$, and 1,000 for $N = 6$. As the number of collocation points increases the stiffness ratio increases, also, necessitating longer computation

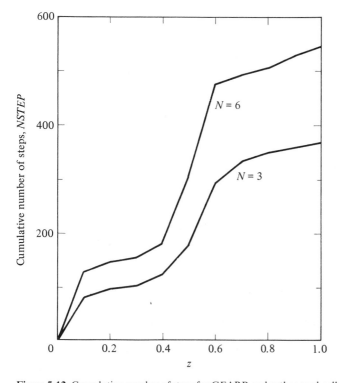

Figure 5-12 Cumulative number of steps for GEARB and orthogonal collocation with $\text{Bi}_w = 20$.

times with explicit methods. This is why Runge–Kutta takes longer than the GEARB method for large N (see Table 5-3). We can apply these ideas to choose the step sizes reported in Table 5-1. For $\alpha = \alpha' = 1$, $|\lambda|_{max} = 70$ for $N = 3$ and the improved Euler method has $p = 2$. Thus $\Delta z \leqslant p/|\lambda|_{max} = 0.029$ and step sizes of 0.01 are stable. The cumulative number of steps needed with GEARB and the problem with $Bi_w = 20$ is shown in Fig. 5-12. Near the region of the sharp rise in temperature more steps (smaller Δz) are needed.

The next application is the diffusion of heat and mass in a porous catalyst pellet, i.e. the transient version of Eqs. (4-120) to (4-123). The specific problem is the one solved by Ferguson and Finlayson.[5]

$$M_1 \frac{\partial T}{\partial t} = \nabla^2 T + \phi^2 \beta R(c, T)$$

$$M_2 \frac{\partial c}{\partial t} = \nabla^2 c - \phi^2 R(c, T)$$

(5-176)

$$\frac{\partial c}{\partial x}\bigg|_{x=0} = 0 \qquad \frac{\partial T}{\partial x}\bigg|_{x=0} = 0$$

(5-177)

$$-\frac{\partial T}{\partial x}\bigg|_{x=0} = Bi[T(1, t) - g_1(t)]$$

$$-\frac{\partial c}{\partial x}\bigg|_{x=0} = Bi_m[c(1, t) - g_2(t)]$$

(5-178)

The functions $g_1(t)$ and $g_2(t)$, which can vary in time, represent the external temperature and concentration surrounding the pellet. We can apply orthogonal collocation to these equations to obtain

$$M_1 \frac{dT_j}{dt} = \sum_{i=1}^{N} B'_{ji} T_i + \phi^2 \beta R_j + \frac{B_{j,N+1} Big_1(t)}{Bi + A_{N+1,N+1}}$$

$$M_2 \frac{dc_j}{dt} = \sum_{i=1}^{N} B'_{ji} c_i - \phi^2 R_j + \frac{B_{j,N+1} Bi_m g_2(t)}{Bi_m + A_{N+1,N+1}}$$

(5-179)

$$T(x, 0) = T_0(x) \qquad c(x, 0) = c_0(x)$$

(5-180)

where B'_{ji} is given by Eq. (5-166). We use Bi for heat transfer and Bi_m for mass transfer. We must integrate the $2N$ equations, Eqs. (5-179), subject to the initial conditions given in Eqs. (5-180). These equations are similar to Eqs. (5-165), and the same method can be used by replacing length z by time t. The major difference is that as time proceeds mass can be added to the system through $g_2(t)$, whereas in the reactor problem $Bi_m = 0$ and no mass is added down the length of the reactor. If the reaction is very fast compared to radial dispersion in the pellet, the mass injected at $r = 1$ can be rapidly consumed, leading to steep gradients in the radial direction. In the reactor, on the other hand, the reaction proceeds somewhat uniformly at all radii since no mass is injected at $r = 1$. If the temperature does not vary radially, there would be no concentration change radially at all; $N = 1$ would suffice. Consequently, low N orthogonal collocation approximation is more likely to suffice for the reactor and the transient pellet problem with low ϕ, but not for the transient pellet problem with high ϕ.

We next integrate Eqs. (5-179) for two cases. Both use a first-order, irreversible reaction with

$$R(c, T) = c\, e^{\gamma - \gamma/T} \tag{5-181}$$

where $\gamma = 20$, $\beta = 0.6$, and $\phi^2 = 0.25$. The values of M_1 and M_2 are 176 and 199, respectively. The first case corresponds to boundary conditions of the first kind $(\mathrm{Bi} \to \infty, \mathrm{Bi}_m \to \infty)$. The initial conditions for temperature and concentration are

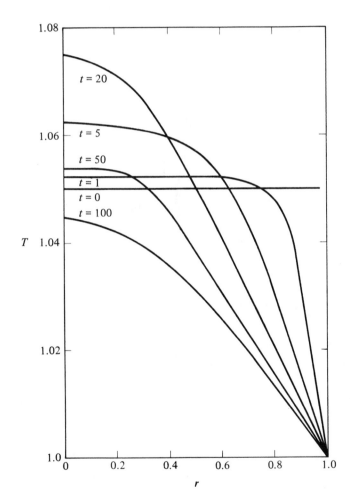

Figure 5-13 Radial temperature profiles in catalyst pellet based on Eqs. (5-176), (5-177), (5-182), and (5-183) with $N = 6$.

$$T(x, 0) = 1.05 \qquad c(x, 0) = 1.0 \tag{5-182}$$

At time zero the boundary value of temperature is changed to

$$T(1, t) = 1.0 \qquad c(1, t) = 1.0 \tag{5-183}$$

Thus the problem represents an approach to steady state when the boundary temperature is lowered. For $\gamma = 20$ and an activation energy of 20 kcal/mole the reference temperature is 1000 K, so that a decrease by 5 percent is a 50 K change. The temperature profile is shown in Fig. 5-13. Here we see that the center temperature first rises and then falls to the steady-state value. By using GEARB to integrate to steady state we can integrate quickly. As the steady state is approached the profiles change slowly and large time steps can be taken by the

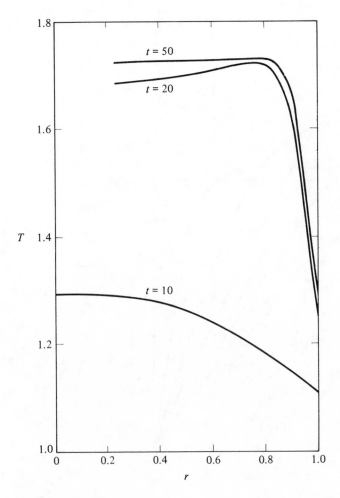

Figure 5-14 Radial temperature profiles in catalyst pellet based on Eqs. (5-176) and (5-177) with Bi $= 27.65$ and Bi$_m = 33.25$.

implicit method. An explicit method would be less economical as small time steps would have to be taken all the way to steady state, since the stable step size is controlled by the stability limit.

For the next example we take the same equations, but now for boundary conditions of the third kind with $Bi = 27.65$ and $Bi_m = 33.25$. The initial temperature and concentration profiles are taken as the $N = 2$ approximation to the middle steady-state solution with boundary temperature and concentration of one. At time zero the boundary temperature is raised to 1.1 and held there. This 10 percent temperature perturbation is sufficient to drive the solution from the intermediate to the upper steady state (with the highest temperature). The temperature solution using $N = 6$ and GEARB is shown in Fig. 5-14, and we see that very steep gradients are present. Consequently, a large N is necessary. This problem has also been solved with the improved Euler method. In this case the stable step size given by Eq. (5-173) is 0.23. Calculations made with $\Delta t = 0.1$ are stable, indicating that Eq. (5-173) gives a reliable estimate of step length.

This demonstrates the application of orthogonal collocation as a numerical method for solving the problem of transient diffusion and reaction in a pellet. The orthogonal collocation method is also useful in a first approximation for studying the stability of a set of equations. We illustrate that advantage here. The steady-state version of Eq. (5-176) can have three steady-state solutions under certain conditions. The second solution is unstable and the upper steady state is stable for large Lewis numbers, where $Le = M_1/M_2$, but unstable for small Lewis numbers (i.e. $Le < 1$). Indeed, for Lewis numbers of less than one a limit cycle develops, and the concentration and temperature profiles in the catalyst change in a periodic fashion with time. We apply a one-term orthogonal collocation method to study this phenomenon. The equations become

$$\frac{dT_1}{d\tau} = \frac{B_{11}}{Le}(T_1 - T_2) + \frac{\beta}{Le}R_1$$

$$\frac{dc_1}{d\tau} = B_{11}(c_1 - c_2) - R_1 \tag{5-184}$$

$$\frac{t}{M_2} = \tau$$

$$-A_{22}(-T_1 + T_2) = Bi(T_2 - 1)$$
$$-A_{22}(-c_1 + c_2) = Bi_m(c_2 - 1) \tag{5-185}$$

where B_{11} and A_{22} are obtained from Table 4-6 for the appropriate geometry. The boundary conditions are combined with the differential equations to obtain

$$\frac{dT_1}{d\tau} = d(1 - T_1) + \frac{\beta}{Le}R \qquad d = \frac{-B_{11}}{Le}\frac{Bi}{A_{22} + Bi}$$

$$\frac{dc_1}{d\tau} = b(1 - c_1) - R_1 \qquad b = -B_1\frac{Bi_m}{A_{22} + Bi_m} \tag{5-186}$$

We call the steady-state solutions T_0 and c_0, define $T = T_1 - T_0$ and $c = c_1 - c_0$, and then linearize Eqs. (5-186) about T_0 and c_0 to get

$$\frac{dT}{d\tau} = -dT + \frac{\beta}{\text{Le}}(R_T T + R_c c)$$

$$\frac{dc}{d\tau} = -bc - (R_T T + R_c c) \tag{5-187}$$

$$R_T = \frac{\partial R}{\partial T}\bigg|_{T_0, c_0} \qquad R_c = \frac{\partial R}{\partial c}\bigg|_{T_0, c_0} \tag{5-188}$$

We solve these linear equations by assuming an exponential solution $T = A\, e^{\mu t}$, $c = B^{\mu t}$ giving rise to a set of homogeneous equations in A and B. A solution exists only for certain values of μ, which are the roots to the determinant

$$\begin{vmatrix} \mu + d - \dfrac{\beta}{\text{Le}}R_T & -\dfrac{\beta}{\text{Le}}R_c \\[2ex] R_T & \mu + b + R_c \end{vmatrix} = 0 \tag{5-189}$$

This gives a quadratic in μ^2

$$\mu^2 + \mu Q + c = 0 \tag{5-190}$$

$$Q = d + b - \frac{\beta}{\text{Le}}R + R_c \qquad c = db + dR_c - \frac{\beta b R_T}{\text{Le}}$$

The quadratic has the solution $2\mu = -Q \pm (Q^2 - 4c)^{1/2}$. If Q is negative then one eigen value has a positive real part and the solution is unstable. By expressing the R_T and R_c in terms of the reaction rate at steady state R_1 we can write Q in the form

$$Q = d\left(1 + \gamma\frac{1 - T_1}{T_1^2} + \text{Le}\frac{\kappa}{c_1}\right)$$

$$\kappa = \frac{b}{d\text{Le}} = \frac{\text{Bi}_m}{\text{Bi}}\frac{A_{22} + \text{Bi}}{A_{22} + \text{Bi}_m} \tag{5-191}$$

Clearly if $T_1 < 1$ then Q is positive. If $T_1 > 1$ then we obtain the following condition for negative Q:

$$\text{Le}\kappa \leqslant c_1\left(-1 - \gamma\frac{1 - T_1}{T_1^2}\right) \tag{5-192}$$

The right-hand side is a numerical value, which depends on the steady-state solution. If the Lewis number is below a critical value, the steady state is unstable. For the example in Fig. 5-14, the upper steady state is $c_1 = 0.02075$ and $T_1 = 1.5989$, and we get from Eq. (5-192) $\text{Le}\kappa \leqslant 0.077$. Since $\text{Le}\kappa = 0.90$ the upper solution is stable. Further calculation reveals that the lower steady state is also stable (both μ are negative), while the intermediate steady state is unstable (one μ is positive). For realistic values of parameters occurring in chemical engineering

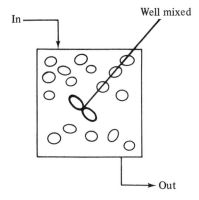

In

Well mixed

Out   **Figure 5-15** Schematic stirred tank with catalyst pellets.

the Lewis number is usually much greater than one. Thus this problem represents an interesting mathematical one, although not of much importance for industrial chemical reactions. Problem 3-8 is a particular case of Eqs. (5-184). For the small Lewis number used there no steady-state solutions are stable, and the solution is a limit cycle.

For the next example we take the case of a stirred reactor filled with catalyst pellets. Then we must solve the equations governing the reactor as well as the diffusion–reaction problem in the pellet. The stirred tank is shown in Fig. 5-15 and a typical mass balance is

$$V\frac{dc}{dt} = F(c_{\text{in}} - c) + VR(c) \tag{5-193}$$

where F is the volumetric flow rate in and out, V is the volume of reactor, and R is the rate of reaction per reactor volume. We consider such a reactor for the reduction of nitric oxide by the two reactions

$$NO + CO \rightarrow CO_2 + \tfrac{1}{2}N_2$$
$$NO + H_2 \rightarrow H_2O + \tfrac{1}{2}N_2$$

The reaction rate is given by an expression of the form

$$R = \frac{kk_i k_{\text{NO}} p_i p_{\text{NO}}}{(1 + k_i p_i + k_{\text{NO}} p_{\text{NO}})^2} \tag{5-194}$$

where p_i is the partial pressure of carbon monoxide and hydrogen for the two reactions, and k and k_i are the reaction rate parameters. The dimensionless equations are for the concentration of ith species in the pellet c_i, and, the concentration in the reactor c_i'.

$$\tau_1 \frac{\partial c_i}{\partial t} = \nabla^2 c_i - d_1(r_1 + r_2)$$

$$\frac{\partial c_i}{\partial r}(0, t) = 0 \qquad -\frac{\partial c_i}{\partial r}\bigg|_{r=1} = \text{Bi}_i[c_i(1, t) - c_i'] \tag{5-195}$$

$$\tau_2 \frac{\partial T}{\partial t} = \nabla^2 T + d_4(-\Delta H_1 r_1 - \Delta H_2 r_2)$$

$$\frac{\partial T}{\partial r}(0, t) = 0 \qquad -\frac{\partial T}{\partial r}\bigg|_{r=1} = \mathrm{Bi}[T(1, t) - T']$$

(5-196)

$$\tau_3 \frac{dc_i'}{dt} = c_{i,\mathrm{in}}' - c_i' + d_5[c_i(1, t) - c_i']$$

$$\tau_3 \frac{dT'}{dt} = T_{\mathrm{in}}' - T' + d_6[T(1, t) - T']$$

(5-197)

$$d_1 = \frac{\rho_s R_p^2}{D_e} \qquad d_4 = \frac{\rho_s R_p^2}{k_e} \qquad d_5 = \frac{6k_{mi}(1-\varepsilon)V}{F2R_p} \qquad d_6 = \frac{6h(1-\varepsilon)V}{F2R_p \rho_f c_{pf}}$$

The catalyst density is ρ_s, the particle radius is R_p, and the effective diffusivity and thermal conductivity are D_e and k_e, respectively. The mass transfer coefficient between the fluid in the stirred tank and the pellet is k_{mi}, while the heat transfer coefficient is h. The void fraction is ε, the heat capacity of the pellet is $\rho_s C_{ps}$, while that of the fluid is $\rho_f C_{fs}$. We wish to solve these equations under transient conditions when the flow rate, and the concentration and temperature of the entering stream vary. We can easily apply orthogonal collocation to Eqs. (5-195) and (5-196) and reduce them to a set of ordinary differential equations to be solved along with Eqs. (5-197). The time constants associated with different phenomena are

$$\tau_1 = \frac{\varepsilon_s R_p^2}{D_e} \qquad \tau_2 = \frac{(\rho C_p)_s R_p^2}{k_e} \qquad \tau_3 = \frac{\varepsilon V}{F}$$

(5-198)

For typical cases $\tau_1 = 0.3$ sec, $\tau_2 = 21$ sec, and $\tau_3 = 0.003$ sec. With time constants this different we know that the problem is stiff and implicit methods are needed. This is called the complete model.

We can make a physical approximation and say that really fast phenomena occur instantaneously. Then we can neglect the time derivatives in those equations. We thus obtain the quasistatic model by neglecting time derivatives in Eqs. (5-197) and in the diffusion of mass given by Eqs. (5-195). The slowest phenomenon is the heat transfer, and we retain the time derivative in the heat conduction equation for the pellet, Eqs. (5-196).

Still another simplification can be made in this case, and that is to recognize that the carbon monoxide is usually far in excess and that the reaction rate is essentially a first-order reaction in relation to nitric oxide concentration. The major temperature drop occurs outside the pellet; the temperature profile inside the pellet is relatively uniform. Thus the problem of diffusion and reaction of mass in the pellet reduces to a steady-state boundary-value problem and, in fact, a linear one that can be solved analytically. Doing so then relieves us from solving the coupled equations for the three species and the temperature in the pellet. This model we refer to as the simple model. All three models are integrated with an improved Euler method, Eqs. (3-78) and (3-79).

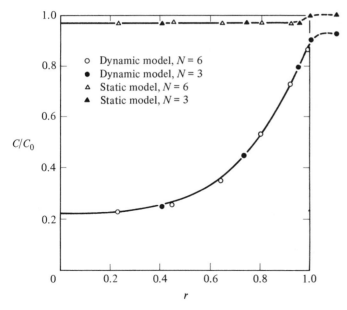

Figure 5-16 Concentration in catalyst pellet. (*After Ferguson and Finlayson.*[6])

A typical concentration profile for carbon monoxide is shown in Fig. 5-16 for the dynamic and quasistatic models. This solution occurs 0.01 sec after a step change in carbon monoxide concentration. The quasistatic model overestimates the concentration by 500 percent, but the error in the carbon monoxide coming out of the stirred tank is only 8 percent, and 0.1 sec later is correct to 0.5 percent. The effect of assuming a quasistatic model on the nitric oxide is similarly small. The original article gives other reasons why the quasistatic model is appropriate.[6] We are interested here primarily in the computation time for the various models. These are listed in Table 5-5. The dynamic model used an explicit, improved Euler method. It is apparent that an implicit scheme would have been best, but at the time of the study the useful implicit integration packages were not available. The quasistatic model and simple model used the same integration routine, but since the problem was not so stiff larger time steps could be used. The results in Table 5-5 indicate that small time steps of 0.0004 sec were necessary for the dynamic model, whereas the quasistatic model could use time steps of 0.5 sec and was consequently much faster. For the same time step $N = 3$, $\Delta t = 4 \times 10^{-4}$, the quasistatic model was more expensive than the dynamic model, because in the dynamic model the right-hand side was evaluated at each time step, but in the quasistatic model a two-point boundary-value problem (with $4N$ unknowns) was solved at each time step, and this involved iteration. The simple model was about seven times as fast as the quasistatic model as a result of not having to solve the boundary-value problem at each time step. In that case the effectiveness factor was

Table 5-5 Computation times for different models

N	Δt	Ratio of real time to computation time	Stable step size by Eq. (5-173)
		Dynamic model	
6	5×10^{-5}	700	0.00018
3	4×10^{-4}	41	0.0018
	Quasistatic model, $\tau_1 = \tau_3 = 0$, Eqs. (5-196)		
1	4×10^{-4}	100	
1	1×10^{-2}	6	
1	5×10^{-1}	0.14	17
	Simple model, $\tau_1 = \tau_3 = 0$, linear reaction at average temperature		
	5×10^{-1}	0.02	17

given as an analytic function. Thus the major part of the computation time is associated with solving the equations for diffusion in the pellet. This means that having an efficient tool, such as orthogonal collocation, is particularly welcome for problems of this type, because it may be necessary to solve such two-point boundary-value problems thousands of times in a simulation. Estimates of stable time steps are obtained from Eq. (5-173) for each equation individually, and the stability limits are listed in Table 5-5. The time steps used were about half of the maximum permissible, and the much smaller step size needed for $N = 6$ is correctly indicated.

In conclusion, the orthogonal collocation method is effective when applied to parabolic partial differential equations in one space dimension. The first term in the solution may give useful qualitative information and higher approximations are very accurate. Effective integration methods include the fixed step, improved Euler method (a second-order Runge–Kutta), and variable step methods: RKINIT (a fourth-order, explicit Runge–Kutta) and GEARB (a multiple-order, implicit Adams method).

5-5 FINITE DIFFERENCE

We illustrate the finite difference method by application to Eq. (5-74) without the reaction rate term. If we let $c_i(t) = c(x_i, t)$, the equation is

$$\frac{dc_i}{dt} = \frac{1}{\Delta x^2} \left[D(c_{i+1/2})(c_{i+1} - c_i) - D(c_{i-1/2})(c_i - c_{i-1}) \right] \qquad (5\text{-}199)$$

We can write this in the general form of Eq. (5-155)

$$\frac{d\mathbf{c}}{dt} = \mathbf{AA(c)c} \tag{5-200}$$

The only difference between orthogonal collocation and finite difference is the form of the matrix $\mathbf{AA}$ and the number of terms needed for good accuracy. The integration method can be any of the methods presented in Chapter 3. A simple Euler method gives

$$\frac{c_i^{n+1} - c_i^n}{\Delta t} = \frac{1}{\Delta x^2}[D(c_{i+1/2}^n)(c_{i+1}^n - c_i^n) - D(c_{i-1/2}^n)(c_i^n - c_{i-1}^n)] \tag{5-201}$$

where

$$c_i^n = c(x_i, t_n) \tag{5-202}$$

We expect the truncation error to be $0(\Delta t)$ and $0(\Delta x^2)$ since these are the truncation errors of the respective parts in Chapters 3 and 4. We can check the truncation error by expanding c^{n+1} in a Taylor series about c^n

$$c_i^{n+1} = c_i^n + \frac{\partial c}{\partial t}\bigg|_i^n \Delta t + \frac{\partial^2 c}{\partial t^2}\bigg|_i^n \frac{\Delta t^2}{2!} + 0(\Delta t^3) \tag{5-203}$$

The spatial derivatives are also expanded by Eqs. (4-29) and (4-30). Substitution into Eq. (5-201) gives

$$\frac{\partial c}{\partial t}\bigg|_i^n + \frac{\Delta t}{2}\frac{\partial^2 c}{\partial t^2}\bigg|_i^n + 0(\Delta t^2) = D(c_{i+1/2}^n)\left(\frac{1}{\Delta x}\frac{\partial c}{\partial x}\bigg|_i^n + \frac{\partial^2 c}{\partial x^2}\bigg|_i^n \frac{1}{2} + \frac{\Delta x}{6}\frac{\partial^3 c}{\partial x^3}\bigg|_i^n\right)$$

$$- D(c_{i-1/2}^n)\left(\frac{1}{\Delta x}\frac{\partial c}{\partial x}\bigg|_i^n - \frac{\partial^2 c}{\partial x^2}\bigg|_i^n \frac{1}{2} + \frac{\Delta x}{6}\frac{\partial^3 c}{\partial x^3}\bigg|_i^n\right) \tag{5-204}$$

If $D(c)$ is constant then the truncation error of the right-hand side is $0(\Delta x^2)$. If D depends on c then we can use

$$\frac{\partial D}{\partial x}\bigg|_i^n = \frac{D(c_{i+1/2}^n) - D(c_{i-1/2}^n)}{\Delta x} \tag{5-205}$$

and the scheme is second-order provided

$$\tfrac{1}{2}[D(c_{i+1/2}^n) + D(c_{i-1/2}^n)] = D(c_i^n) + 0(\Delta x^2) \tag{5-206}$$

This is the case for the two approaches in Eqs. (4-40), (4-41), and (4-51) given in Sec. 4-2. The time truncation error is $0(\Delta t)$. As $\Delta t \to 0$ and $\Delta x \to 0$ we get the original equation evaluated at the ith spatial grid point and the nth time level.

The stability of Eq. (5-199) can be studied using the methods of Chapter 3. Let us do this for the case $D = $ constant. Then Eq. (5-199) reduces to

$$\frac{dc_i}{dt} = \frac{D}{\Delta x^2}(c_{i+1} - 2c_i + c_{i-1}) = \frac{D}{\Delta x^2}\sum_{j=1}^{n+1} B_{ji}c_i \tag{5-207}$$

where the matrix $\mathbf{B}$ is tridiagonal. Suppose the boundary conditions are the first kind at both $i = 1$ and $i = n+1$. The stability of Eq. (5-207) is governed by the largest eigen value in absolute magnitude (see Sec. 3-7)

$$\alpha \equiv \Delta t \frac{D}{\Delta x^2} \leqslant \frac{p}{|\lambda|_{\text{max}}} \tag{5-208}$$

The largest eigen value is bounded by

$$|\lambda|_{\text{max}} \leqslant \max_{2 < j < n} \sum_{i=2}^{n} |B_{ji}| \tag{5-209}$$

and this gives $|\lambda|_{\text{max}} = 4$. Substitution into Eq. (5-208) gives stability for

$$\frac{\Delta t D}{\Delta x^2} \leqslant \frac{p}{4} \tag{5-210}$$

Now if we do a calculation with finite n and actually calculate the eigen values we find they are smaller than the upper bound, Eq. (5-209). For $n = 2$, 3, and 5 we get $|\lambda|_{\text{max}} = 2$, 3, and 3.41, respectively. For $n \to \infty$, $|\lambda|_{\text{max}} = 4$ and the upper bound is in fact the exact value. This means that for any finite n we can calculate with a step size Δt larger than Eq. (5-210) by some small amount and retain stability. For the Euler method of integration $p = 2$ and Eq. (5-210) gives the value $\frac{1}{2}$ on the right-hand side.

There is another, more common, way to study stability of the equations, using the following theorem:

Theorem If

$$c_i^{n+1} = A c_{i+1}^n + B c_i^n + C c_{i-1}^n \tag{5-211}$$

and A, B, and C are positive and $A + B + C \leqslant 1$, then the scheme is stable and the errors die out.

PROOF Apply absolute values to Eq. (5-211) and make the right-hand side larger by replacing each term by its absolute value

$$|c_i^{n+1}| \leqslant |A c_{i+1}^n| + |B c_i^n| + |C c_{i-1}^n| = A|c_{i+1}^n| + B|c_i^n| + C|c_{i-1}^n| \tag{5-212}$$

Replace each $|c_i^n|$ by $\max_k |c_k^n|$ where k ranges from 2 to n. Thus

$$|c_i^{n+1}| \leqslant (A + B + C) \max_k |c_k^n| = \max_k |c_k^n| \tag{5-213}$$

But this equation holds for each i and so it holds if we replace the left-hand side by the maximum over i

$$\max_i |c_i^{n+1}| \leqslant \max_k |c_k^n| \tag{5-214}$$

This equation is applied to each time level to obtain

$$\max_i |c_i^{n+1}| \leqslant \max_k |c_k^0| \tag{5-215}$$

Thus the solution is bounded as $n \to \infty$. If the problem is

$$\frac{\partial c}{\partial t} = D \frac{\partial^2 c}{\partial x^2} \qquad (5\text{-}216)$$

$$c(x,0) = f(x) \qquad c(0,t) = g_1(t) \qquad c(1,t) = g_2(t) \qquad (5\text{-}217)$$

then the error $u = c - c_{\text{ex}}$ is governed by

$$\frac{\partial u}{\partial t} = D \frac{\partial^2 u}{\partial x^2} \qquad (5\text{-}218)$$

$$u(x,0) = u(0,t) = u(1,t) = 0 \qquad (5\text{-}219)$$

If error is introduced at time t, then this error decreases in time, according to Eqs. (5-214) and (5-215), provided that the calculations are done exactly. Now Eq. (5-201) with $D = $ constant can be rearranged to the form of Eq. (5-211)

$$c_i^{n+1} = \alpha c_{i+1}^n + (1-2\alpha)c_i^n + \alpha c_{i-1}^n \qquad (5\text{-}220)$$

$$\alpha = \frac{D\Delta t}{\Delta x^2} \qquad (5\text{-}221)$$

If A, B, and C are positive and $A+B+C = 1$ then the scheme is stable. The coefficients add up to one and A and C are always positive. B is positive when

$$1 - 2\alpha \geqslant 0 \qquad \alpha \leqslant \tfrac{1}{2} \qquad (5\text{-}222)$$

thus

$$\frac{D\Delta t}{\Delta x^2} \leqslant \tfrac{1}{2} \qquad (5\text{-}223)$$

If Δt obeys Eq. (5-223) then the calculations are stable. [Compare to Eq. (5-210).] We have not proved that a larger Δt is unstable. Such a proof is more difficult although true for $n \to \infty$. We have seen by actually calculating the eigen values that larger Δt are in fact stable for finite n. The implications of Eq. (5-223) are that if Δx is decreased by a factor of two then the time step must be decreased by a factor of four. Richtmyer and Morton[12] suggest for nonlinear problems $D(c)$ that the same criterion be used as a guide using the current local value of $D(c_i^n)$. Other geometries are examined in problem 5-19.

Other methods than Euler's can be used to integrate Eq. (5-199). The Crank–Nicolson method was developed by evaluating the right-hand side at the average of its values at time levels n and $n+1$. We illustrate this classical method for the case with $D = $ constant. We evaluate the right-hand side at the two time levels and weight them by $1 - \beta$ and β, respectively, giving

$$\frac{c_i^{n+1} - c_i^n}{\Delta t} = D(1-\beta)\left(\frac{c_{i+1}^n - 2c_i^n + c_{i-1}^n}{\Delta x^2}\right) + D\beta\left(\frac{c_{i+1}^{n+1} - 2c_i^{n+1} + c_{i-1}^{n+1}}{\Delta x^2}\right) \qquad (5\text{-}224)$$

Now the equations are of the form

$$-\alpha\beta c_{i-1}^{n+1} + (1+2\alpha\beta)c_i^{n+1} - \alpha\beta c_{i+1}^{n+1} = c_i^n + \alpha(1-\beta)(c_{i+1}^n - 2c_i^n + c_{i-1}^n) \qquad (5\text{-}225)$$

These can be written as

$$\mathbf{AAc} = \mathbf{f(c)} \tag{5-226}$$

where the matrix $\mathbf{AA}$ is tridiagonal. The case $\beta = \frac{1}{2}$ is the classical Crank–Nicolson method, $\beta = 0$ gives the Euler method, and $\beta = 1$ gives the backward Euler method. We have seen that the implicit methods with $\beta = \frac{1}{2}$ and 1 are suitable for stiff problems and that the choice $\beta = 1$ provides an L stable method. The choice $\beta = \frac{1}{2}$ is second-order with error $O(\Delta t^2)$; the other choices give $O(\Delta t)$. The stability of the equations is governed by

$$\alpha = \frac{D\Delta t}{\Delta x^2} \leqslant \frac{0.5}{1 - 2\beta} \tag{5-227}$$

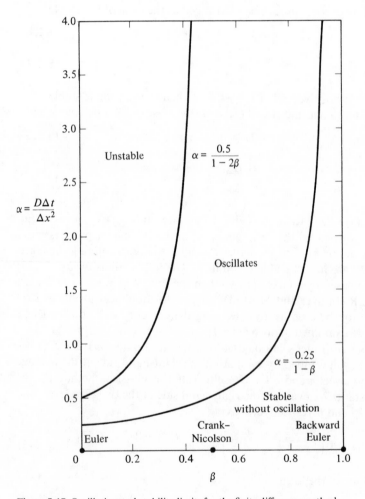

Figure 5-17 Oscillation and stability limits for the finite difference method.

For $\beta = \frac{1}{2}$ the upper bound on Δt is infinite indicating A stability. Only the case $\beta = 1$ gives L stability. A chart of the stability regions for different β is shown in Fig. 5-17. The oscillation limit is derived using the analysis of Eqs. (3-125) to (3-127) to give

$$|\lambda|_{max} \frac{D\Delta t}{\Delta x^2} \leqslant \frac{1}{1-\beta} \qquad (5\text{-}228)$$

or

$$\alpha = \frac{D\Delta t}{\Delta x^2} \leqslant \frac{0.25}{1-\beta} \qquad (5\text{-}229)$$

When using the Euler method ($\beta = 0$) we must use $\alpha < 0.5$ for stability but $\alpha < 0.25$ to avoid oscillations. When using the Crank–Nicolson method we can use large α and have a stable scheme, but it oscillates unless $\alpha < 0.5$. In that case we should compare the extra work to solve the tridiagonal equations of Crank–Nicolson ($\beta = 0.5$) with the work needed to solve the equations using Euler's method ($\beta = 0$) and a slightly smaller Δt: $\alpha < 0.25$. The reason Crank–Nicolson is so popular compared with the Euler method is that for many problems the former solution oscillates. Such oscillations are small, however, since the accuracy goes as Δt^2, and may be smaller than the errors in the Euler solution. For harder problems with large $|\lambda|_{max}$ in Eq. (5-208) [or D in Eq. (5-227) replaced by $|\lambda|_{max}/4$] the tendency to oscillate may be very noticeable because α is taken so large. Then the enhanced stability of the backward Euler method is needed.

The effect of a reaction rate term on the stability limitation is easy to deduce if the reaction rate is linear in concentration. The equations are then linear

$$\frac{c_i^{n+1} - c_i^n}{\Delta t} = \frac{D}{\Delta x^2}(c_{i+1}^n - 2c_i^n + c_{i-1}^n) - kc_i^n \qquad (5\text{-}230)$$

They rearrange to give

$$c_i^{n+1} = \alpha c_{i+1}^n + (1 - 2\alpha - k\Delta t)c_i^n + \alpha c_{i-1}^n \qquad (5\text{-}231)$$

Now $A + B + C < 1$, and A and C are positive. B is positive and the calculations are stable if

$$\alpha = \frac{D\Delta t}{\Delta x^2} \leqslant \frac{0.5}{1 + k\Delta x^2/2D} \qquad (5\text{-}232)$$

As the reaction rate increases the stable step size is decreased, although this effect is mitigated by a grid with small Δx.

The boundary and initial conditions are also important in the calculations. If the boundary condition and initial condition are incompatible at their common points, the analyst must make an arbitrary choice. Suppose the conditions are: $c(t,0) = 0$, $c(0,x) = 1$. What value do we assign to $c(0,0)$? Whatever choice we make introduces errors which, if the scheme is stable, will decay in successive time levels. The recommendation is to use the boundary-condition value and make $c(0,0) = 1$, based on the comparison by Wilkes.[15]

When the equation is nonlinear we have several options to solve it. The Euler method is illustrated in Eq. (5-201). The modified backward Euler method can be employed by evaluating the right-hand side at the $n+1$ level, except for the $D(c)$ terms which are evaluated at the nth time level. The same approach can be used with the Crank–Nicolson method, giving

$$\frac{c_i^{n+1} - c_i^n}{\Delta t} = \frac{1}{2\Delta x^2} \{ D(c_{i+1/2}^n)(c_{i+1}^{n+1} + c_{i+1}^n) - [D(c_{i+1/2}^n) + D(c_{i-1/2}^n)]$$

$$\times (c_i^{n+1} + c_i^n) + D(c_{i-1/2}^n)(c_{i-1}^{n+1} + c_{i-1}^n) \} \quad (5\text{-}233)$$

An improvement is made if the diffusivity is evaluated at the $n+\frac{1}{2}$ time level. The value of c there is obtained by applying a Euler method to step forward $\Delta t/2$. This approach is a combination of a second-order Runge–Kutta method

$$\frac{y^{n+1/2} - y^n}{\Delta t/2} = f(y^n) \qquad \frac{y^{n+1} - y^n}{\Delta t} = f(y^{n+1/2}) \quad (5\text{-}234)$$

and the trapezoid rule

$$\frac{y^{n+1} - y^n}{\Delta t} = \tfrac{1}{2}[f(y^n) + f(y^{n+1})] \quad (5\text{-}235)$$

In all of the implicit methods the tridiagonal matrix must be decomposed twice for each time step. Problem 5-20 shows that the truncation error of this scheme is Δt^2.

We can also just take the equations in the form

$$\frac{dc_i}{dt} = F_i(c) \quad (5\text{-}236)$$

In this form we can apply any of the methods of Chapter 3. Below we use the improved Euler method (or second-order Runge–Kutta method), with fixed step size, the fourth-order Runge–Kutta method with a variable step size (RKINIT), and the implicit GEARB system. In the latter cases the nonlinear system of equations is solved with Newton–Raphson, and the derivatives are calculated numerically. Thus in the last two cases, the user just has to provide a main program and a subroutine to calculate the right-hand side F given c.

We first apply these methods to Eqs. (5-156) to (5-158) for a packed bed reactor with radial dispersion. The radial direction is divided into $n-1$ equal intervals using n points. The equation for concentration at an internal grid point is obtained from Eq. (5-156) by applying the difference formula for the laplacian

$$\frac{dc_i}{dt} = \alpha \left[\frac{c_{i-1} - 2c_i + c_{i+1}}{\Delta r^2} + \left(\frac{a-1}{r_i} \right) \frac{c_{i+1} - c_{i-1}}{2\Delta r} \right] + \beta R_i \quad (5\text{-}237)$$

Here we use a to denote geometry, thus for planar geometry $a = 1$, cylindrical $a = 2$, or spherical $a = 3$. At either boundary we employ a false boundary point, giving the second-order equations for the boundary conditions

$$-\frac{c_0 - c_2}{2\Delta r} = 0 \qquad \frac{c_{n+1} - c_{n-1}}{2\Delta r} = \text{Bi}_w(c_n - c_w) \quad (5\text{-}238)$$

For the boundary node we must then employ Eq. (5-237) with the boundary conditions

$$\frac{dc_1}{dt} = \frac{2a}{\Delta r^2}(c_2 - c_1) + \beta R_1 \qquad (5\text{-}239)$$

$$\frac{dc_n}{dt} = \alpha\left[\frac{2(c_{n-1} - c_n) - 2\mathrm{Bi}_w\Delta r(c_n - c_w)}{\Delta r^2} + (a-1)\mathrm{Bi}_w(c_w - c_n)\right] + \beta R_n \qquad (5\text{-}240)$$

The set of n equations—Eq. (5-239) for $i = 1$, Eq. (5-237) for $i = 2$ to $n-1$, and Eq. (5-240) for $i = n$—is then solved using the different methods to integrate in time.

Results for the first case with $\mathrm{Bi}_w = 1$ and $T_w = 0.92$ are shown in Table 5-6. Notice that as the step length is decreased in the second-order Runge–Kutta method the answer for the average concentration approaches the value obtained with the variable step, fourth-order Runge–Kutta method. This suggests that the

Table 5-6 Average concentration for Eq. (5-156) with $\mathrm{Bi}_w = 1$ and $T_w = 0.92$

							Number of	
							Function	Jacobian
				Error	Computation		evalua-	evalua-
Δz	$n+1$	Δr	$\langle c \rangle_{0.4}$	$\times 10^4$	time, sec	Steps	tions	tions
			Second-order Runge–Kutta, fixed Δz					
0.01	3	0.5	0.17533	24	0.37			
0.005	3	0.5	0.17536	25	0.74			
0.0025	3	0.5	0.17537	25	1.4			
0.01	5	0.25	0.17316	2.6	0.57			
0.005	5	0.25	0.17318	2.8	1.1			
0.0025	5	0.25	0.17319	2.9	2.3			
0.01	9	0.125	Unstable	—	—			
0.005	9	0.125	0.1729509	0.48	Unstable			
0.0025	9	0.125	0.1729562	0.53	3.9			
			Fourth-order Runge–Kutta, variable Δz, $\varepsilon = 10^{-5}$					
	3	0.5	0.175371	25	1.1			
	5	0.25	0.1731905	2.9	1.6			
	9	0.125	0.1729587	0.56	4.6			
Extrapolated Δr^2			0.1728814	0.22	6.1			
			GEARB, variable Δz, $MF = 22$, $\varepsilon = 10^{-5}$					
	3	0.5	0.1753993	25	1.2	195	374	30
	5	0.25	0.1732321	3.3	1.8	176	377	22
	7	0.1666	0.1730533	1.5	2.6	175	427	24
	9	0.125	0.1730161	1.1	3.1	171	393	22
	11	0.10	0.1729811	0.78	4.0	166	411	23
	13	0.0833	0.1730104	1.0	5.1	188	460	27

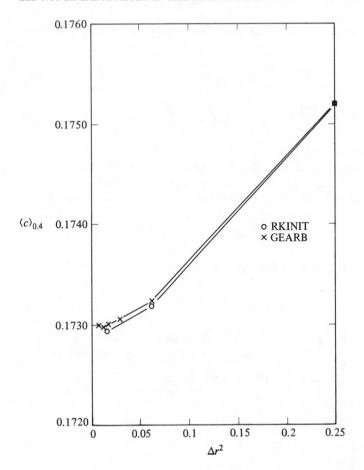

Figure 5-18 Average concentration at $z = 0.4$, $Bi_w = 1$, and $T_w = 0.92$, using the finite difference method.

fourth-order Runge–Kutta method gives results that have very little truncation error of Δz; the only error is due to Δr. The results can be interpolated to give

$$\langle c \rangle_{0.4} = 0.1728814 + 0.004945\Delta r^2 - 0.3945\Delta z^2 \qquad (5\text{-}241)$$

For the fourth-order Runge–Kutta method we use $\Delta z = 0$ in the above formula. The results with $n = 3$ or $\Delta r = 0.5$ do not follow this formula, so that in this case Δr is not small enough that the truncation error formulas are followed. (Remember that the truncation error applies only as $\Delta r \to 0$.) This is demonstrated in Fig. 5-18. Results obtained with GEARB are less accurate than those obtained with Runge–Kutta, but for large n the GEARB results require less computation time. We recall that the explicit method must use a smaller Δz when Δr becomes small, due to the stability limit, Eq. (5-223). An implicit method, however, has no such limitation by stability and larger Δz can be used for small Δr; Δz is then chosen by the accuracy requirement. The results from the second-order Runge–

Kutta method shows that a step size that is stable for large Δr is not stable for smaller Δr. For other geometry the stability limit comparable to Eq. (5-223) is

$$\frac{D\Delta t}{\Delta r^2} \leqslant \frac{1}{2a} \qquad (5\text{-}242)$$

for a single linear equation with boundary conditions of the first kind. Using the values $\Delta z = 0.005$ and $\Delta r = 0.125$ we get a value of $\Delta z/\Delta r^2 = 0.32$, and with $\Delta z = 0.0025$ we get 0.16. The stability limit gives $\Delta z/\Delta r^2 \leqslant 0.25$ and the results in

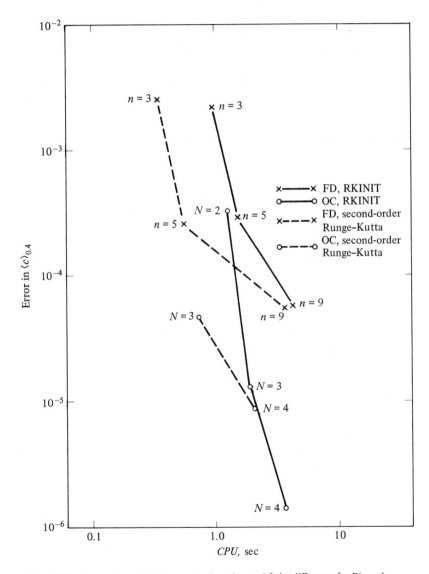

Figure 5.19 Comparison of orthogonal collocation and finite difference for $\mathrm{Bi}_w = 1$.

Table 5-6 indicate this limit is close to that achieved even with the reaction rate term included.

Results obtained by orthogonal collocation are compared in Fig. 5-19 to those using finite difference. The collocation results are very much more accurate than the finite difference results. We can use Eq. (5-241) to determine how many finite difference grid points are necessary to achieve the same accuracy as obtained with three collocation points, 1.3×10^{-5},

$$1.3 \times 10^{-5} = 0.004945 \Delta r^2 \tag{5-243}$$

We need $\Delta r = 0.05$ or 21 grid points to give equivalent accuracy to only three collocation points. To make the temporal error the same size with a second-order Runge–Kutta method requires $\Delta z = 0.006$, which is within the range of the step lengths used.

We next compare the finite difference and orthogonal collocation solutions when both methods use the fourth-order Runge–Kutta method to integrate in

Table 5-7 Average concentration for Eq. (5-156) with $Bi_w = 20$ and $T_w = 1.0$

| | | | | | | Number of | |
| | | | Error | Computation | | Function | Jacobian |
Δz	$n+1$	$\langle c \rangle_{0.6}$	$\times 10^4$	time, sec	Steps	evaluations	evaluations
				Second-order Runge–Kutta, fixed Δz			
0.01	3	0.81644	1000	0.37			
0.005	3	0.87564	440	0.71			
0.0025	3	0.87619	431	1.4			
0.01	5	0.89975	195	Unstable			
0.005	5	0.91630	30	1.1			
0.0025	5	0.920154	8.9	2.3			
0.00125	5	0.920465	12	4.5			
0.0025	9	0.9193089	0.49	3.9			
				Fourth-order Runge–Kutta, variable Δz, $\varepsilon = 10^{-5}$			
	3	0.87634	430	0.96			
	5	0.920564	13	1.8			
	7	0.915718	35	3.4			
	9	0.919566	3.1	6.1			
	11	0.920657	14	9.2			
	13	0.921070	18	13			
				GEARB, variable Δz, $\varepsilon = 10^{-5}$			
	3	0.87655	427	1.1	192	387	31
	5	0.921292	20	1.8	180	403	25
	9	0.920027	7.7	3.2	154	408	23
Extrapolated Δr^2		0.919605	3.5				

time. The collocation solution with $N = 2$ has about the same accuracy as the finite difference solution with $n = 5$, and the collocation solution is only about 10 percent faster. For more accurate solutions, however, the collocation method is plainly superior. If 4 sec are spent in the calculation, the collocation results are 40 times more accurate than the finite difference results. The difference between the methods is accentuated for small errors: the orthogonal collocation method is better the smaller the desired error.

Consider next the same problem with $Bi_w = 20$ and $T_w = 1$, which exhibits more severe spatial gradients (compare Figs. 5-6 and 5-11). We might expect at the outset that a method that uses finite differences or finite elements might be more suited to this problem. Finite difference results are shown in Table 5-7. We note that the errors are generally larger than those in Table 5-6; this illustrates the fact that this problem has more severe gradients and more radial points are needed. As before the fourth-order Runge–Kutta method is more accurate than GEARB for the same Δr, but GEARB takes less computation time, at least for small Δr. The errors in the second-order Runge–Kutta method are not proportional to Δz, so that the Δz values are not small enough for this problem to achieve the limiting behavior as small Δz. The formula for the truncation error is

$$\langle c \rangle_{0.6} = 0.91923 + 0.0213 \Delta r^2 - 64 \Delta z^2 \qquad (5\text{-}244)$$

The coefficient of Δz^2 is uncertain but illustrates the fact that small Δz are necessary for a fixed-step method.

The computation times are illustrated in Fig. 5-20. Calculations made with a Crank–Nicolson method but with the reaction rate term evaluated at the nth time level are generally less accurate than the solutions reported in Table 5-6. These solutions are also compared to the collocation results. Now the Runge–Kutta method is better for finite difference and the GEARB method is better for orthogonal collocation. In this case the finite difference method is superior for low errors. For an error of 0.01 the finite difference method takes about half the computation time, while for an error of 0.001 the two methods are equivalent, and for smaller errors the trend is that the collocation method takes less time. For 4 sec computation time we need six interior collocation points and about eight finite difference grid points; both calculations give the same error. This illustrates again the fact that this problem has steep gradients in the radial direction and more points are needed in this direction.

The computation times for these methods are proportional to the total number of grid points at which the reaction rate is evaluated. It is interesting to examine the computation times for fixed-step methods and compare them with the work estimates. In the orthogonal collocation method we need $2N^2$ multiplications to evaluate the right-hand side involving the dispersion terms and mN multiplications to evaluate the reaction rate, where m is the number of multiplications needed to evaluate one rate. Thus the computation time CPU should be a function of N and N^2. The data are fit to a quadratic and the best least squares fit is

$$\text{CPU} = (\text{number of axial steps}) \, 0.0025N \qquad (5\text{-}245)$$

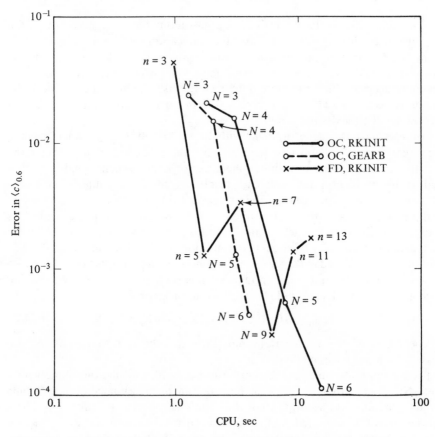

Figure 5-20 Comparison of orthogonal collocation and finite difference for $Bi_w = 20$.

which shows that the computation time is proportional to the number of collocation points. This suggests that the major computation time is associated with the evaluation of the reaction rate term. The finite difference method gives

$$CPU = (\text{number of axial steps}) \, 0.00060N \qquad (5\text{-}246)$$

but in this case the number of calculations needed to evaluate the right-hand side is proportional to the number of terms. In both cases the computation time is proportional to the total number of grid points in both axial and radial directions. We can thus arrive at an estimate of the cost per grid point for the two methods

$$
\begin{aligned}
\text{Cost for orthogonal collocation} &\simeq 0.0025 \text{ sec/collocation point/axial point} \\
\text{Cost for finite difference} &\simeq 0.00060 \text{ sec/grid point}
\end{aligned}
\qquad (5\text{-}247)
$$

Thus for these methods the finite difference method uses only 24 percent of the computation time per grid compared with the collocation method. The collocation method must achieve its dramatic improvement because it can use many fewer terms.

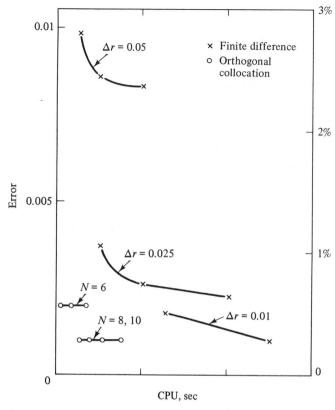

Figure 5-21 Error in the temperature surface flux versus computation time, with $Bi = Bi_m = \infty$.

Next we apply the finite difference method to the problem treated with orthogonal collocation, Eqs. (5-176) to (5-178), for diffusion and reaction in a catalyst pellet. This problem was solved earlier by Ferguson and Finlayson.[5] The collocation results were obtained with an improved Euler method in time, and the finite difference method used Crank–Nicolson. Errors in heat flux are shown in Fig. 5-21. The collocation method can use 6 collocation points and achieve the accuracy of a finite difference solution found using 100 grid points. Alternatively, the collocation solution is obtained with only 5 percent of the computation time needed for the finite difference method. Part of this speed advantage comes from being able to take a larger step size in the collocation method ($\Delta t = 0.05$ instead of 0.01) and the rest comes from having many fewer unknowns to represent the spatial variations of the solution (6 instead of 100).

In summary, for many diffusion–reaction problems the orthogonal collocation method gives very accurate answers, but as the gradients become larger the finite difference method may be competitive. The fourth-order Runge–Kutta method and GEARB both are good schemes to use to integrate in the time-like direction. RKINIT is more accurate and usually faster for a small number of terms, whereas GEARB is faster for a larger number of terms when the equations are stiffer due to the spatial approximation.

5-6 ORTHOGONAL COLLOCATION ON FINITE ELEMENTS

The diffusion–reaction problems treated above are all well suited to orthogonal collocation. The parameters can be changed, however, to ones that give steep profiles during the transients. This is done in a fashion similar to the steady-state case. For example take Bi small (5) and Bi_m large (250), even for $\beta = 0.02$ steep profiles are obtained. If that happens it is necessary to resort to a method that approximates the solution piecewise (i.e. we must use the finite difference method, orthogonal collocation on finite elements, or Galerkin finite elements). The details of applying the method of orthogonal collocation on finite elements to transient problems are similar to those for steady-state problems, so we need only outline them.

Considering Eqs. (5-176) to (5-178) we apply orthogonal collocation on finite elements. The domain is discretized as shown in Fig. 4-12. First we apply lagrangian interpolation. The residual is evaluated at the collocation points. The equations for concentration only are given since the ones for temperature are similar.

$$M_2 \frac{dc_I}{dt} = \frac{1}{h_k^2} \sum_{J=1}^{N+2} B_{IJ} c_J + \frac{a-1}{x_i} \frac{1}{h_k} \sum_{J=1}^{N+2} A_{IJ} c_J - \phi^2 R(c_I, T_I) \qquad (5\text{-}248)$$

whereas before

$$i = (k-1)(N+1) + I \qquad (5\text{-}249)$$

We require continuity of the derivative across the elements, Eq. (4-269). The boundary conditions are Eqs. (4-270) and (4-271). The system of equations can be written in the form

$$\mathbf{CC} \frac{d\mathbf{c}}{dt} = \mathbf{AAc} - \mathbf{f(c)} \qquad (5\text{-}250)$$

where the matrices $\mathbf{CC}$ and $\mathbf{AA}$ have the structure shown in Fig. 4-13. The matrix $\mathbf{CC}$ multiplying the time derivative has the value M_2 on the diagonal of each equation representing a residual and zero elsewhere, since the boundary conditions and flux continuity conditions have no time derivatives in them. The matrix $\mathbf{AA}$ is identical to that given in Sec. 4-9. The vector $\mathbf{f}$ includes the reaction rate term in the residual equations and either zero or some value for the boundary conditions and flux continuity conditions.

If we apply a backward Euler method or Crank–Nicolson method in time then we get equations in the form

$$\mathbf{CC} \frac{\mathbf{c}^{n+1} - \mathbf{c}^n}{\Delta t} = \mathbf{AA}[(1-\beta)\mathbf{c}^n + \beta \mathbf{c}^{n+1}] + (1-\beta)\mathbf{f(c^n)} + \beta \mathbf{f(c^{n+1})} \qquad (5\text{-}251)$$

$$(\mathbf{CC} - \beta \Delta t \mathbf{AA})\mathbf{c}^{n+1} = [\mathbf{CC} + (1-\beta)\Delta t \mathbf{AA}]\mathbf{c}^n + (1-\beta)\Delta t \mathbf{f}(\mathbf{c}^n) + \beta \Delta t \mathbf{f}(\mathbf{c}^{n+1}) \quad (5\text{-}252)$$

The left-hand side has the structure of Fig. 4-13. The reaction rate term must be linearized, Eq. (4-278), to solve Eq. (5-252). The solution then proceeds as in any method, except that the LU decomposition of the matrix is done taking into account the structure of Fig. 4-13.

Explicit methods are possible if desired. Each residual equation can be calculated using whatever method is chosen. After doing this we have values of the solution at each interior collocation point at the new time. We also need the boundary values and the solution at the points between elements. We can obtain these by solving the algebraic equations for boundary conditions, Eqs. (4-270) and (4-271), and the element continuity equation, Eq. (4-269), using the solution at the interior collocation points as known. Then the matrices are in tridiagonal form

$$\mathbf{AAc} = \mathbf{f} \quad (5\text{-}253)$$

with

$$\frac{1}{h_1}(A_{11}c_1 + A_{1,N+2}c_{N+2}) = -\frac{1}{h_1}\sum_{J=2}^{N+1} A_{1J}c_J \quad (5\text{-}254)$$

$$AA_{ij} = \begin{cases} j = i-1 & \dfrac{1}{h_{k-1}}A_{N+2,1} \\[2mm] j = i & \dfrac{1}{h_{k-1}}A_{N+2,N+2} - \dfrac{1}{h_k}A_{11} \\[2mm] j = i+1 & -\dfrac{1}{h_k}A_{1,N+2} \end{cases} \quad (5\text{-}255)$$

$$f_i = \frac{1}{h_k}\sum_{J=2}^{N+1} A_{1J}c_J - \frac{1}{h_{k-1}}\sum_{J=2}^{N+1} A_{N+2,J}c_J \quad (5\text{-}256)$$

$$-\frac{1}{h_{NE}}(A_{N+2,1}c_{NT-N-1} + A_{N+2,N+2}c_{NT}) - \text{Bi}_m c_{NT}$$

$$= \frac{1}{h_{NE}}\sum_{J=2}^{N+1} A_{N+2,J}c_J - \text{Bi}_m \quad (5\text{-}257)$$

This part of the solution is referred to as smoothing. It takes the solution at the interior collocation points and provides the solution at the element boundaries in such a way that the final solution is continuous and has continuous first derivatives (or fluxes).

One disadvantage of the equations in the form of Eq. (5-250) is that they involve coupled differential equations and algebraic equations. Most integration packages are not suitable for such systems, so that the analyst must provide a satisfactory package. This is not true of the finite difference method, however, which can easily be applied using the integration packages.

The cumbersome algebraic equations can be eliminated by using Hermite instead of lagrangian interpolation. Doing this for Eqs. (5-176) to (5-177) gives the residuals

$$M_2 \sum_{I=1}^{4} H_{JI} \frac{da_I}{dt} = \frac{1}{h_k^2} \sum_{I=1}^{4} B_{JI} a_I + \frac{a-1}{x_j} \sum_{I=1}^{4} A_{JI} a_I - \phi^2 R \left(\sum_{I=1}^{4} H_{JI} a_I \right) \quad (5\text{-}258)$$

along with the boundary conditions of Eqs. (4-302) and (4-303). The last two algebraic equations are easily solved and the results can be introduced into the residuals by eliminating two unknowns. The residuals can be written in the form

$$\mathbf{CC} \frac{d\mathbf{a}}{dt} = \mathbf{AAa} - \mathbf{f(a)} \quad (5\text{-}259)$$

where the $\mathbf{CC}$ matrix has the structure of Fig. 4-19 without the boundary conditions and is not a diagonal matrix because of $\mathbf{H}$. The $\mathbf{AA}$ matrix has the same structure and is the same matrix as that derived in Sec. 4-9. Explicit methods cannot be applied to this system of equations easily because of the nondiagonal nature of the $\mathbf{CC}$ matrix. Even if the right-hand side is evaluated using known information, a set of coupled equations still remains due to the $\mathbf{CC}$ matrix. An LU decomposition must be performed. For this problem the $\mathbf{CC}$ matrix is constant in time, so that it can be decomposed once per problem, and an explicit method used with successive fore-and-aft sweeps to solve for the successive right-hand sides. This is not a disadvantage for implicit methods since the LU decomposition must be performed anyway. The nondiagonal nature of the $\mathbf{CC}$ matrix also presents problems when using integration packages, which are often written for equations in the form

$$\frac{d\mathbf{a}}{dt} = \mathbf{f(a)} \quad (5\text{-}260)$$

rather than in the form of Eq. (5-259). The analyst must then devise a suitable package.

5-7 GALERKIN FINITE ELEMENT METHOD

Galerkin finite elements are also useful when the transient solution has steep gradients. We apply them here to the diffusion–reaction equations, Eqs. (5-156) to (5-158). The domain is divided into finite elements as shown in Figs. 4-20 to 4-22. The Galerkin method is applied in the same way: a trial function is assumed on each element, it is substituted into the differential equation to form the residual, and the weighted residual is set to zero with the weighting function as one of the trial functions. The equations are the same as Eqs. (4-320) to (4-325) with the time-dependent term added

$$\sum_e \sum_I C_{JI}^e \frac{dc_J^e}{dt} = \sum_e \sum_I B_{JI}^e c_I^e - \sum_e F_J^e \quad (5\text{-}261)$$

with the mass matrix

$$C_{JI}^e = \Delta x_e \int_0^1 N_J(u) N_I(u) r^{a-1} \, du \quad (5\text{-}262)$$

and $\mathbf{B}^e$ and $\mathbf{F}^e$ are defined by Eqs. (4-320) and (4-321). We must then integrate these in time. The nonlinear reaction rate term must be evaluated using numerical quadrature each time step or it can be interpolated. The form of the equation is Eq. (5-250), and the structure of the equations is tridiagonal for linear basis functions. Quadratic basis functions give the structure shown in Fig. 4-23.

The time integration has the same difficulties which the Hermite polynomials have on finite elements: explicit methods involve at least one LU decomposition each problem and integration packages are not readily available. The equations present no problem though if an implicit method is applied and the analyst provides a suitable package.

There is an alternative to the time integration problem that has been called lumping. In this procedure each row of the **CC** matrix is added up and put on the diagonal. The off-diagonal terms are set to zero. With this approximation the equations can now be solved with explicit methods and the integration packages are suitable. The accuracy can be degraded, however, as described for the convective diffusion equation treated below.

The Galerkin method has this disadvantage for time-dependent problems (a nondiagonal mass matrix), and it has the added disadvantage of extensive computation time to evaluate nonlinear integrals each iteration and each time step. Each of these problems can be overcome, but at the expense of degrading the accuracy or increasing the computation time. While the Galerkin method is not too promising for one-dimensional problems, it proves to be a superior method in two dimensions (see Chapter 6).

5-8 CONVECTIVE DIFFUSION EQUATION

Several methods are applied to the convective diffusion equation because it has a sharp front but is a linear problem with an exact solution. We examine the steady-state problem first to assess the difficulty of the problem.

In steady state we wish to solve the ordinary differential equation

$$\frac{d^2c}{dx^2} - \text{Pe}\frac{dc}{dx} = 0 \tag{5-263}$$

with boundary conditions

$$c(0) = 1 \qquad c(1) = 0 \tag{5-264}$$

The exact solution is

$$c = \frac{e^{\text{Pe}} - e^{\text{Pe}x}}{e^{\text{Pe}} - 1} \tag{5-265}$$

Successive derivatives of the exact solution are

$$c^{(n)} = \frac{d^n c}{dx^n} = \frac{-1}{e^{\text{Pe}} - 1}\text{Pe}^n e^{\text{Pe}x} \tag{5-266}$$

and the norm of the nth derivative is

$$||c^{(n)}|| = \left[\int_0^1 (c^{(n)})^2 \, dx \right]^{1/2} = \left[\frac{1}{2\text{Pe}} (e^{2\text{Pe}} - 1) \right]^{1/2} \frac{\text{Pe}^n}{e^{\text{Pe}} - 1} \qquad (5\text{-}267)$$

We wish to investigate the interpolation error if a finite difference or finite element method is used to interpolate the exact solution. The error estimates are provided by Prenter.[10] For lagrangian interpolation of order n the errors are

$$||c - p|| \leqslant \frac{||c^{(n+1)}|| h^{n+1}}{4(n+1)} \qquad (5\text{-}268)$$

where $p(x)$ is the lagrangian interpolation. For Hermite cubic polynomials the error estimate is

$$||c^{(j)} - p^{(j)}|| \leqslant \frac{||c^{(n+1)}|| n! h^{n+1-j}}{(j-1)!(n+1-j)!} \qquad 1 \leqslant j \leqslant n \qquad (5\text{-}269)$$

$$||c^{(j)} - p^{(j)}|| \leqslant \frac{||c^{(4)}|| h^{4-j}}{(4-j)12^{4-[j]}} \qquad 0 \leqslant j \leqslant 3 \qquad (5\text{-}270)$$

where $[j] = j$ if j is even and $[j] = j+1$ if j is odd. Using the properties of the exact solution gives an estimate for the mean square error, and we wish to make this less than a prescribed value ε

$$||c - p|| \leqslant \frac{\text{Pe}^{n+1/2} h^{n+1}}{4\sqrt{2}(n+1)} \leqslant \varepsilon \qquad \text{lagrangian interpolation} \qquad (5\text{-}271)$$

$$||c - p|| \leqslant \frac{\text{Pe}^{3.5} h^4}{4 \quad 12^4} \leqslant \varepsilon \qquad \text{Hermite interpolation} \qquad (5\text{-}272)$$

This limitation requires a certain number of intervals in each method. The number of intervals is given in Table 5-8. We note that for a large error ($\varepsilon = 0.1$) the piecewise constant lagrangian interpolations ($n = 0$) require the fewest points. This is because the number of intervals needed in the higher-order methods depends on higher-order derivatives of the function, and the derivatives and their norms increase with each differentiation. As the desired error is reduced (say $\varepsilon = 0.001$), the linear lagrangian interpolation requires the fewest intervals of the lagrangian interpolants of various orders. For even more stringent error criteria ($\varepsilon = 10^{-6}$) the third-order interpolants require the fewest intervals. The Hermite polynomials, which have continuous first derivatives, do a better job than the lagrangian interpolants, which are only continuous. (We should recall though that the orthogonal collocation with lagrangian interpolation is made to have continuous first derivatives and hence is equivalent to the Hermite interpolation.) Indeed for errors lower than 0.01, the Hermite cubics require the fewest intervals.

Suppose we use these error estimates for the interpolant and combine them with the work necessary to solve Eq. (5-263). The error of the approximate solution is not necessarily that given by the interpolation error, but it cannot be any smaller than the interpolant error. The work necessary to apply the methods is

Table 5-8 Number of intervals required to satisfy the error criteria of Eqs. (5-271) and (5-272), Pe $= 10^5$

			Number of intervals		
			Lagrangian		Hermite
ε	$n = 0$	$n = 1$	$n = 2$	$n = 3$	$n = 3$
10^{-1}	560	5,300	12,000	19,000	2,500
10^{-2}	5,600	17,000	27,000	34,000	4,400
10^{-3}	56,000	53,000	57,000	61,000	7,900
10^{-4}	560,000	170,000	120,000	110,000	14,000
10^{-5}	5,600,000	530,000	270,000	190,000	25,000
10^{-6}	56,000,000	1,700,000	570,000	340,000	44,000

			Number of operations*		
	FD, $5n$	GFEM–2, $16NE$	GFEM–3, $23NE$	OCFE–L, $37.33NE$	OCFE–H, $12NE$
10^{-1}	0.026	0.19	0.44	0.093	0.030
10^{-2}	0.084	0.43	0.78	0.16	0.054
10^{-3}	0.26	0.91	1.4	0.29	0.094
10^{-4}	0.84	1.9	2.5	0.52	0.17
10^{-5}	2.6	4.3	4.4	0.93	0.30
10^{-6}	8.4	9.1	7.8	1.6	0.54

* Number of operations $\times 10^{-6}$ needed to solve a linear system of size specified above.

also given in Table 5-8. For large errors ($\varepsilon > 0.1$) the finite difference method requires less work. The work estimates for the Hermite cubics are given for both low-order and high-order orthogonal collocation on finite elements, with the latter clearly preferable. The interpolation error is plotted versus the number of multiplications necessary to solve the problem in Fig. 5-22. This graph clearly portrays the guidelines: if low accuracy is desired in a solution with steep gradients the finite difference method is best, but if high accuracy is desired higher-order methods are preferable.

The transient problem has the same difficulties if the solution has large gradients. Consider the convective diffusion equation with a steep change of concentration at $x = 0$

$$\frac{\partial c}{\partial t} + \text{Pe}\frac{\partial c}{\partial x} = \frac{\partial^2 c}{\partial x^2} \tag{5-273}$$

$$c(x,0) = 0 \qquad c(0,t) = 1 \qquad \frac{\partial c}{\partial x}(1,t) = 0 \tag{5-274}$$

We do not apply orthogonal collocation (with a global polynomial) because the front is steep for large Pe. The finite difference, orthogonal collocation on finite

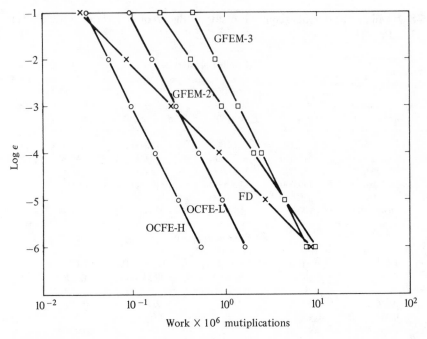

Figure 5-22 Interpolation error as a function of work.

elements, and Galerkin finite element methods are all applied in turn. For the finite element methods the backward Euler method is used in time, with a small time step, so that oscillations are not introduced by the time integration, whereas GEARB is used with the finite difference method.

The finite difference method applied to Eq. (5-273) gives the following equation, accurate to $0(\Delta x^2)$,

$$\frac{dc_i}{dt} + \text{Pe}\frac{c_{i+1} - c_{i-1}}{2\Delta x} = \frac{c_{i+1} - 2c_i + c_{i-1}}{\Delta x^2} \tag{5-275}$$

This equation can be represented in the form

$$\mathbf{C}\frac{d\mathbf{c}}{dt} = \mathbf{AAc} \tag{5-276}$$

and for the finite difference method $\mathbf{C} = \mathbf{I}$. Results for Pe = 1,000 and $\Delta x = 0.01$ are shown in Fig. 5-23. Oscillations in the solution are evident. These results were obtained using a time step small enough that the oscillations were not due to the temporal integration. Thus we are left with an unsatisfactory solution. Another solution which uses five times as many grid points, $\Delta x = 0.002$, is shown in Fig. 5-23. The oscillations are now gone, but at a considerable increase in computation time (22 times).

Another approach to improve the solution is to introduce upstream deriva-

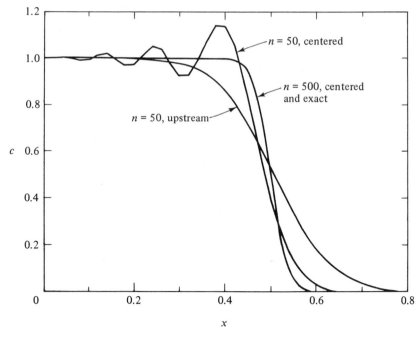

Figure 5-23 Convective diffusion equation with Pe = 1,000 and t = 0.0005.

tives when the Peclet number is large. We use a first-order expression for the first derivative

$$\left.\frac{\partial c}{\partial x}\right|_i = \frac{c_i - c_{i-1}}{\Delta x} \tag{5-277}$$

Equation (5-275) becomes

$$\frac{dc_i}{dt} + \text{Pe}\frac{c_i - c_{i-1}}{\Delta x} = \frac{c_{i+1} - 2c_i + c_{i-1}}{\Delta x^2} \tag{5-278}$$

and

$$\text{Truncation error} = -\frac{\text{Pe}\,\Delta x}{2}\frac{\partial^2 c}{\partial x^2} \tag{5-279}$$

The truncation error is found by inserting a Taylor series for c_i into Eq. (5-278). Thus solving Eq. (5-278) is the same as solving the following equation to second order $0(\Delta x^2)$:

$$\frac{dc_i}{dt} + \text{Pe}\frac{c_{i+1} - c_{i-1}}{2\Delta x} = \frac{1 + \text{Pe}\,\Delta x/2}{\Delta x^2}(c_{i+1} - 2c_i + c_{i-1}) \tag{5-280}$$

The effect of this upstream derivative is to add dispersion to the numerical solution. Figure 5-23 shows the solution for Pe = 1,000 and 51 grid points, while Fig. 5-24 shows the solution for Pe = 87,790 and 301 grid points. With a second-

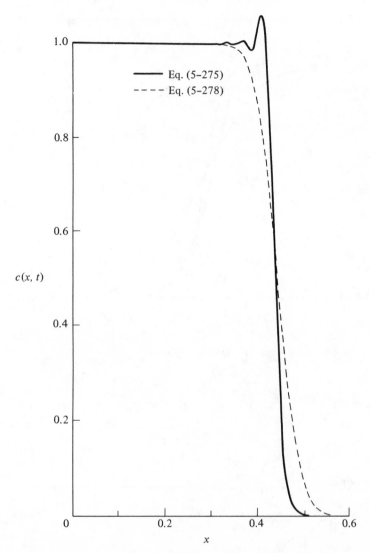

Figure 5-24 Finite difference solution at $t = 5 \times 10^{-6}$, Pe = 87,790, Co = 0.13, $\Delta t = 5 \times 10^{-9}$, and $NT = 301$.

order scheme the solution oscillates. With a first-order scheme, Eq. (5-280) or (5-278), the oscillations disappear and instead we have a smoother solution with the front smoothed out by the numerical dispersion. If this effect is acceptable, numerical dispersion can be introduced.

The method of orthogonal collocation on finite elements with lagrangian interpolation would use

$$\frac{dc_I}{dt} + \frac{\text{Pe}}{\Delta x_k} \sum_{J=1}^{N+2} A_{IJ} c_J = \frac{1}{\Delta x_k^2} \sum_{J=1}^{N+2} B_{IJ} c_J \qquad (5\text{-}281)$$

The problem can be written in the form of Eq. (5-276) and the matrix is block diagonal (see Fig. 4-13). The condition at the element boundaries is still the continuity of the first derivative. If Hermite interpolation is used, the matrix has the structure of Fig. 4-19 and the residuals are

$$\sum_{J=1}^{4} H_{IJ} \frac{da_J}{dt} + \frac{Pe}{\Delta x_k} \sum_{J=1}^{4} A_{IJ} a_J = \frac{1}{\Delta x_k^2} \sum_{J=1}^{4} B_{IJ} a_J \tag{5-282}$$

Results are shown in Fig. 5-25 for the orthogonal collocation method using lagrangian interpolation. Oscillations are very small when 50 elements are used,

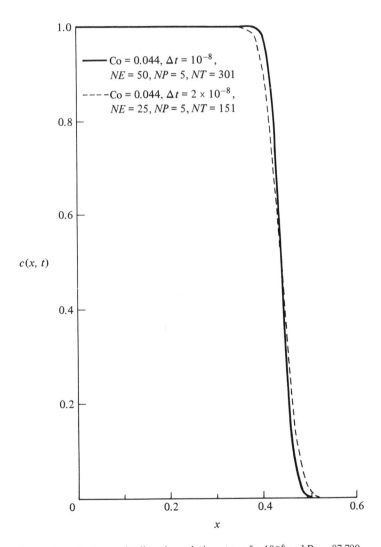

Figure 5-25 Orthogonal collocation solution at $t = 5 \times 10^{-6}$ and Pe $= 87{,}790$.

and the front is smoothed somewhat when only 25 elements are used. The steepness of the front may be correctly modeled but 301 collocation points were needed. The collocation solution with 301 collocation points is more expensive than a finite difference solution with 301 grid points, but the collocation solution does not oscillate nearly as much. (Compare Fig. 5-25 with Fig. 5-24.)

With the collocation method we can introduce numerical dispersion in an ad hoc way by solving

$$\frac{\partial c}{\partial t} + \mathrm{Pe}\frac{\partial c}{\partial x} = \left(1 + \mathrm{Pe}\frac{\Delta x}{2}\right)\frac{\partial^2 c}{\partial x^2} \qquad (5\text{-}283)$$

We rearrange this equation by dividing by $1 + \mathrm{Pe}\,\Delta x/2$ and define new variables

$$\tau = t\left(1 + \mathrm{Pe}\frac{\Delta x}{2}\right) \qquad \mathrm{Pe}' = \frac{\mathrm{Pe}}{(1 + \mathrm{Pe}\,\Delta x/2)} \qquad (5\text{-}284)$$

If we wish to solve for $\mathrm{Pe} = 1{,}000$ using $\Delta x = 0.1$, we simply solve

$$\frac{\partial c}{\partial \tau} + \mathrm{Pe}'\frac{\partial c}{\partial x} = \frac{\partial^2 c}{\partial x^2} \qquad (5\text{-}285)$$

with $\mathrm{Pe}' = 1{,}000/51 = 19.61$ and translate the terms using Eq. (5-284). Such an approach is unsatisfying, but may be necessary in order to solve the problem in a given amount of computer time.

The Galerkin method gives

$$C^e_{IJ}\frac{dc^e_J}{dt} + \frac{\mathrm{Pe}}{\Delta x}A^e_{IJ}c^e_J = \frac{1}{\Delta x^2}B^e_{IJ}c^e_J \qquad (5\text{-}286)$$

The matrices are given in Table 4-9 for the appropriate choice of basis function. The structure of the equations is tridiagonal for linear basis functions and is shown in Fig. 4-23 for quadratic basis functions. The linear basis functions give the explicit equation

$$\frac{1}{6}\frac{dc_{i+1}}{dt} + \frac{2}{3}\frac{dc_i}{dt} + \frac{1}{6}\frac{dc_{i-1}}{dt} + \mathrm{Pe}\frac{(c_{i+1}-c_{i-1})}{2\Delta x} = \frac{c_{i+1}-2c_i+c_{i-1}}{\Delta x^2} \qquad (5\text{-}287)$$

One disadvantage in the Galerkin method is that the left-hand side is not diagonal. Thus the integration codes that are discussed in Chapter 3 are not directly applicable. One approach to this difficulty in the Galerkin method is called lumping. The left-hand side represents the time derivative of the mass in an element, and in lumping the entire mass is "lumped" with the ith node. Equation (5-287) then becomes

$$\frac{dc_i}{dt} + \mathrm{Pe}\frac{c_{i+1}-c_{i-1}}{2\Delta x} = \frac{c_{i+1}-2c_i+c_{i-1}}{\Delta x^2} \qquad (5\text{-}288)$$

This is just the finite difference equation.

If we apply Galerkin finite elements methods we also get oscillations if $\mathrm{Pe}\Delta x$ is not small enough. Thus artificial dispersion needs to be introduced here, too. This is done using a different weighting function. The Galerkin equations are written as

$$\int_0^1 W_J \left(\frac{\partial c}{\partial t} + Pe \frac{\partial c}{\partial x} \right) dx = - \int_0^1 \frac{\partial c}{\partial x} \frac{\partial W_J}{\partial x} dx \qquad (5\text{-}289)$$

and the weighting function is chosen to introduce numerical dispersion. The weighting function (see Fig. 5-26a) is taken as[7]

$$
\begin{aligned}
W_I &= N_I + \alpha F(x) & x_{i-1} \leqslant x \leqslant x_i \\
W_I &= N_I - \alpha F(x) & x_i \leqslant x \leqslant x_{i+1} \\
F(u) &= -3(u^2 - u) & u = \frac{x - x_i}{x_{i+1} - x_i}
\end{aligned}
\qquad (5\text{-}290)
$$

The Galerkin linear element equations become

$$\frac{1}{6} \frac{dc_{i+1}}{dt} + \frac{2}{3} \frac{dc_i}{dt} + \frac{1}{6} \frac{dc_{i-1}}{dt} + \frac{Pe}{2\Delta x} [(1-\alpha)c_{i+1} + 2\alpha c_i - (1+\alpha)c_{i-1}]$$

$$= \frac{1}{\Delta x^2} (c_{i+1} - 2c_i + c_{i-1}) \quad (5\text{-}291)$$

The choice of α can be used to optimize the results. The upstream dispersion as in

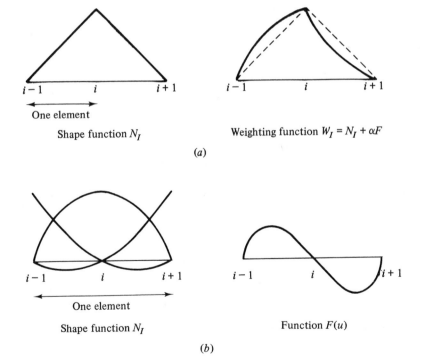

One element

Shape function N_I

Weighting function $W_I = N_I + \alpha F$

(a)

One element

Shape function N_I

Function $F(u)$

(b)

Figure 5-26 Upstream weighting functions. (a) Linear basis functions. (b) Quadratic basis functions.

the finite difference method is given when $\alpha = 1$. The choice

$$\alpha = \coth\left(\frac{\text{Pe}\,\Delta x}{2}\right) - \frac{2}{\text{Pe}\,\Delta x} \tag{5-292}$$

makes the errors zero at the nodes for the steady-state problem. With upstream dispersion the calculation can be made with a larger element size and oscillations are not present provided

$$\text{Pe}\,\Delta x \leqslant \frac{2}{1-\alpha} \tag{5-293}$$

For the quadratic trial functions we choose the weighting function[8] for i as an end node

$$W_I = N_I - \alpha F$$

for i as a midside node

$$W_I = N_I + 4\beta F \tag{5-294}$$

and

$$F(u) = \tfrac{5}{2}u(2u^2 - 3u + 1)$$

$$\beta_0 = \coth\left(\frac{\text{Pe}\,\Delta x}{4}\right) - \frac{4}{\text{Pe}\,\Delta x}$$

$$\alpha_0 = 2\left[\tanh\left(\frac{\text{Pe}\,\Delta x}{2}\right)\right]\left(1 + \frac{3\beta_0}{\text{Pe}\,\Delta x} + \frac{12}{\text{Pe}^2\,\Delta x^2}\right) - \frac{12}{\text{Pe}\,\Delta x} - \beta_0$$

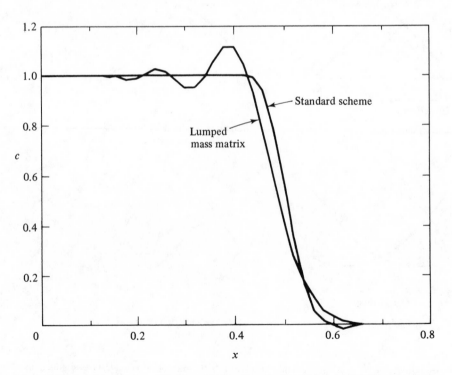

Figure 5-27 Galerkin solutions with linear trial functions for $\text{Pe} = 1{,}000$, $t = 0.0005$, and $n = 50$.

which is also pictured in Fig. 5-26*b*. With upstream weighting functions the solution does not oscillate, but it is damped more than the exact solution (see Fig. 5-27).

For each of the methods then we see that the solution frequently oscillates, but these oscillations can be eliminated by introducing more points/elements or by introducing numerical dispersion in some form. Which method is best? Using several methods Jensen and Finlayson[9] have examined the steady-state equation

$$\text{Pe}\frac{dc}{dx} = \frac{d^2c}{dx^2} \tag{5-295}$$

$$c(0) = 1 \qquad c(1) = 0 \tag{5-296}$$

The steady-state equation can be solved exactly by difference formulas, and the results show that oscillations occur unless Δx is small. In particular, a criterion can be developed that

$$\text{Pe}\,\Delta x \leqslant B \tag{5-297}$$

to eliminate oscillations. The value of B depends on the method, and several values are listed in Table 5-9. We see that all the methods have a limit and that the limits are close to each other. Thus based on this theoretical limit we cannot choose the

Table 5-9 Limits to element size to prevent oscillations. Table value is B in $\text{Pe}\,\Delta x \leqslant B$

Method	Theoretical B (based on monotonicity)	Practical 1 percent	B 10 percent
FD			
Central	2	3	13
Upwind	α		
GFEM			
Linear standard	2	9	70
With upstream weighting function	$\dfrac{2}{1-\alpha}$		
GFEM			
Quadratic standard	2		
With dispersion weighting function	2		
GFEM			
Cubic, C°	4.644		
OCFE–L			
Quadratic	2		
Cubic	3.464		
Quartic	4.644		
OCFE–H			
Cubic	3.464		
Quartic	4.644		

best method. In a numerical solution, however, oscillations can be present, though small—indeed so small that they are not observable. The criterion to meet this standard is not available and Eq. (5-297) only provides a guide. Based on extensive experimentation, Ehlig[4] reported some practical limits as shown in Table 5-9. Settari, et al.[13] reported practical limits for the finite difference method of six, rather than two. For the Galerkin finite element method with Hermite cubics a value of 250 proved a satisfactory limit, rather than about 4.6. It is concluded that for any method a great many intervals are needed to prevent oscillations unless artificial numerical dispersion is introduced into the method.

Computation time for the various methods is illustrated in Fig. 5-28. These results are for a calculation with a time step so small that there is no temporal truncation error. On this basis, the finite difference method is very much more expensive than the finite element method, even when linear basis functions are used. The higher-order methods are preferable on this scale. When using a

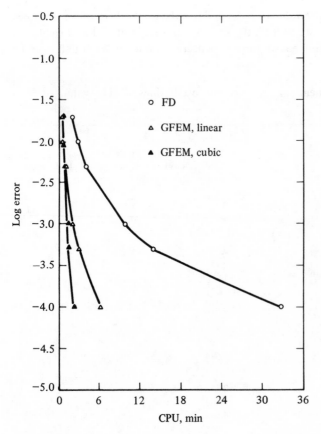

Figure 5-28 Computation time versus error for convective diffusion equation with Pe = 887.9. (*After* Price, et al.[11])

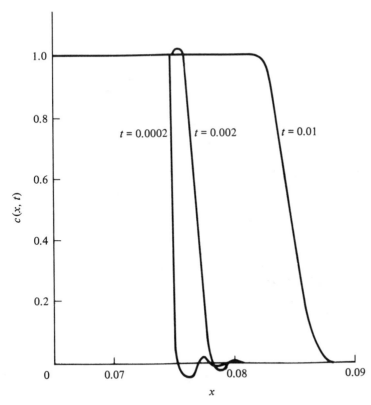

Figure 5-29 Movement of sharp front, color equation, using orthogonal collocation on finite elements with 200 elements. Front initially at $x = 0.075$. $NP = 7$, $\Delta t = 2 \times 10^{-4}$, and Co = 0.04.

practical time step, rather than a very small one, the differences may not be so marked, and the relative position of the curves may change.

We next consider the color equation, which eliminates the dispersion term altogether, such that

$$\frac{\partial c}{\partial t} + \frac{\partial c}{\partial x} = 0 \tag{5-298}$$

$$c(x,0) = 0 \qquad c(0,t) = 1 \tag{5-299}$$

This equation is hyperbolic and has an exact solution since the inlet concentration is propagated with velocity 1.0 without change in shape. Solutions at an early time are shown in Fig. 5-29. Initially, there are oscillations because the front is vertical. As the calculation proceeds these oscillations are damped, but the steep front is smoothed, too.

The properties of the exact solution are dependent on the Courant number

$$\text{Co} = \frac{\Delta t}{\Delta x} \tag{5-300}$$

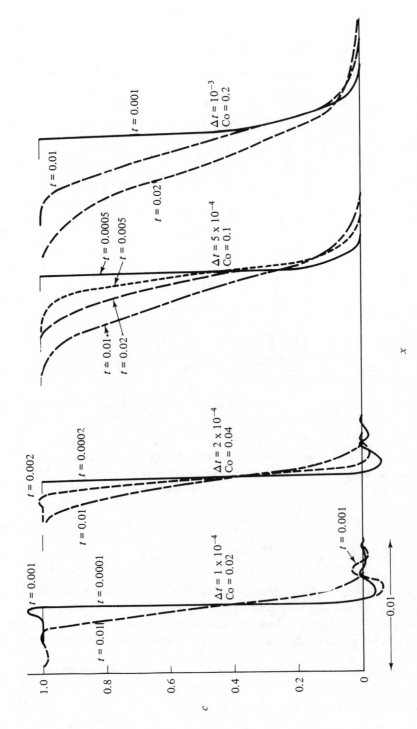

Figure 5-30 Effect of time step on sharpness of front and oscillations, using color equation, orthogonal collocation on finite elements with $NE = 200$, $NP = 7$, and $NT = 1,201$.

It can be shown that a finite difference solution with upstream convective terms and an explicit integration scheme is stable if the Courant number is less than one and that an implicit method is stable for all Courant numbers. With centered convective terms, the explicit scheme is unstable for any Δt, while the implicit scheme is stable for any Δt. (See problem 5-32.)

The influence of different Courant numbers while keeping the number of elements fixed is shown in Fig. 5-30. As the Courant number is increased the oscillations are damped but the profiles are less steep. As the Courant number is decreased by decreasing the time step oscillations are introduced. We thus have a dilemma: if we use too large a Δt we get numerical dispersion and the front is damped, if we use too small a Δt we get the oscillations inherent in the spatial approximation. Only by a judicious choice of Δx and Δt, and by some relaxation of our standards, can we solve such a problem numerically. In the convective diffusion problem the Courant number is

$$\text{Co} = \frac{\text{Pe}\,\Delta t}{\Delta x} \qquad (5\text{-}301)$$

We see in Figs. 5-24 and 5-25 that the Courant number is in the range 0.04–0.13 for good solutions.

The convective diffusion problem is an interesting and useful one because it gives guidelines that are relevant if a problem has steep fronts that move in time. We see that all the numerical methods give a solution that oscillates unless enough elements or grid points are used. For very large Peclet numbers, and very steep fronts, we can introduce numerical dispersion into all the methods to make the calculation possible with a larger Δx, but the solution is less accurate. If we must model the sharp front then sufficient points must be employed.

5-9 FLOW THROUGH POROUS MEDIA

An important area of numerical analysis is the simulation of the flow of fluids through porous media. Applications exist in the flow of water in underground aquifers and the flow of oil–water mixtures in oil fields. Here we consider the flow of water in dry soils by solving the partially saturated equations.

The mass balance for liquid water is

$$\frac{\partial}{\partial t}(\rho\phi S) + \frac{\partial}{\partial x}(\rho q) = 0 \qquad (5\text{-}302)$$

where ρ is the water density ($\rho = 1\,\text{g/cm}^3$), ϕ is the porosity of the rock, and S is the saturation defined as the fraction of free space that is occupied by water. The combination $\rho\phi S$ is the mass density per system volume. The mass flux is ρq in units of mass of water per unit time per unit total cross sectional area, and q is the volumetric flux. Darcy's law is used to relate the mass flux to pressure gradients, which drive the flow,

$$\rho q = -\frac{\rho k}{\mu}\left(\frac{\partial p}{\partial x} - \rho g\right) \qquad (5\text{-}303)$$

Table 5-10 Typical soil properties

k_0 (microns^2)	0.50
ϕ	0.485
S_r	0.32
A (cm)	231
B (cm)	146
η	3.65
λ	6.65

where k is the permeability, μ the viscosity of water ($\mu = 0.00894$ g/cm/sec), p the water pressure, and g the acceleration of gravity ($g = 980$ cm/sec^2). Here we are going to assume that the density, porosity, and viscosity are constant. The saturation and permeability depend on the capillary pressure

$$p_c = p_{\text{air}} - p \tag{5-304}$$

Since we are allowing the air to fill the void spaces not filled with water the air pressure is taken as constant, which we take as zero. Then the capillary pressure is

$$p_c = -p \tag{5-305}$$

The saturation time derivative can then be expressed as

$$\frac{\partial S}{\partial t} = -\frac{dS}{dp_c}\frac{\partial p}{\partial t} \tag{5-306}$$

This equation is nondimensionalized using the definitions

$$p' = \frac{p}{\rho g L} \qquad k_r = \frac{k}{k_0} \qquad x' = \frac{x}{L} \qquad t' = \frac{t}{t_c} \qquad t_c = \frac{\phi L \mu}{k_0 \rho g} \tag{5-307}$$

Thus

$$-\frac{dS}{dp'_c}\frac{\partial p'}{\partial t'} = \frac{\partial}{\partial x'}\left(k_r \frac{\partial p'}{\partial x'}\right) - \frac{\partial k_r}{\partial x'} \tag{5-308}$$

where k_0 is the absolute permeability and is the permeability when the soil is completely filled with water. The last term is not used if gravity is neglected. We use the following relations for the dependence of S and k_r on p_c:

$$k_r = \frac{1}{1 + (p'_c L/B)^{\lambda}}$$

$$\frac{(S - S_r)}{1 - S_r} = \frac{1}{1 + (p'_c L/A)^{\eta}} \tag{5-309}$$

Typical soil properties are listed in Table 5-10. Equation (5-308) is solved subject to the boundary conditions

$$p' = \text{BP1} = \text{constant} \qquad \text{at } x' = 0 \tag{5-310}$$

$$\frac{\partial p'}{\partial x'} = 0 \qquad \text{at } x' = 1$$

and the initial conditions

$$p' = BP0 = \text{constant} \qquad \text{at } t' = 0 \tag{5-311}$$

For convenience we now drop the primes in Eqs. (5-308) to (5-311).

We anticipate the solution somewhat by recognizing that when the initial conditions and boundary conditions are very different the solution will proceed with a sharp front. Suppose the soil is initially dry and at time zero the boundary at $x = 0$ is brought in contact with water, which begins to infiltrate the soil. If the front is sharp initially it is located near $x = 0$, and only as time proceeds does the front move into the entire region to $x = 1$. Thus for small time, at least, we can ignore the boundary condition at $x = 1$ and solve the problem on an infinite domain. We then try a similarity transformation as derived in Sec. 5-1, and we find that it works when gravity is neglected. Making the change of variable $\eta = x/t^{1/2}$ we obtain the equation

$$\frac{d}{d\eta}\left[k_r(p)\frac{dp}{d\eta}\right] - \frac{\eta}{2}\frac{dS}{dp_c}\frac{dp}{d\eta} = 0 \tag{5-312}$$

with the boundary conditions

$$p(0) = BP1 \qquad p(\infty) = BP0 \tag{5-313}$$

This equation can be solved using any of the methods for boundary-value problems. It has been solved using the time-dependent method by adding a time derivative and integrating to steady state (using finite differences with a variable grid spacing) and by orthogonal collocation on finite elements (using a variable element spacing). The solutions for different boundary conditions are shown in

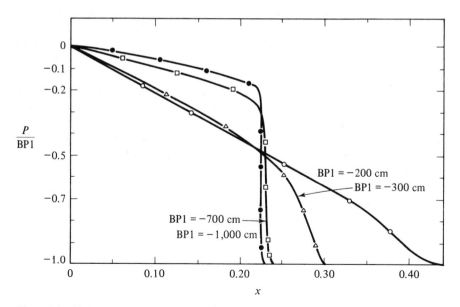

Figure 5-31 Similarity solution to flow through porous media, when gravity is neglected.

Fig. 5-31 and we see that the profile is very steep. The steepness of the profile can be changed by changing the initial condition or initial soil dryness. The grid spacing in the finite difference method for BP1 $= -1,000$ cm was $\Delta\eta = 0.005$ near the front and $\Delta\eta = 0.625$ near the origin, giving a ratio of 125 for the largest to smallest element size.

These solutions have the interesting behavior that for very dry soils the front is very steep and a convective-type solution results. The nature of this solution can be seen if we differentiate Eq. (5-308) and divide by k_r to get

$$-\frac{1}{k_r}\frac{dS}{dp_c}\frac{\partial p}{\partial t} - \frac{1}{k_r}\frac{\partial k_r}{\partial x}\frac{\partial p}{\partial x} = \frac{\partial^2 p}{\partial x^2} \tag{5-314}$$

Comparing this equation to the convective diffusion equation we find the Peclet number is equivalent to $\partial \ln k_r/\partial x$. In the very dry regions of the soil the k_r varies with $p_c^{-\lambda}$ and the term is

$$-\frac{1}{k_r}\frac{\partial k_r}{\partial x} = -\frac{\partial \ln k_r}{\partial x} = \frac{d \ln k_r}{dp_c}\frac{\partial p}{\partial x} = \frac{\lambda}{p}\frac{\partial p}{\partial x} \tag{5-315}$$

When the pressure gradient is large (see Fig. 5-31) this term is very large, too, giving a problem with a large convective term.

We can also define a pseudo-Courant number based on a parallel with the convective diffusion equation. If we measure the actual velocity of the front v_f, then accurate simulations have been obtained for $v_f \Delta t/\Delta x \simeq 0.1$–$0.4$, as in the convective diffusion equation. Such a guideline is not useful until one knows v_f, which is usually not known until one solution has been determined, but it does provide some guidance for Δt when Δx is changed or if v_f is known a priori.

In addition, the coefficient of the time derivative changes many orders of magnitude. It is a function of $p_c = -p$, and for $p = 0$, -200, -300, and $-1,000$ cm it takes the value 0, 0.29, 0.17, and 0.0012, respectively, for $L = 100$ cm. Thus the problem is stiff because the coefficient of the time derivative varies over several orders of magnitude for nodes from the boundary through the front. We thus expect all the difficulties apparent in the solution of the convective diffusion equation. Namely, small elements are needed to eliminate oscillations, and if large elements are used some form of numerical dispersion must be introduced. Typical solutions obtained by solving Eq. (5-308) are shown in Fig. 5-32, and the oscillations are apparent in all the methods. It is also clear that the solutions do not agree with each other and the front is moving with different velocities depending on the method used to solve the problem.

The stiffness of the system of equations is quantified by actually calculating the eigen values. Several less severe cases are integrated and at specific times the jacobian of the right-hand side is evaluated. If the equations are written in the forms

$$C_j \frac{dp_j}{dt} = \frac{1}{\Delta x^2} f_j(\mathbf{p}) \tag{5-316}$$

$$\frac{dp_j}{dt} = \frac{1}{\Delta x^2} F_j(\mathbf{p}) = \frac{1}{\Delta x^2}\frac{f_j}{C_j} \tag{5-317}$$

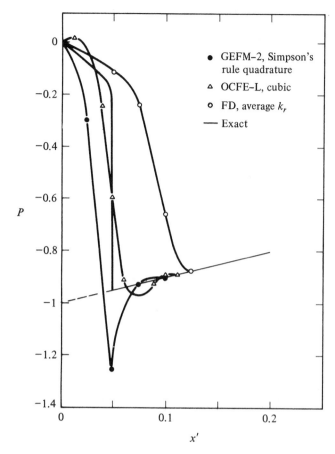

Figure 5-32 Solutions at $t = 0.005$ and $L = 1,000$ cm.

The jacobian is

$$\frac{1}{\Delta x^2} \frac{\partial F_j}{\partial p_i} = \frac{1}{\Delta x^2} \frac{1}{C_j} \left(\frac{\partial f_j}{\partial p_i} - \frac{f_j}{C_j} \frac{\partial C_j}{\partial p_i} \right) \tag{5-318}$$

This can be done for any of the methods. The method used in this case is orthogonal collocation on finite elements. We are interested in the ratio of the largest to the smallest eigen value, which gives the stiffness ratio SR

$$SR = \frac{\max_k |\lambda_k|}{\min_k |\lambda_k|} \tag{5-319}$$

We find numerically that the lowest eigen value is characteristic of the problem, and using different Δx and degree of polynomial gives the same value for the lowest eigen value. If the solution has some oscillations in it, one or two of the

Table 5-11 Eigen values for diffusion problem and orthogonal collocation on finite elements

NP	NE ($\Delta x = 1/NE$)	$\min\lvert\lambda_i\rvert$	$\max\lvert\lambda_i\rvert$		Maximum eigen value of modified **B** matrix*	Stiffness ratio $\max\lvert\lambda_i\rvert/\min\lvert\lambda_i\rvert$
4 (Cubic polynomials)	5	2.48	$885 =$	$35.41/\Delta x^2$		360
	10	2.47	$3,585 =$	$35.85/\Delta x^2$	36.00	1,450
	20	2.48	$14,384 =$	$35.96/\Delta x^2$		5,800
5	5	2.48	$243 =$	$97.35/\Delta x^2$		980
	10	2.47	$9,801 =$	$98.01/\Delta x^2$	98.23	4,000
	20	2.48	$39,268 =$	$98.17/\Delta x^2$		16,000
6	5	2.48	$5,524 =$	$220.96/\Delta x^2$		2,200
	10	2.47	$22,188 =$	$221.88/\Delta x^2$	222.20	9,000
	20	2.48	$88,848 =$	$222.12/\Delta x^2$		36,000

* **B** matrix in orthogonal collocation is modified by deleting the first and last rows and columns.

eigen values are imaginary, with an occasional one with a positive real part (indicating a growing error). For a good numerical solution, however, all the eigen values must be real and negative. The largest eigen value depends on the Δx used and the method chosen to approximate the spatial dependence. Results for the diffusion problem ($dS/dp_c = -1, k_r = 1$) are given in Table 5-11.

When the equation is written in the form of Eq. (5-314) it suggests that the stiffness may come from the coefficient in front of the time derivative. Thus we define a coefficient ratio CR as

$$CR = \frac{2 \leqslant i \leqslant NT-1 \left| \frac{1}{k_r} \frac{dS}{dp_c} \right|_i}{2 \leqslant i \leqslant NT-1 \left| \frac{1}{k_r} \frac{dS}{dp_c} \right|_i} \tag{5-320}$$

and we correlate the stiffness ratio versus the coefficient ratio in Fig. 5-33. For a variety of different problems there is a reasonable correlation. Thus we can look at the coefficient ratio before solving a problem and determine the difficulty of the problem. One unpleasant fact is that the coefficient ratio must be evaluated only for nodes in the domain (i.e. not the boundary conditions) since the boundary conditions do not involve the time derivative. In fact, before calculating the eigen values of the jacobian it is necessary to solve the linear algebraic equations and insert them into the differential equations. The solution at the first node in from the boundary condition may not be known so some guess is necessary. If the boundary condition has a positive head, an internal node can have $p = 0$, in which case the coefficient of the time derivative is zero. This means that the equation for this node is also algebraic, and this equation must be eliminated before calculating

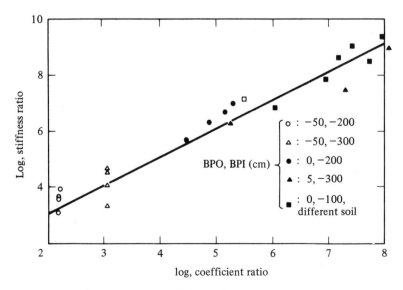

Figure 5-33 Stiffness ratio versus coefficient ratio.

the jacobian. In spite of these difficulties, which apply mainly to the actual calculation of the eigen values, it is feasible to choose a representative value for the pressure at the first node and evaluate the coefficient ratio, Eq. (5-320), using that value. The result is only a guide to the analyst anyway. We know that for problems with large coefficient ratios the stiffness ratio is large and the problem is stiff. This means we must use an implicit method to integrate the equations, and since the eigen value is so large either a small time step is necessary or the method must be L stable. We have found in Chapter 3 that methods that are not L stable tend to give oscillatory errors for large $\lambda \Delta t$. In particular, the Crank–Nicolson method in time requires $\lambda \Delta t \leqslant 2$ to avoid oscillations. By contrast, for the solutions shown in Fig. 5-33 the largest eigen value may be 7×10^8 or 6×10^{10} with a time step of $\Delta t = 10^{-4}$ or 5×10^{-5} for solutions with BP1 $= -200$ and -300 cm, respectively. Then $\lambda \Delta t$ is 7×10^4 or 3×10^6, which is very large.

We next see how to introduce numerical dispersion into the solutions. In the finite difference method, the equation is

$$-\frac{dS}{dp_{c,i}}\frac{dp_i}{dt} = \frac{1}{\Delta x^2}[k_{i+1/2}(p_{i+1}-p_i)-k_{i-1/2}(p_i-p_{i-1})] \qquad (5\text{-}321)$$

We can introduce upstream dispersion by evaluating the permeability in either of two ways.

$$k_{i-1/2} = k(p_{i-1/2}) \qquad \text{exact} \qquad (5\text{-}322)$$

$$k = k_{i-1} \qquad \text{upstream when } k_{i-1} > k_i \qquad (5\text{-}323)$$

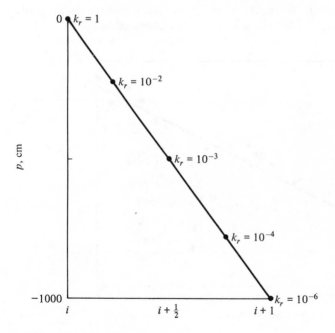

Figure 5-34 Permeabilities for a sharp front.

$$k = \tfrac{1}{2}(k_i + k_{i-1}) \qquad \text{average} \qquad (5\text{-}324)$$

Here we assume the front is moving from the $(i-1)$th node to the ith node, so that the permeabilities are evaluated at the upstream position. The second of these equations defines upstream permeability and the third uses average permeabilities. If the front is very steep as shown in Fig. 5-34, the pressure at the ith node may be 0, at the $(i+1)$th node $-1,000$ cm, and at the $(i+\tfrac{1}{2})$th node -500 cm. This can give rise to permeabilities of 1, 10^{-3}, and 10^{-6}, respectively, at the three nodes i, $i+\tfrac{1}{2}$, and $i+1$. The upstream permeability found using Eq. (5-323) is 1 and the average permeability using Eq. (5-324) is 0.5, and these are both very different from the true value at the $(i+\tfrac{1}{2})$th node of 10^{-3}. Taking $p_{i+1} = -1,000$, too, gives for Eq. (5-321)

$$-\frac{dS}{dp_c}\bigg|_i \frac{dp_i}{dt} \simeq -\frac{10^{-3}}{\Delta x^2}(p_i - p_{i-1}) \qquad \text{exact} \qquad (5\text{-}325)$$

$$\simeq -\frac{1}{\Delta x^2}(p_i - p_{i-1}) \qquad \text{upstream} \qquad (5\text{-}326)$$

$$\simeq -\frac{1}{2\Delta x^2}(p_i - p_{i-1}) \qquad \text{average} \qquad (5\text{-}327)$$

The truncation error of the method using an average permeability is $0(\Delta x^2)$ as we have determined before. The use of upstream permeabilities gives

$$\text{RHS}_i = \frac{1}{\Delta x^2}[k_i(p_{i+1} - p_i) - k_{i-1}(p_i - p_{i-1})] \qquad (5\text{-}328)$$

when $p_{i-1} > p_i$ and $k_{i-1} > k_i$, or $\partial k/\partial x < 0$. Using Eqs. (4-29) and (4-30) we get

$$\text{RHS}_i = \frac{\partial}{\partial x}\left(k\frac{\partial p}{\partial x}\right)\bigg|_i - \frac{\Delta x}{2}\frac{\partial}{\partial x}\left(\frac{\partial k}{\partial x}\frac{\partial p}{\partial x}\right) + 0(\Delta x^2) \qquad (5\text{-}329)$$

Thus the use of upstream permeability gives a method that is only accurate to $0(\Delta x)$. The errors have the form of a dispersion term and correspond to solving the original problem with the dispersion coefficient

$$k_r' = k_r - \frac{\Delta x}{2}\frac{\partial k}{\partial x} \qquad (5\text{-}330)$$

Since $\partial k/\partial x < 0$ the numerical permeability is positive, and we are adding numerical dispersion to the equations. This increases the errors but dampens the oscillations. The use of average permeabilities introduces numerical dispersion, too, but not as much as is introduced by the upstream permeability.

For the Galerkin method, we can introduce an upstream weighting function into the Galerkin equations to bring in numerical dispersion. It turns out, however, that the interpolation of k_r onto the trial function space introduces numerical dispersion, too. To see this let us look at the Galerkin term

$$-\sum_J \frac{1}{\Delta x^2}\int_0^1 k_r \frac{dN_I}{du}\frac{dN_J}{du}\,du\,P_J \qquad (5\text{-}331)$$

We evaluate the relative permeability at the nodes and write it as

$$k_r = \sum_K k_K N_K \tag{5-332}$$

Then the evaluation of (5-331) is straightforward yielding

$$\int_0^1 k_r \frac{dN_I}{du} \frac{dN_J}{du} du = \sum_k k_K \int_0^1 N_K \frac{dN_I}{du} \frac{dN_J}{du} du \tag{5-333}$$

If we use linear basis functions the integrals are

$$\int_0^1 N_K \frac{dN_I}{du} \frac{dN_J}{du} du = \begin{pmatrix} \frac{1}{2} & -\frac{1}{2} \\ -\frac{1}{2} & \frac{1}{2} \end{pmatrix} \qquad K = 1, 2 \tag{5-334}$$

Then a typical right-hand side for the ith node RHS_i is, remembering we have to assemble two terms of the form of Eq. (5-333),

$$\mathrm{RHS}_i = \frac{1}{2\Delta x^2} [(k_{i+1} + k_i)(p_{i+1} - p_i) - (k_i + k_{i-1})(p_i - p_{i-1})] \tag{5-335}$$

This is the same as the finite difference equation with average permeabilities. (The left-hand sides are different though.) Since the finite difference equation introduces numerical dispersion so does the interpolation of permeabilities, Eq. (5-332).

If we evaluate Eq. (5-331) using two gaussian quadrature points at $u = 0.211\ldots, 0.788\ldots$ in the element we get

$$-\frac{1}{\Delta x^2} \sum_J \sum_{k=1} W_k k_r(u_k) \left(\frac{dN_I}{du} \frac{dN_J}{du} \right)_{u_k} P_J \tag{5-336}$$

For the same pressure profile in Fig. 5-34 typical values of the k_r at the two quadrature points are 10^{-2} and 10^{-4}. Using these values gives

$$\mathrm{RHS}_i \simeq \frac{-10^{-2}}{2\Delta x^2} p_{i-1} \tag{5-337}$$

This is much closer to the original finite difference equation with $k_{i+1/2} = k_r(p_{i+1/2})$ and has little dispersion. Consequently, the interpolation of the relative permeability, Eq. (5-332), rather than using a more exact quadrature, introduces numerical dispersion into the equations.

Instead of using gaussian quadrature to evaluate (5-331) let us use the trapezoid rule. In this case we get $W_k = \frac{1}{2}$ in (5-336), with $u_1 = 0$, and $u_2 = 1$. The result is

$$\int_0^1 \left(-\frac{dS}{dp_c} \right) N_J N_I \, du = \begin{cases} -\frac{1}{2} \dfrac{dS}{dp_c}(u_k) & J = I \\ 0 & J \neq I \end{cases} \tag{5-338}$$

The equation for the right-hand side is given by Eq. (5-335) since the trapezoid rule integrates Eq. (5-333) exactly. If we combine the equations for the ith node we get

$$-\frac{dS}{dp_{c,i}} \frac{dp_i}{dt} = \frac{1}{2\Delta x^2} [(k_{i+1} + k_i)(p_{i+1} - p_i) - (k_i + k_{i-1})(p_i - p_{i-1})] \tag{5-339}$$

which is exactly the same as the finite difference method with averaged permeabilities. Thus the Galerkin method, with linear trial functions and the trapezoid rule for quadratures, is equivalent to the finite difference method with averaged permeabilities. This method has some damping features compared to the exact equation but is still second order, $O(\Delta x^2)$.

The weighted Galerkin equation is

$$\int_0^1 W_I \left(-\frac{dS}{dp_c} \right) \frac{\partial p}{\partial t} = -\int_0^1 k_r \frac{\partial p}{\partial x} \frac{\partial (W_I)}{\partial x} dx \tag{5-340}$$

which leads to equations of the form

$$\sum_e C_{IJ}^e \frac{dp_J^e}{dt} = \sum_e B_{IJ}^e p_J^e \tag{5-341}$$

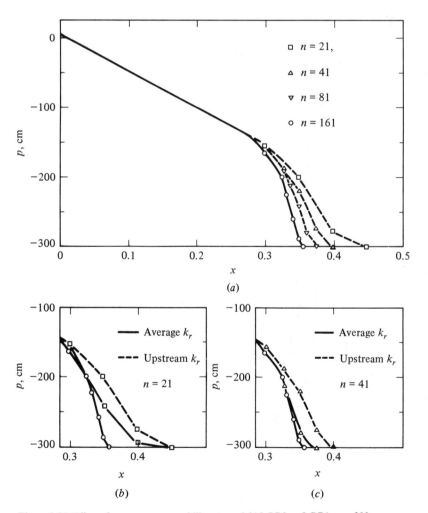

Figure 5-35 Effect of upstream permeability at $t = 0.015$, BP0 $= 5$, BP1 $= -300$ cm.

with the element matrices

$$C^e_{IJ} = \Delta x \int_0^1 W_I \left(-\frac{dS}{dp_c} \right) N_J \, du$$

$$B^e_{IJ} = \frac{-1}{\Delta x} \int_0^1 k_r \frac{dN_J}{du} \frac{dW_I}{du} \, du \tag{5-342}$$

If we interpolate the relative permeability we introduce more numerical dispersion. The weighting functions are taken as Eq. (5-290) or (5-294).

There has not been any systematic way developed to introduce numerical

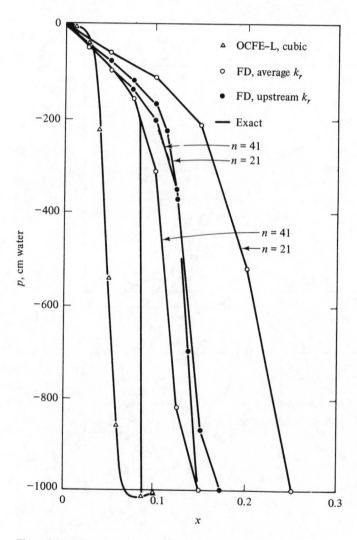

Figure 5-36 Effect of upstream permeability at $t = 0.015$, $L = 1,000$.

dispersion into the collocation method. One way would be to use the relative permeability given by Eq. (5-330), and another would be to evaluate the relative permeability at the pressure one node upstream. These approaches would make it possible to solve the problems with the collocation method and large elements, but the high-order accuracy of the collocation method would then be lost. The author's philosophy is that if one needs to introduce numerical dispersion into the problem one might as well use the simplest low-order method, which is the finite difference method.

The effect of the numerical dispersion is illustrated in Figs. 5-35 and 5-36. Figure 5-35a shows the accurate solution obtained with the finite difference method with average permeabilities and $n = 161$. The other solutions are obtained when upstream permeabilities and fewer grid points are used. For this case, which is relatively mild, the solution (not shown) with the exact permeabilities $k_i = k_{i+1/2} = k(p_{i+1/2})$ is very poor using a large Δx (i.e. $n = 21$ and 41). For $n = 80$ the solution is better, and for $n = 161$ the same solution is obtained as with average permeabilities. Of course, as $\Delta x \to 0$ the permeability does not vary much from one node to another, so that the averaged permeability is a closer and closer approximation to the exact value. Figure 5-36 gives results for a much steeper profile. The curves for the finite difference method and averaged permeabilities show that the front is at different positions depending on Δx, for 21, 41, and 81 intervals. The use of upstream permeability makes the front location depend less on Δx, but its position is still incorrect. The method of orthogonal collocation on finite elements gives the location of the front, which is incorrect, but as close as the finite difference method. Unfortunately the solution oscillates, too. It is clear that small elements are necessary if we wish to model this phenomenon. All these simulations are with the backward Euler method in time. The boundary condition at $x = 0$ is a fixed pressure.

Let us compare solutions obtained with different numerical methods for the case of a soil column with $L = 100$ cm, an initial dryness of $p = -300$ cm, and a boundary condition of $p = +5$ cm. The finite difference method is applied using averaged permeabilities and the collocation method is applied without numerical dispersion. The backward Euler method is used in time with a fixed step size. In both methods we use permeabilities at the known time

$$\frac{\partial}{\partial x}\left(k_r^n \frac{\partial p^{n+1}}{\partial x}\right) \tag{5-343}$$

Typical errors and computation times are shown in Fig. 5-37. We see that as the computation time is increased in any method by taking smaller and smaller Δt the error decreases and eventually approaches an error dictated entirely by the spatial discretization. Then as we decrease Δx the spatial discretization error decreases. The time steps for the first point shown for each discretization are for $NT = 21$ 5×10^{-4}, for $NT = 41$ 2×10^{-4}, and for $NT = 81$ 2×10^{-4}. Using orthogonal collocation on finite elements, the values are for $NE = 10$ 2×10^{-4}, and for $NE = 20$ 10^{-4}. It is clear that an even smaller time step must be used for the finite difference method ($n = 80$) in order to reach the spatial discretization error. The

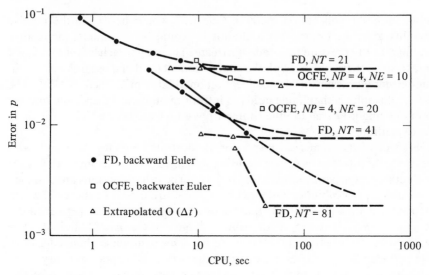

Figure 5-37 Accuracy versus computation time. Error at $x = 0.25$, $t = 0.015$, fixed Δt, and $P_{ex} = -1.2221$.

collocation results are not as good as the finite difference results because no numerical dispersion is introduced, because many nodes are needed to approximate the steep profile, and because larger time steps than the finite difference method are not possible for this problem.

In Fig. 5-38 we compare the fixed time step backward Euler method with a

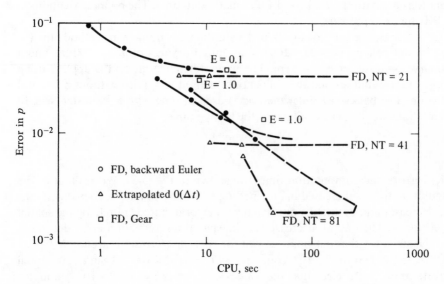

Figure 5-38 Comparison of backward Euler (fixed Δt) and GEAR (variable Δt) for finite difference method.

variable time step method using Gear's algorithm. Gear's algorithm achieves a very accurate answer for the time-dependent problem in that most of the error is spatial truncation error. Unfortunately, we have no need to make the temporal truncation error smaller than the spatial truncation error and, in this sense, Gear's algorithm is inefficient because it solves the time-dependent equations more accurately than is necessary, with an appropriate increase in work. For this problem the finite difference answers follow the extrapolation

$$\text{FD error} = 124\Delta t + 12.9\Delta x^2 \qquad (5\text{-}344)$$

and the computation time is

$$\begin{aligned}
\text{CPU}_{\text{FD}} &= 0.0012 \text{ sec/time step/grid point} \\
\text{CPU}_{\text{OCFE}} &= 0.0044 \text{ sec/time step/collocation point}
\end{aligned} \qquad (5\text{-}345)$$

using a fixed time step and the backward Euler method.

It is of interest to explore why Gear's algorithm does not achieve better results for this problem. In this problem the variable coefficient of the time derivative takes large and small values as we go from the boundary to the interior region, thus making the problem stiff. As the front moves, however, the range of values is always the same, since the boundary points are fixed at each end of the front. Thus there are always nodes with small coefficients and nodes with large coefficients in front of their time derivatives and the problem is always stiff. It is not possible for large time steps to be taken while the front is moving, in contrast to other problems for which Gear's algorithm has proved useful. Since the eigen values are so large, however, $\lambda \Delta t$ is large and an A and L stable method is required.

The next set of comparisons all use a fixed time step but use different methods of integrating the equations. The temporal methods of integration are:

(1) Crank–Nicolson with a variable weighting β. $\beta = 1$ is the backward Euler method. The matrices are evaluated at an estimated time and solution, according to β. Either one or more iterations may be taken each time step, with the matrix reevaluated each iteration. The Crank–Nicolson method ($\beta = \frac{1}{2}$) is $0(\Delta t^2)$, and A but not L stable.
(2) Modified backward Euler method with the matrices evaluated at the known time and solution. This method is $0(\Delta t)$, and both A and L stable.
(3) Nørsett method, either second- or third-order, using the same algorithm for nonlinear problems as derived for linear ones, with the matrices evaluated at the known solution. This is high-order [$0(\Delta t^2)$ or $0(\Delta t^3)$], and both A and L stable.

The spatial approximations are made for the collocation method on finite elements using polynomials of orders two through six. The Galerkin method is applied using polynomials of order one, two, and three. Quadratures are evaluated using a variety of quadrature schemes. $NQ = 2$ uses the trapezoid rule, $NQ = 3$ uses the two end points and the midpoint (Simpson's rule), and $NQ = 4$ and $NQ = 5$ use gaussian quadrature involving two and three quadrature points.

Using linear polynomials and $NQ = 2$ gives the finite difference method with average permeabilities, although the program was written in the finite element scheme.

Table 5-12 Numerical errors for different methods for $5 \rightarrow -300$ cm. Values of $p(x = 0.25, t = 0.015)$, exact $= -1.222$

(a) **Effect of integration method**

Orthogonal collocation on finite elements, $NP = 4$, $NE = 20$, $NT = 61$					
Method	Δt	p	Error	Computation time, sec	
Crank–Nicolson $\beta = 0.5$, ITER $= 1$	2.5 (-4)	-1.426	0.206	4.1	
Backward Euler $\beta = 1.0$, ITER $= 1$	2.5 (-4)	-1.269	0.047	4.2	
	1.25 (-4)	-1.241	0.019	8.1	
	0, ext.	-1.213	0.009	12.3	
Nørsett					
Second-order	2.5 (-4)	-1.247	0.025	5.3	
Third-order	2.5 (-4)	-1.242	0.020	6.1	
Galerkin method, quadratic, $NQ = 3$, $NT = 61$					
Crank–Nicolson					
$\beta = 0.5$, ITER $= 1$	2.5 (-4)	-1.228	0.006	5.5	
ITER $= 2$	5.0 (-4)	-1.330	0.108	5.6	
Backward Euler $\beta = 1.0$, ITER $= 1$	2.5 (-4)	-1.260	0.038	5.4	
	1.25 (-4)	-1.241	0.019	11.6	
	0, ext.	-1.222	0.000	17.0	
Nørsett					
Second-order	2.5 (-4)	-1.238	0.016	6.3	
Third-order	1 (-3)	-1.988	0.766	1.8	
	5 (-4)	-1.270	0.048	3.9	
	2.5 (-4)	-1.233	0.011	7.1	
	1 (-4)	-1.227	0.005	18.5	

(b) **Galerkin method, different polynomials and quadrature using Nørsett integration, third-order, $\Delta t = 2.5 \times 10^{-4}$**

NP	NE	NT	NQ	p	Error	Computation time, sec
2	40	41	2	-1.243	0.021	5.4
2	40	41	3	-1.246	0.024	6.3
3	20	41	2	-3.000	1.778	4.4
3	20	41	3	-1.236	0.014	4.9
3	20	41	4	-1.251	0.029	5.5
4	10	31	5	-1.260	0.038	5.7
4	20	61	5	-1.258	0.036	11.5

Typical errors for the problem with an initial condition of -300 cm and a boundary condition of $+5$ cm are listed in Table 5-12. Looking at the orthogonal collocation results we see that the Crank–Nicolson method does not do a good job of integrating the equations, but both the backward Euler and Nørsett methods are good, with no clear preference. For the Galerkin method, the Crank–Nicolson method is best, although the Nørsett method is nearly as good. Using extrapolation techniques we estimate the spatial truncation error to be 0.009 for the collocation method with 20 elements and 61 nodes, and 0 for the Galerkin method using quadratic trial functions with 30 elements and 61 nodes. The collocation method is less accurate and slightly faster for the same number of nodes. Comparing Galerkin methods with different degrees of polynomials we get the following comparisons: take linear ($NP = 2$) with three quadrature points ($NQ = 3$), quadratic ($NP = 3$) with four quadrature points ($NQ = 4$), and cubic ($NP = 4$) with five quadrature points ($NQ = 5$). This ensures that the numerical quadratures are relatively accurate. Results in Table 5-12b indicate that when the number of elements is changed to keep the total number of nodes fixed the errors and computation times are comparable. Thus we cannot make a case in this example for high- or low-order elements.

Looking next at the effect of quadrature we see that when using linear elements the use of two quadrature points (giving the finite difference method with average permeabilities) is slightly better than that using three quadrature points, but for this case the importance of the dispersion is not manifest. The low-order quadrature with quadratic trial functions is not suitable. We can also see the effect of taking several iterations in the Crank–Nicolson method. The results in Table 5-12 and other calculations indicate that if we take two iterations with a step size Δt the error is greater than if we take one iteration and two steps, each with a size step $\Delta t/2$, giving equivalent computation times. Thus it may not always pay to iterate on the nonlinear terms. Therefore, it is concluded that the Nørsett methods are a viable alternative to the backward Euler (L stable) or Crank–Nicolson (A stable) method. None of the spatial approximations consistently proves itself best.

In summary, we see that the soil problem can be a very difficult one and gives very steep solutions. The best method in space is the finite difference method with average permeabilities, while the best method in time is the backward Euler method or a Nørsett method. Small elements are required or numerical dispersion can be introduced to make the solution less steep and allow larger elements, but the accuracy is degraded, too.

5-10 COMPARISON

We are now in a position to compare the methods and include in our comparison orthogonal collocation, finite difference, orthogonal collocation on finite elements, and Galerkin finite elements. The spatial discretization errors decrease with smaller Δx as listed in Table 4-11, while the temporal truncation errors have the truncation error Δt^m depending on the method chosen in Chapter 3. If we choose to apply an implicit method involving the decomposition of a matrix, then the work required to do this is shown in Table 4-11.

Different spatial approximations allow different time steps and we have seen that smaller Δx requires smaller Δt. This effect can be quantified for the diffusion equation

$$\frac{\partial c}{\partial t} = \frac{\partial^2 c}{\partial x^2} \tag{5-346}$$

When we apply a spatial approximation to this equation we get equations of the form

$$C_{ji} \frac{dc_i}{dt} = AA_{ji}C_i \tag{5-347}$$

and the difficulty of integration depends on the eigen values of the matrix. For collocation and finite difference methods the matrix $\mathbf{C}$ is the identity matrix and we just need the eigen values of the matrix $\mathbf{AA}$. The lowest eigen value is the one corresponding to the physical problem, i.e. the first eigen value to the eigen function problem arising in separation of variables,

$$\frac{d^2 X}{dx^2} + \lambda_1 X = 0 \qquad \lambda_1 = \pi^2 \tag{5-348}$$

The highest eigen value is generally dependent on Δx

$$\lambda_{max} = \frac{LB}{\Delta x^2} \tag{5-349}$$

This has been determined by computations similar to those reported in Table 5-11.

Values of LB are given in Table 5-13 for the different methods. For the Galerkin methods and the Hermite collocation method the eigen values are the solution to the equation

$$\det|C_{ji}\lambda - A_{ji}| = 0 \tag{5-350}$$

Table 5-13 Value of LB in Eq. (5-349)

Method	LB
FD	4
OCFE–L, $NP = 4$	36
$NP = 5$	98
$NP = 6$	222
OCFE–H, $NP = 4$	36
GFEM–1	12
GFEM–2	60
GFEM–1, lumped	4
GFEM–2, lumped	24

and these values behave similarly to Eq. (5-349). What this means is that the stiffness ratio for a linear diffusion problem is given by

$$SR = \frac{LB}{\lambda_1 \Delta x^2} \tag{5-351}$$

and thus as Δx decreases the problem becomes more stiff, and the explicit time step must decrease, since

$$\Delta t \leqslant \frac{p}{|\lambda|_{max}} \leqslant \frac{p \Delta x^2}{LB} \tag{5-352}$$

We take first an "easy" problem of the type

$$\frac{\partial c}{\partial t} = \frac{\partial}{\partial x}\left[D(c)\frac{\partial c}{\partial x} \right] + R(c) \tag{5-353}$$

Suppose we apply an explicit method of integration. Then the work estimates are just the time needed to calculate the right-hand side for each time step, multiplied by the number of time steps, which can be determined from Table 5-13. Suppose we integrate to a time such that the total number of time steps is $LB/\Delta x^2$. We can then calculate the work necessary to derive the solution. We choose the number of elements and grid points such that the results should give comparable accuracy, based on the experiences related in Sec. 5-5. We use collocation with three interior grid points, finite difference with ten nodes, orthogonal collocation on finite elements with cubic polynomials and three elements, and Galerkin linear polynomials with ten elements and five Galerkin quadratic elements. Also we use the same estimates of $m_1 = 15$ multiplications to evaluate the rate of reaction and $m_2 = 4$ to evaluate the nonlinear diffusivity. The work per time step for each of the methods is given in Table 5-14 for the cases of large numbers of points. Using Tables 5-13 and 5-14 and the number of time steps given by $LB/\Delta x^2$ gives the results in Table 5-14 for the number of multiplications required to solve this problem, when each method has comparable accuracy.

The fastest methods are the orthogonal collocation method and the collocation on finite element methods. This is primarily due to the low number of terms and the large time step allowed. The finite difference method is the next best method, with the Galerkin methods being the slowest. These conclusions, of course, depend on the number of terms needed in the various methods, which have been chosen here to correspond with experimental results and the error terms in Table 4-11. Table 5-14 just confirms the conclusion of Sec. 5-5 in which we found the orthogonal collocation method is far superior to the other methods for diffusion–reaction problems.

Next we consider a problem that has steep profiles. Now we solve an LU decomposition at each time step so the work estimates are derived from Table 4-11. We suppose the number of time steps is the same for each method and choose the element size so that each method has the same number of unknowns. Take orthogonal collocation with 60 collocation points, orthogonal collocation with 67 elements, finite difference with 200 grid points, and the Galerkin methods with 200

Table 5-14 Operation count for explicit and implicit methods

Method	Formula	Explicit				Implicit			
		Values	Evaluate RHS once	Number of time steps	Total count $\times 10^{-6}$	Values	One time step	Number of time steps	Total count $\times 10^{-6}$
OC	$NP(2NP+20)$	$NP=3$	78	184	0.014	$NP=60$	73,000	1,000	73
OCFE-H	$NE(10NP+33)$	$NP=4, NE=3$	219	324	0.071	$NP=4, NE=67$	6,400	1,000	6.4
OCFE-L	$NE[NP(2NP+21)-27]$	$NP=4, NE=3$	267	324	0.086	$NP=4, NE=67$	6,400	1,000	6.4
GFEM-2	$103NE$	$NE=5$	515	1,500	0.77	$NE=100$	9,200	1,000	9.2
GFEM-2*	$84NE$	$NE=5$	420	1,500	0.63	$NE=100$	7,200	1,000	7.2
GFEM-1	$80NE$	$NE=10$	800	1,200	0.96	$NE=200$	15,000	1,000	15
GFEM-1*	$36NE$	$NE=10$	360	1,200	0.43	$NE=200$	6,400	1,000	6.4
FD	$23n$	$n=10$	230	400	0.092	$NE=200$	5,400	1,000	5.4

* Interpolate $D(c)$ and $R(c)$.

and 100 elements for linear and quadratic trial functions. Then all methods have the same number of unknowns, 200, except for orthogonal collocation. The number of time steps is the same for each method and let us assume it is 1,000. The summary of results is given in Table 5-14. The finite difference method is best by a small margin. This result is in accord with the experience of Sec. 5-9. Notice that finite element or finite difference methods have to be used; global orthogonal collocation is too expensive.

In summary, if a problem is not too difficult and does not have a solution with steep gradients, the orthogonal collocation with a global polynomial is the preferred method of spatial approximation. If the problem has steep gradients then a finite difference or finite element method is preferred. For large errors the finite difference method is quite suitable, whereas for small errors the higher-order finite element methods are preferred. This is the case for interpolation of a steep solution (see Sec. 5-8). It may be that the solution is so steep that numerical dispersion must be introduced. In that case the simplest method is probably the finite difference method, as we found in Sec. 5-9.

STUDY QUESTIONS

1. How to apply the following methods to parabolic partial differential equations
   a. Method of Weighted Residuals
   b. Finite difference method
   c. Galerkin finite elements method
   d. Orthogonal collocation on finite elements
2. Similarity transformation
   a. Application
   b. General limitations
   c. Clues to when one exists
3. Treatment of semi-infinite domain
4. Separation of variables
   a. Limitations
   b. Relationship to
      i Method of Weighted Residuals
      ii Galerkin finite elements method
5. Stability limitations
   a. Orthogonal collocation
   b. Finite difference method
6. Time integration methods
   a. Application
   b. Stable step size—influence of
      i Degree of polynomials
      ii Element size
      iii Geometry

7. Application of global polynomials compared with
   a. Finite difference method
   b. Finite elements methods
8. Prediction of stable step size for explicit methods and any spatial approximation

PROBLEMS

Similarity transformation

5-1 Derive a similarity transformation, if possible, for the nonlinear problem

$$\frac{\partial c}{\partial t} = \frac{\partial^2 c}{\partial x^2} - kc^2$$

$$c(x, 0) = 0 \qquad c(\infty, t) = 0 \qquad c(0, t) = 1$$

5-2 Derive a similarity transformation, if possible, to the problem

$$\frac{\partial c}{\partial t} = \frac{1}{x^{a-1}} \frac{\partial}{\partial x}\left(x^{a-1} \frac{\partial c}{\partial x}\right)$$

$$c(x, 0) = 0 \qquad c(\infty, t) = 0 \qquad c(1, t) = 1$$

Make a conclusion about the different geometries for $a = 1, 2$, and 3.

5-3 Derive Eq. (5-312). Consider an initial-value method of solving this equation by adding a time derivative. What difference is there between solving that time-dependent equation and solving the original time-dependent equation, Eq. (5-308)?

5-4 The boundary-layer equations for flow past a flat plate are

$$\frac{\partial u}{\partial x} + \frac{\partial v}{\partial y} = 0$$

$$u\frac{\partial u}{\partial x} + v\frac{\partial u}{\partial y} = v\frac{\partial^2 u}{\partial y^2}$$

$$u(x, 0) = 0 \qquad u(x, \infty) = u(0, y) = U$$

Is the problem an initial-value or boundary-value problem in the x variable and/or in the y variable? Write

$$v = -\int_0^y \frac{\partial u}{\partial x} dy$$

and derive a similarity transformation for the problem $u = \phi(\eta)$. Derive the equation for $f(\eta)$, where $df/d\eta = \phi$ and $f(0) = 0$. Write down the equations to solve the resulting boundary-value problem for f using a shooting method.

Separation of variables

5-5 Equations (5-48), (5-49), and (5-50) have been solved twice, once using separation of variables, Eq. (5-105), and once using a similarity transformation, Eq. (5-42), as an approximation for small time. Evaluate the solution at $x = 0.25, 0.5, 0.75$, and 1.0 for times $t = 0.01, 0.1$, and 1.0. Notice that as the time approaches zero more and more of the expansion given by Eq. (5-105) is needed for a good approximation. Comment on the relative ease of using the two solutions.

5-6 Minimize the least squares residual in Eq. (5-103) and show that the least squares method gives the same value of A_n as the Galerkin criterion.

5-7 Derive a solution to the following problem using separation of variables:

$$\frac{\partial c}{\partial t} = D\frac{\partial^2 c}{\partial r^2}$$

$$c(r, 0) = 0 \qquad c(1, t) = 1$$

$$\frac{\partial c}{\partial r} = 0 \qquad \text{at } r = 0$$

Derive the eigen value problem

$$\frac{d^2 R}{dr^2} + \lambda R = 0$$

$$R(1) = 0 \qquad \frac{dR}{dr}(0) = 0$$

The eigen functions are

$$R_n = \cos(2n+1)\frac{\pi r}{2}$$

Method of Weighted Residuals

5-8 Derive the equations that when solved give an approximation valid for large time for the problem

$$\frac{\partial c}{\partial t} = \frac{\partial}{\partial x}\left(D\frac{\partial c}{\partial x}\right) \qquad D = 1 + \alpha c$$

$$c(0, t) = 1 \qquad c(x, 0) = 0$$

$$\frac{\partial c}{\partial x} = 0 \qquad \text{at } x = 1$$

Do not solve them, but use any Method of Weighted Residuals to derive them.

5-9 (*a*) Apply the integral method to the boundary-layer equations in problem 5-4. Use the trial function

$$u = U_\infty \tfrac{1}{2}\eta(3 - \eta^2)$$

$$\eta = \frac{x}{\delta(t)}, \ \delta(t) = \left(\frac{vx}{U_\infty}\right)^{1/2}$$

and integrate over $0 \leqslant \eta \leqslant 1$. Show that this solution is the same as found if the equation for $df/d\eta$ in problem 5-4 is solved by the integral method using the same trial function for $\phi(\eta)$.

(*b*) Derive the polynomial that has the following properties to be used to solve the boundary-layer equations:

(i) $\phi(0) = 0$, $\phi(1) = 1$
(ii) conditions (i) plus $\phi'(1) = 0$
(iii) conditions (ii) plus $\phi''(0) = 0$
(iv) conditions (iii) plus $\phi''(1) = 0$
What do these conditions mean in words? What criteria are we applying?

Orthogonal collocation

5-10 Write down the orthogonal collocation equations for problem 5-8.

5-11 Consider diffusion and reaction in a spherical catalyst with a first-order, irreversible, non-isothermal reaction. Initially, the temperature is 1.0 and the concentration $c = 1 - 0.4725(1 - r^2)$, and the boundary conditions are

$$T(1) = c(1) = 1.0$$

Derive the equations for a one-term orthogonal collocation solution. Check these against problem 3-8.

5-12 Check the entries in Table 5-4 for N = 1 and spherical geometry.

5-13 For the example in Fig. 5-14 determine the three steady-state solutions and their eigen values μ. Determine the limits on Lewis number that make the upper steady state unstable.

5-14 Integrate the transient version of problems 4-7a to 4-7c to steady state from the initial condition $c(r) = 0$. Use N = 2.

5-15 Integrate the transient version of problem 4-8 to steady state from the initial condition $c(r) = 0$.

Similarity transformation with orthogonal collocation

5-16 The following problem represents diffusion into a fluid flowing down an inclined plane:

$$\tfrac{3}{2}(1-x^2)\frac{\partial c}{\partial t} = \frac{\partial^2 c}{\partial x^2}$$

$$c(0, z) = 1 \qquad c(x, 0) = 0$$

$$\frac{\partial c}{\partial x} = 0 \qquad \text{at } x = 1$$

Apply a similarity transformation and solve this problem for a solution valid for small z. Apply a one-term orthogonal collocation method to derive a solution for large time. Derive the equations that one would solve if the expansion were the similarity solution plus the orthogonal polynomial expansion used in orthogonal collocation, and orthogonal collocation is applied.

Finite difference

5-17 Write a finite difference equation for the following diffusion–reaction problem using a variable grid spacing:

$$\frac{\partial c}{\partial t} = \frac{1}{r^2}\frac{\partial}{\partial r}\left(r^2\frac{\partial c}{\partial r}\right) - R(c)$$

$$c(r, 0) = 0$$

$$\frac{\partial c}{\partial r} = 0 \qquad \text{at } r = 0$$

$$-\frac{\partial c}{\partial r}\bigg|_{r=1} = \text{Bi}_m[c(1, t) - 1]$$

5-18 Apply the finite difference method to Eq. (5-199) with D constant and $\Delta x = 0.5$ and 0.3333. Calculate the eigen value of the right-hand side if an explicit method is used to integrate in time. How does this compare to the upper bound of four?

5-19 Apply the finite difference method to the problem of diffusion in a sphere and a cylinder. Deduce the limit on step size if an explicit method is used, by employing the positivity rule. What can you say about the stable step size for the three geometries: planar, cylindrical, and spherical?

5-20 Determine the truncation error of Eqs. (5-233) and (5-280).

5-21 Integrate the transient version of problems 4-7a to 4-7c to steady state from the initial conditions $c(r) = 0$.

5-22 Integrate the transient version of problem 4-8 to steady state from the initial condition $c(r) = 0$.

5-23 Determine the truncation error of Eq. (5-321) when using upstream permeabilities.

5-24 Determine the stability limits for the finite difference method applied to the color equation. Consider explicit and implicit methods, and upstream and centered difference expressions. To do this write the difference equation and then substitute the solution

$$c_j^n = \zeta^n \exp \text{im}\sqrt{-1}$$

Solve for ζ and find the conditions under which $|\zeta| < 1$, which ensures the solution decays and the method is stable.

Orthogonal collocation on finite elements

5-25 Integrate the transient version of problems 4-7a to 4-7c to steady state from the initial condition $c(r) = 0$.

5-26 Integrate the transient version of problem 4-8 to steady state from the initial condition $c(r) = 0$.

Galerkin finite element method

5-27 Derive the Galerkin equations for a variable element spacing for linear trial functions for problem 5-17. Compare with the finite difference equations.

5-28 Derive Eq. (5-261).

5-29 Integrate the transient version of problems 4-7a to 4-7c to steady state from the initial conditions $c(r) = 0$.

5-30 Integrate the transient version of problem 4-8 to steady state from the initial condition $c(r) = 0$.

5-31 Derive Eq. (5-335).

General

5-32 Consider the transient problem whose steady-state equations are given in problem 4-8. The initial condition is the steady-state solution derived in problem 4-8. At time zero the boundary concentration (now 1) is changed to 1.1. Identify an appropriate method for both space and time for the five cases in problems 4-8a to 4-8e. The final steady state can be determined using the same program employed for problem 4-8 but with the new boundary concentration.

5-33 A more accurate way to integrate the convective diffusion equation was given by Van Genuchten.[14] Crank–Nicolson is applied but the diffusion term is augmented. For the term at the $(n+1)$th level use the coefficient $(1 - \text{Pe}\,\Delta t/6)$ times the second derivative. For the term at the nth level use the coefficient $(1 + \text{Pe}\Delta t/6)$ times the second derivative. Show that such a scheme is third-order correct in time for any method.

5-34 The diffusion equation in plane geometry is to be integrated using a variety of methods in space and the improved Euler method in time. Determine the largest stable step size for: OC; OCFE, $NP = 4, NP = 5$; FD, $n = 40$; GFEM–1, $NE = 40$; GFEM–2, $NE = 20$.

BIBLIOGRAPHY

A complete treatment of separation of variables is given by
>Weinberger, H. F.: *A First Course in Partial Differential Equations with Complex Variables and Transformation Methods*, Blaisdell, New York, 1965.

The application of orthogonal collocation to chemically reacting partial differential equations was pioneered by Ferguson and Finlayson,[5] and by
>Finlayson, B. A.: "Packed Bed Reactor Analysis by Orthogonal Collocation," *Chem. Eng. Sci.*, vol. 26, pp. 1081–1091, 1971.

The method is reviewed in
>Finlayson, B. A.: "Orthogonal Collocation in Chemical Reaction Engineering," *Cat. Rev. Sci. Eng.*, vol. 10, pp. 69–138, 1974.

A careful mathematical analysis of the finite difference method applied to partial differential equations is made by Richtmyer and Morton.[12] Many applications are discussed in
>Carnahan, B., H. A. Luther, and J. O. Wilkes: *Applied Numerical Methods*, John Wiley & Sons Inc., New York, 1969.

The ability to apply the Galerkin finite element method to the convective diffusion equation with strong convection was first shown by Price, *et al.*[11] The application to the color equation (with no dispersion) was made by
>Gresho, P. M., R. L. Lee, and R. L. Sani: "Advection-Dominated Flows, with Emphasis on the Consequences of Mass Lumping," in *Preprints of Second International Symposium on Finite Element Methods in Flow Problems*, S. Margherita Ligure, Italy, June 14–18, 1976, pp. 745–756.

A Fourier analysis is provided there as well as in
>Pinder, G. F., and W. G. Gray: *Finite Element Simulation in Surface and Subsurface Hydrology*, Academic Press, New York, 1977.

An improved method for solving problems with large convection terms is provided by Price, *et al.*,[11] who move the small elements with the front, and by
>Jensen, O. K., and B. A. Finlayson: "Solution of the Transport Equations Using a Moving Coordinate System", *Adv. Water Resources*, vol. 3, pp. 9–18, 1980.

who move the coordinate system so that the sharp front is fixed in time. The concept of stiffness applied to the equations for movement of water in dry soils was first expressed by

> Finlayson, B. A.: "Water Movement in Desiccated Soils," in W. G. Gray, *et al.* (ed.) *Finite Elements in Water Resources*, Pentech Press, London, 1977, pp. 3.91–3.106.

Applications to water movement in wetter soils are described in the book by Pinder and Gray (1977) cited above. The finite difference and orthogonal collocation methods are compared in the paper by Finlayson (1971) cited above. The following paper compares methods used for integrating in time when the spatial approximation is by finite difference:

> Kurtz, L. A., R. E. Smith, C. L. Parks, and L. R. Boney: "A Comparison of the Method of Lines to Finite Difference Techniques in Solving Time-Dependent Partial Differential Equations," *Computers and Fluids*, vol. 6, pp. 49–70, 1978.

A comparison of several methods of spatial approximation is provided by

> Douglas, J., Jr.: "A Survey of Numerical Methods for Parabolic Differential Equations," *Adv. Comp.*, vol. 2, pp. 1–54, 1961.
>
> Wexler, A.: "Computation of Electromagnetic Fields," *IEEE Trans.*, *Microwave Theory Tech.*, vol. 17, pp. 416–439, 1969.
>
> Walsh, J.: "Finite Difference and Finite Element Methods of Approximation," *Proc. Roy. Soc. (London)*, vol. A323, pp. 155–165, 1971.
>
> Gourlay, A. R.: "Some Recent Methods for the Numerical Solution of Time-Dependent Partial Differential Equations," *Proc. Roy. Soc. (London)*, vol. A323, pp. 219–235, 1971.
>
> Madsen, N. K., and R. F. Sincovec: "General Software for Partial Differential Equations," in L. Lapidus, *et al.* (ed.) *Numerical Methods for Differential Systems; Recent Developments in Algorithms, Software and Applications*, Academic Press, New York, 1976, pp. 229–242.
>
> Hopkins, T. R., and R. Wait: "A Comparison of Galerkin, Collocation and the Method of Lines for Partial Differential Equations," *Int. J. Num. Methods Eng.*, vol. 12, pp. 1081–1107, 1978.

REFERENCES

1. Ames, W. F.: "Recent Developments in the Nonlinear Equations of Transport Processes," *Ind. Eng. Chem. Fund.*, vol. 8, pp. 522–536, 1969.
2. Ames, W. F.: *Nonlinear Partial Differential Equations in Engineering*, Academic Press, New York, 1965.
3. Crank, J.: *The Mathematics of Diffusion*, Oxford University Press, 1956.
4. Ehlig, C.: "Comparison of Numerical Methods for Solution of the Diffusion–Convection Equation in One- and Two-Dimensions," in W. G. Gray, *et al.* (ed.) *Finite Elements in Water Resources*, Pentech Press, London, 1977, pp. 1.91–1.102.
5. Ferguson, N. B., and B. A. Finlayson: "Transient Chemical Reaction Analysis by Orthogonal Collocation," *Chem. Eng. J.*, vol. 1, pp. 327–336, 1970.
6. Ferguson, N. B., and B. A. Finlayson: "Transient Modeling of a Catalytic Converter to Reduce Nitric Oxide in Automobile Exhaust," *A.I.Ch.E.J.*, vol. 20, pp. 539–550, 1974.
7. Heinrich, J. C., P. S. Huyakorn, O. C. Zienkiewicz, and A. R. Mitchell: "An 'Upwind' Finite Element Scheme for Two-Dimensional Convective Transport Equation," *Int. J. Num. Methods Eng.*, vol. 11, pp. 131–143, 1977.
8. Heinrich, J. C., and O. C. Zienkiewicz: "Quadratic Finite Element Schemes for Two-Dimensional Convective-Transport Problems," *Int. J. Num. Methods Eng.*, vol. 11, pp. 1831–1844, 1977.
9. Jensen, O. K., and B. A. Finlayson: "Oscillation Limits for Weighted Residual Methods," *Int. J. Num. Methods Eng.*, vol. 15, 1980.
10. Prenter, P. M.: *Splines and Variational Methods*, John Wiley & Sons, Inc., New York, 1975.
11. Price, H. S., J. C. Cavendish, and R. S. Varga: "Numerical Methods of Higher-Order Accuracy for Diffusion–Convection Equations," *Soc. Pet. Eng. J.*, vol. 8, pp. 293–303, 1968.
12. Richtmyer, R. D., and K. W. Morton: *Difference Methods for Initial-Value Problems*, Interscience, New York, 1967.

13. Settari, A., H. S. Price, and T. Dupont: "Development and Application of Variational Methods for Simulation of Miscible Displacement in Porous Media," in *Fourth Soc. Pet. Eng. Symposium on Numerical Simulation of Reservoir Performance,* Los Angeles, Calif., Feb. 19–20, 1976, pp. 43–67.
14. Van Genuchten, M. T.: "On the Accuracy and Efficiency of Several Numerical Schemes for Solving the Convective-Dispersion Equation," in W. G. Gray, *et al.* (ed.) *Finite Elements in Water Resources,* Pentech Press, London, 1977, pp. 1.71–1.90.
15. Wilkes, J. O.: "In Defense of the Crank–Nicolson Method," *A.I.Ch.E.J.,* vol. 16, p. 501, 1970.

PARTIAL DIFFERENTIAL EQUATIONS IN TWO SPACE DIMENSIONS

In Chapter 3 we treated evolution problems in time, which were ordinary differential equations. Chapter 4 deals with ordinary differential equations in space, which give boundary-value problems. There it is found that the solutions near one boundary are influenced by the boundary condition at the other end of the interval. Chapter 5 combines these problems to give evolution problems in time or a time-like variable together with one spatial coordinate. We now extend to two dimensions in space. Then the problems have a boundary-value character in the two spatial coordinates and, in addition, the problem may be evolutionary in time.

6-1 INTRODUCTION

The steady-state diffusion problem

$$\frac{\partial^2 c}{\partial x^2} + \frac{\partial^2 c}{\partial y^2} = 0 \qquad 0 \leqslant x, y \leqslant 1 \tag{6-1}$$

$$\begin{aligned} c(x,0) = c_1 \qquad c(x,1) = c_2 \qquad 0 \leqslant x \leqslant 1 \\ c(0,y) = c_3 \qquad c(1,y) = c_4 \qquad 0 \leqslant y \leqslant 1 \end{aligned} \tag{6-2}$$

is called an elliptic boundary-value problem, and we see that there are boundary conditions at each end of the x and y intervals. The corresponding time-dependent problem is

$$\frac{\partial c}{\partial t} = D\left(\frac{\partial^2 c}{\partial x^2} + \frac{\partial^2 c}{\partial y^2}\right) \tag{6-3}$$

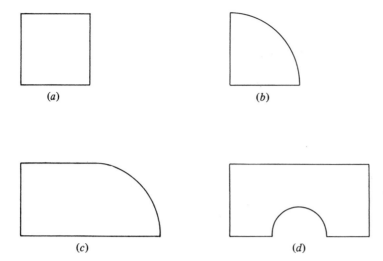

Figure 6-1 Domains with regular and irregular shapes.

$$c = c_1 \text{ on } y = 0 \qquad c = c_2 \text{ on } y = 1$$
$$c = c_3 \text{ on } x = 0 \qquad c = c_4 \text{ on } x = 1 \qquad (6\text{-}4)$$
$$c = f(x, y) \qquad \text{at } t = 0$$

This is a parabolic partial differential equation, and the evolutionary methods of Chapter 3 can be applied to it. There are some special techniques, though, which are only applicable to problems of the type shown above, and it is these techniques which are to be considered.

The addition of the second dimension is seemingly straightforward, but can have far-reaching consequences. Consider the two domains in Figs. 6-1a and 6-1b. Each of these has a boundary on one of the coordinates lines, such as $x = $ constant or $y = $ constant. Separation of variables is then a suitable technique for solution if the problem is linear, and the Method of Weighted Residuals is suitable if the problem is nonlinear, since the boundary conditions can be easily met by the trial function. If the two domains are combined, however, to obtain the domain in Fig. 6-1c then this feature is lost. Now the boundaries are not a coordinate line. Holes in the region, as shown in Fig. 6-1d, are even worse. Yet regions like these arise in the analysis of engineering problems, and methods must be developed to solve them.

Another difficulty that can arise is when the boundary data are not continuous. For example in Eq. (6-1) it may be that $c_2 = c_3 = c_4 = 0$ but $c_1 = 1$. Then at the corners $y = 0$ and $x = 0$ or $y = 0$ and $x = 1$, the boundary condition is not defined. The solution will have a discontinuity there, and the derivatives will be infinite. Such a problem is not well posed in a mathematical sense, and even in regular geometries, like Fig. 6-1a, the solution to Eq. (6-1) is not uniformly convergent under these boundary conditions.[15]

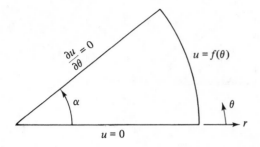

Figure 6-2 Boundary conditions for Eqs. (6-5) to (6-8).

Difficulties can also arise even when boundary data are continuous. Consider the following problem for the domain pictured in Fig. 6-2

$$\frac{1}{r}\frac{\partial}{\partial r}\left(r\frac{\partial u}{\partial r}\right)+\frac{1}{r^2}\frac{\partial^2 u}{\partial \theta^2}=0 \tag{6-5}$$

$$u = 0 \qquad \text{along } \theta = 0 \tag{6-6}$$

$$\frac{\partial u}{\partial n}=\frac{\partial u}{\partial \theta}=0 \qquad \text{along } \theta = \alpha \tag{6-7}$$

$$u = f(\theta) = \sin\frac{\pi\theta}{2\alpha} \qquad \text{along } r = 1 \tag{6-8}$$

Separation of variables can be applied to the differential equation by assuming a trial function in the form

$$u = R(r)T(\theta) \tag{6-9}$$

and substituting into Eq. (6-5). The resulting solution is of the form

$$u = \frac{a_0}{2}+\Sigma(a_n r^n \cos n\theta + b_n r^n \sin n\theta) \tag{6-10}$$

We take $n \geqslant 0$ for finite solutions. For the boundary conditions $u = 0$ on $\theta = 0$ we must have $a_0 = 0$ and $a_n = 0$, while the conditions that $\partial u/\partial \alpha = 0$ on $\theta = \alpha$ gives

$$b_n \cos n\alpha = 0 \tag{6-11}$$

or

$$n\alpha = \text{odd multiple of }\frac{\pi}{2}$$

The first term has $n = \pi/(2\alpha)$ and gives $\sin \pi\theta/(2\alpha)$. Then the boundary condition at $r = 1$ eliminates all but the first term. The solution is then

$$u(r,\theta) = r^{\pi/(2\alpha)}\sin\frac{\pi\theta}{2\alpha} \tag{6-12}$$

The radial derivative of the solution is

$$\frac{\partial u}{\partial r}=\frac{\pi}{2\alpha}r^{\pi/2\alpha-1}\sin\frac{n\theta}{2\alpha} \tag{6-13}$$

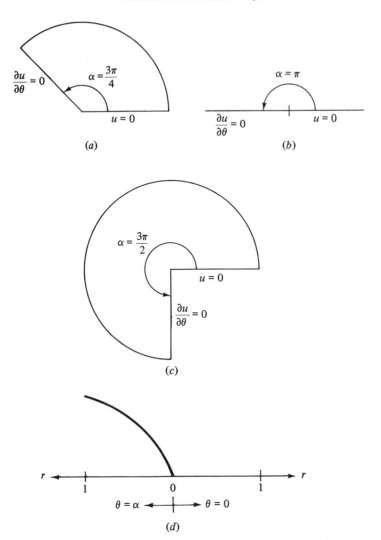

Figure 6-3 Problems with singularities. (a) $u \simeq r^{2/3}$, $du/dr \simeq r^{-1/3}$. (b) $u \simeq r^{1/2}$, $du/dr \simeq r^{-1/2}$. (c) $u \simeq r^{1/3}$, $du/dr \simeq r^{-2/3}$. (d) Solution along $\theta = 0$ or $\theta = \alpha$ for (b).

Whenever $\alpha > \pi/2$ the power of r in the solution is less than one, and the power of r in the radial derivative is negative. This means that the solution has an infinite radial derivative at the center. Three cases and their behavior are given in Figs. 6-3a, 6-3b, and 6-3c. The solution along $\theta = 0$ or $\theta = \alpha$ for case (b) is plotted in Fig. 6-3d. The derivative is infinite at $r = 0$. These results occur even though the problem is seemingly benign: no discontinuities are prescribed, the domain is regular, and separation of variables gives the exact solution.

What is the impact of such results? In Sec. 4-13 we see that the convergence of different methods is different: the finite difference method converges as Δx^2, Galerkin quadratic as Δx^3, and cubic collocation as Δx^4. There is a proviso, however, that the exact solution is continuous and has a certain number of continuous derivatives. The same results apply to two-dimensional problems, except now we see that the restriction of continuous derivatives may be a sleeper: an otherwise nice problem does not give continuous derivatives. We see in Sec. 4-13 that the higher-order methods, such as Galerkin finite elements with quadratic trial functions or orthogonal collocation on finite elements with cubic trial functions, require more work per element than the low-order methods, such as finite difference or Galerkin linear finite elements. The extra work per element must be made up by having to use fewer elements to achieve the same accuracy. Otherwise the high-order methods are not competitive. Yet in the examples given above the high-order methods do not converge rapidly, because the rate of convergence, or the power m in Δx^m, is set by the properties of the exact solution rather than the method. In such cases the low-order methods may prove more economical. Thus the things we have learned in one dimension may or may not help us in two dimensions. The whole question is very problem-dependent.

6-2 FINITE DIFFERENCE

We can easily write down the finite difference equation for Eq. (6-3) by using the same techniques developed above and applying them in the x, y, and t variables. Let

$$c_{ij}^n = c(x_i, y_j, t_n) \tag{6-14}$$

and an explicit method applied to Eq. (6-3) gives

$$\frac{c_{ij}^{n+1} - c_{ij}^n}{\Delta t} = \frac{D}{\Delta x^2}(c_{i+1,j}^n - 2c_{i,j}^n + c_{i-1,j}^n) + \frac{D}{\Delta y^2}(c_{i,j+1}^n - 2c_{i,j}^n + c_{i,j-1}^n) \tag{6-15}$$

This can be arranged to give

$$c_{ij}^{n+1} = \frac{D\Delta t}{\Delta x^2}(c_{i+1,j}^n + c_{i-1,j}^n) + \frac{D\Delta t}{\Delta y^2}(c_{i,j+1}^n + c_{i,j-1}^n)$$

$$+ c_{ij}^n \left(1 - \frac{2D\Delta t}{\Delta x^2} - \frac{2D\Delta t}{\Delta y^2}\right) \tag{6-16}$$

By the positivity rule, extended to involve five terms, the calculation is stable provided the coefficients are all positive and add up to one or less. This gives the stability limit

$$1 > \left(\frac{2D}{\Delta x^2} + \frac{2D}{\Delta y^2}\right)\Delta t \tag{6-17}$$

$$\Delta t \leqslant \frac{1}{2D}\frac{\Delta x^2 \Delta y^2}{\Delta x^2 + \Delta y^2} \tag{6-18}$$

If $\Delta x = \Delta y$ we get

$$\Delta t \leqslant \frac{\Delta x^2}{4D} \tag{6-19}$$

or a time step half as large as in the one-dimensional problem. Thus the stable step size goes down as the number of dimensions increases.

An important consideration in multidimensional problems is that the scale of the problem increases dramatically with the number of dimensions. To illustrate this point let us consider solving a transient diffusion equation, Eq. (6-3), in one, two, and three dimensions. We assume that the domain is a square in two dimensions and a cube in three dimensions. We divide each dimension into n equal intervals and take the domain extending from zero to L in each applicable coordinate x, y, or z. We then have

$$\Delta x = \Delta y = \Delta z = \frac{L}{n} \tag{6-20}$$

Steady state is reached at about $t = L^2/(Ds)$, where s is the number of dimensions. The equations in one, two, and three dimensions, respectively, are then

$$u_i^{n+1} = \alpha(u_{i+1}^n + u_{i-1}^n) + (1 - 2\alpha)u_i^n \tag{6-21}$$

$$u_{ij}^{n+1} = \alpha(u_{i+1,j}^n + u_{i-1,j}^n + u_{i,j+1}^n + u_{i,j-1}^n) + (1 - 4\alpha)u_{ij}^n \tag{6-22}$$

$$u_{ijk}^{n+1} = \alpha(u_{i+1,jk}^n + u_{i-1,jk}^n + u_{i,j+1,k}^n + u_{i,j-1,k}^n + u_{ij,k+1}^n + u_{ij,k-1}^n)$$
$$+ (1 - 6\alpha)u_{ijk}^n \tag{6-23}$$

To calculate the solution at one grid point at the new time requires two multiplications in one, two, or three dimensions. The stable step sizes are

$$\Delta t \leqslant \frac{\Delta x^2}{2Ds} \tag{6-24}$$

and we need to solve for enough steps to reach $t = L^2/(Ds)$

$$\text{Number of steps} = \frac{L^2}{Ds} \times \frac{2Ds}{\Delta x^2} = 2n^2 \tag{6-25}$$

Take a multiplication time of $\frac{1}{6}$ μsec, which applies to the CDC 7600 computer, and which is very fast. The total computation time is then

$$\text{Multiplication time} = 2 \times n^s \times 2n^2 \times \frac{1}{6} \text{ μsec} \tag{6-26}$$

Table 6-1 summarizes the results for different numbers of mesh points and dimensions. If we need 100 mesh points in each direction the three-dimensional calculation takes about one hour. With computer time sold at about $1,000 per hour on this machine, this is clearly too expensive to do many calculations unless they are very important. Of course more complicated problems take longer. Such stringent computing requirements are relieved in the one-dimensional case by using an implicit method, which allows larger step sizes and which works in two and three dimensions as well.

Table 6-1 Multiplication times for diffusion problems

Number of dimensions s	Multiplication time			
	sec $n = 10$	sec $n = 20$	min $n = 100$	hr $n = 1,000$
	Explicit, Eq. (6-26)			
1	0.0007	0.005	0.01	0.2
2	0.007	0.1	1	2×10^2
3	0.07	2	1×10^2	2×10^5
	Implicit, Eq. (6-28)			
1	0.0002	0.001	0.003	5
2	0.08	5	1×10^3	20
3	80	4×10^4	1×10^9	2×10^{16}

n = number of grid intervals in each direction.

Let us apply an implicit method to Eq. (6-3) to give

$$\frac{c_{ij}^{n+1} - c_{ij}^n}{\Delta t} = (1 - \lambda)D\left(\left.\frac{\partial^2 c}{\partial x^2}\right|_{ij}^n + \left.\frac{\partial^2 c}{\partial y^2}\right|_{ij}^n\right) + \lambda D\left(\left.\frac{\partial^2 c}{\partial x^2}\right|_{ij}^{n+1} + \left.\frac{\partial^2 c}{\partial y^2}\right|_{ij}^{n+1}\right) \tag{6-27}$$

The unknowns are now c_{ij}^{n+1}, $c_{i,j+1}^{n+1}$, $c_{i,j-1}^{n+1}$, $c_{i+1,j}^{n+1}$, and $c_{i-1,j}^{n+1}$. If the unknowns are numbered with the i index set to one, letting j go from one to n, then increasing i, etc., a typical grid is shown in Fig. 6-4a and the corresponding matrix structure is shown in Fig. 6-4b. The matrix is now pentadiagonal, but unfortunately the bandwidth is large since it extends from the diagonal at least n entries away in each direction. Such matrices can be decomposed using a banded LU decomposition, but alternative methods have been developed because the bandwidth is so large. The number of multiplications needed to solve the implicit equations (just for the LU decomposition) is $BW(BW + 1)N$, where BW is the half-bandwidth and N is the total number of terms. The half-bandwidth is one, n, and n^2 in one, two, and three dimensions, while the total number of equations is n, n^2, and n^3. Take a step size four times as large as that needed for the explicit method (this provides an accurate solution but one that oscillates only slightly). The multiplication time is then

$$\text{Multiplication time} = BW(BW + 1)n^s \times \frac{n^2}{2} \times \tfrac{1}{6} \, \mu\text{sec} \tag{6-28}$$

Values are given in Table 6-1. For one-dimensional problems the implicit method is cheaper than the explicit method, but not for two or three dimensions. Indeed, because of the large bandwidth the calculations become prohibitive if many grid points are used. Clearly, another method of solution is needed.

Iterative methods have been developed to solve the equations that arise in dimensions higher than one. (Sometimes these are used for one-dimensional

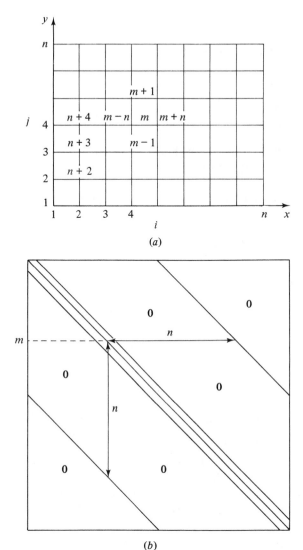

Figure 6-4 Pentadiagonal matrix structure. (*a*) Numbering system. (*b*) **AA** matrix structure.

problems, too.) These iterative methods are illustrated by application to a heat conduction equation, or the equivalent mass transfer equation,

$$\mathbf{\nabla}^2 T - Q = 0 \tag{6-29}$$

$$\text{in } 0 \leqslant x \leqslant 1 \qquad 0 \leqslant y \leqslant 1$$

We consider two types of boundary conditions: Dirichlet where

$$T = \text{given on boundary} \tag{6-30}$$

and Neumann where

$$\frac{\partial T}{\partial n} = \text{given on boundary}$$

The finite difference formulation of the equation is

$$\frac{1}{\Delta x^2}(T_{i+1,j} - 2T_{ij} + T_{i-1,j}) + \frac{1}{\Delta y^2}(T_{i,j+1} - 2T_{ij} + T_{i,j-1}) - Q_{ij} = 0 \qquad (6\text{-}31)$$

The iterative methods are classified as point or line methods depending on whether iterations are performed simultaneously on single points or on lines of points.

The first point-iterative method is the Jacobi method. We rearrange Eq. (6-31) to the form

$$2\left(1 + \frac{\Delta x^2}{\Delta y^2}\right)T_{ij} = T_{i+1,j} + T_{i-1,j} + \frac{\Delta x^2}{\Delta y^2}(T_{i,j+1} + T_{i,j-1}) - \Delta x^2 Q_{ij} \qquad (6\text{-}32)$$

and then write it in the generic formula

$$2\left(1 + \frac{AY}{AX}\right)T_{ij}^{s+1} = [\ ]^s \equiv T_{i+1,j}^s + T_{i-1,j}^s + \frac{AY}{AX}(T_{i,j+1}^s + T_{i,j-1}^s) - \Delta x^2 Q_{ij}^s \qquad (6\text{-}33)$$

where

$$AX = \frac{1}{\Delta x^2} \qquad AY = \frac{1}{\Delta y^2} \qquad (6\text{-}34)$$

We thereby have assumed that the grid spacing is uniform in each direction, but it may be different in the x and y directions. Furthermore anisotropies in the material can be included. These can affect the final equations in the same way as the ratio of grid spacings. Thus the ratio AX/AY can be different from one either due to different grid spacing in the x and y directions or due to anisotropy in the material parameters (here thermal conductivity), or both. The Jacobi method can be improved by noting that if the points are calculated in a definite order, namely from small to large i and then from small to large j, some of the values on the right-hand side are known for the $(s+1)$th iteration. We replace them by those values to obtain the Gauss–Seidel method

$$2\left(1 + \frac{AX}{AY}\right)T_{ij}^{s+1} = \{\ \}^s \equiv T_{i+1,j}^s + T_{i-1,j}^{s+1} + \frac{AX}{AY}(T_{i,j+1}^s + T_{i,j-1}^{s+1}) - \Delta x^2 Q_{ij}^s \qquad (6\text{-}35)$$

The Gauss–Seidel method generally converges twice as fast as the Jacobi method. Both methods converge provided

$$\sum_{i \neq j} |A_{ji}| < A_{jj} \qquad \text{for } \sum_i A_{ji} T_i = f_j \qquad (6\text{-}36)$$

For the equation used above $<$ can be replaced by $\leqslant$, and Eq. (6-36) is true.

Still another improvement is possible. Rather than using Eq. (6-35) to

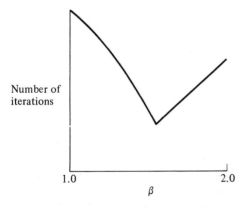

Figure 6-5 Optimal relaxation parameter.

calculate T^{s+1} we use it to calculate an estimate for T^{s+1}

$$2\left(1 + \frac{AX}{AY}\right) T_{ij}^* = \{\ \}^s \tag{6-37}$$

Then using the values at T^s and T^* we extrapolate to find T^{s+1}

$$T_{ij}^{s+1} = T_{ij}^s + \beta(T_{ij}^* - T_{ij}^s) \tag{6-38}$$

The parameter β is called the relaxation parameter. If $\beta = 1$ we have just the Gauss–Seidel method. If $\beta < 1$ we use underrelaxation (i.e. we use only part of the estimated value), while for $\beta > 1$ we call it overrelaxation. The relaxation parameter must be chosen. Values between one and two are usually best, but some experimentation is necessary to find the best value. Crichlow[3] suggests making several test calculations: First we compute the number of iterations necessary to reduce the residuals to a specified small value. The residuals are just the value of the original equation, Eq. (6-31), with the approximate solution substituted in. We then do this calculation for a variety of β. The number of iterations is plotted versus β and a graph similar to Fig. 6-5 results. Clearly we wish to choose the β that results in the minimum number of iterations. An alternative is to compute a fixed number of iterations and look for the maximum residual. A plot of the maximum residual versus β looks similar to Fig. 6-5 and provides a similar choice.

The rates of convergence can be calculated and a very good summary is in the book by Peaceman.[7] Assume that we have the same number of grid points in both x and y directions. The rate of convergence is expressed by means of the formula for the ratio of the maximum error from one iteration to that from the next

$$\frac{\text{Error}^{s+1}}{\text{Error}^s} = 1 - R \tag{6-39}$$

Values of R close to one are needed for a fast convergence. After N iterations the error is reduced by the factor

$$\frac{\text{Error}^N}{\text{Error}^0} = (1 - R)^N \tag{6-40}$$

where for Neumann boundary conditions

$$R = \frac{\pi^2/2n^2}{[1+\max(AX/AY, AY/AX)]} \tag{6-41}$$

and for Dirichlet boundary conditions

$$R = \frac{\pi^2}{2n^2} \tag{6-42}$$

Note that Dirichlet boundary conditions lead to faster convergence since R is bigger by at least a factor of two. Anisotropy ($\Delta x \neq \Delta y$) causes slower convergence for Neumann boundary conditions but not for Dirichlet conditions. It can be shown that R for the Gauss–Seidel method is twice that for the Jacobi method, leading to faster convergence. In the successive overrelaxation method the optimum β should be chosen. Young[11] gives the formula

$$t^2\beta^2 - 16\beta + 16 = 0 \tag{6-43}$$

where $t = 2\cos(\pi/n)$. Peaceman[7] showed that, in the vicinity of the optimum relaxation parameter,

$$R \simeq -\ln(\beta_{opt} - 1) \tag{6-44}$$

We thus have a way of estimating the rates of convergence. If we wish to make the error decrease by a factor of 100, we need to use N iterations, where N is given by

$$0.01 = (1 - R)^N \tag{6-45}$$

The number of iterations for various n and the different methods is given in Table 6-2. We notice that the simple change of using $\beta \neq 1$ means that many fewer iterations are needed. Also more iterations are needed as the grid is refined by using a higher n.

Table 6-2 Iterative methods for Eq. (6-31) with Neumann boundary conditions

	n		
Method	10	20	100
Jacobi			
$\quad R = \pi^2/4n^2$	0.025	0.0062	0.00025
$\quad 1 - R$	0.975	0.9938	0.99975
$\quad N$ in Eq. (6-45)	180	740	1900
Gauss–Seidel			
$\quad R$	0.05	0.012	0.0005
$\quad N$ in Eq. (6-45)	90	370	940
Successive overrelaxation			
$\quad \beta_{opt}$	1.528	1.728	1.939
$\quad R$	0.639	0.317	0.0629
$\quad N$ in Eq. (6-45)	5	12	71

In line iterative methods we iterate on an entire line at once. In Eq. (6-33) we evaluate all terms involving j, but not $j-1$ or $j+1$, at the $(s+1)$th iteration

$$2\left(1 + \frac{AX}{AY}\right)T_{ij}^{s+1} = T_{i+1,j}^{s+1} + T_{i-1,j}^{s+1} + \frac{AX}{AY}(T_{i,j+1}^{s} + T_{i,j-1}^{s}) - \Delta x^2 Q_{ij}^{s} \tag{6-46}$$

If, in addition, we employ Eq. (6-38) we have line successive overrelaxation LSOR. The rates of convergence for the same limiting cases are given by Peaceman[7] for Neumann boundary conditions

$$R = \frac{2\pi}{n} \tag{6-47}$$

and for Dirichlet boundary conditions

$$R = \frac{2\pi}{n}\left(1 + \frac{AY}{AX}\right)^{1/2} \tag{6-48}$$

Line Jacobi is twice as fast as point Jacobi (when $AX = AY$) and line successive overrelaxation is $\sqrt{2}$ as fast as point successive overrelaxation.

The alternating direction implicit method ADI can be used to solve the transient problem, Eq. (6-3). Let us define the operators

$$\delta_{xx}T_{ij}^n = \frac{T_{i+1,j}^n - 2T_{ij}^n + T_{i-1,j}^n}{\Delta x^2} \tag{6-49}$$

$$\delta_{yy}T_{ij}^n = \frac{T_{i,j+1}^n - 2T_{ij}^n + T_{i,j-1}^n}{\Delta y^2} \tag{6-50}$$

$$\nabla^2 T_{ij}^n = \delta_{xx}T_{ij}^n + \delta_{yy}T_{ij}^n \tag{6-51}$$

Then Eq. (6-15) becomes

$$T_{ij}^{n+1} - T_{ij}^n = D\Delta t(\delta_{xx}T_{ij}^n + \delta_{yy}T_{ij}^n) = \Delta t\nabla^2 T_{ij}^n \tag{6-52}$$

while the implicit equation, Eq. (6-27), is

$$T_{ij}^{n+1} - T_{ij}^n = D\Delta t(1 - \lambda)(\delta_{xx}T_{ij}^n + \delta_{yy}T_{ij}^n) + D\Delta t\lambda(\delta_{xx}T_{ij}^{n+1} + \delta_{yy}T_{ij}^{n+1}) \tag{6-53}$$

Rather than evaluating all the terms on the right-hand side at the $(n+1)$th time level, leading to a large banded matrix, let us first evaluate only the x derivatives, but step forward only $\Delta t/2$, using $\lambda = 0$ for δ_{yy} and $\lambda = 1$ for δ_{xx}

$$T_{ij}^{n+1/2} - T_{ij}^n = \frac{D\Delta t}{2}(\delta_{xx}T_{ij}^{n+1/2} + \delta_{yy}T_{ij}^n) \tag{6-54}$$

Next evaluate the x derivatives at the $n+1/2$ time level and the y derivatives at the $(n+1)$th level

$$T_{ij}^{n+1} - T_{ij}^{n+1/2} = \frac{D\Delta t}{2}(\delta_{xx}T_{ij}^{n+1/2} + \delta_{yy}T_{ij}^{n+1}) \tag{6-55}$$

Equations (6-54) and (6-55) give rise to a tridiagonal matrix, and hence are relatively easy to solve. We must solve Eq. (6-54) n times, once for each i, and then

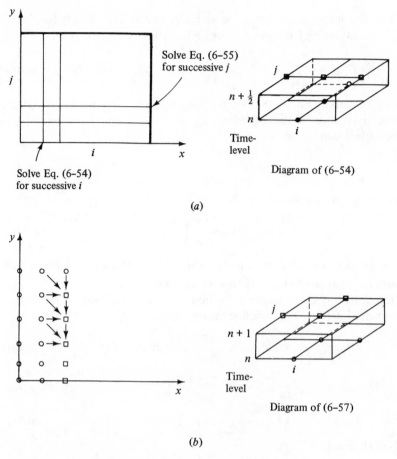

(a)

(b)

Figure 6-6 Alternating direction methods. ○ known value, □ unknown value. (a) Alternating direction implicit method. (b) Alternating direction explicit method.

solve Eq. (6-55) n times, once for each j, assuming an $n \times n$ grid. The operation count is reduced from n^4 for the direct solution of Eq. (6-53) to $4n + 6n^2$ to do one LU decomposition in each of the x and y directions, and n fore-and-aft sweeps for each direction. The computational savings are dramatic, being roughly $n^2/6$. Figure 6-6a illustrates the process.

The temporal truncation error is obtained by adding Eqs. (6-54) and (6-55) to give

$$T_{ij}^{n+1} - T_{ij}^n = D\Delta t \delta_{xx} T_{ij}^{n+1/2} + \frac{D\Delta t}{2} (\delta_{yy} T_{ij}^n + \delta_{yy} T_{ij}^{n+1}) \qquad (6\text{-}56)$$

The x derivative is thus treated using a midpoint rule, which has truncation error Δt^2, and the y derivative is treated using a Crank–Nicolson method, which also has truncation error Δt^2. Thus the overall system has truncation error Δt^2.

There are other alternatives. Equation (6-15) can be written as

$$T_{ij}^{n+1} - T_{ij}^n = D\Delta t \left(\frac{T_{i+1,j}^n - 2T_{ij}^{n+1} + T_{i-1,j}^{n+1}}{\Delta x^2} \right) + D\Delta t \left(\frac{T_{i,j+1}^{n+1} - 2T_{ij}^n + T_{i,j-1}^n}{\Delta y^2} \right) \quad (6\text{-}57)$$

and the value for T_{ij}^{n+1} can be calculated explicitly if the calculations for the $(i-1,j)$ and $(i,j+1)$ points have been done previously. We do the calculations in the order of increasing i and decreasing j, as illustrated in Fig. 6-6b. Such a scheme is an alternating direction explicit method. The truncation error is determined in problem 6-2.

The alternating direction method is also suitable when the equation is elliptic. Let us take Eq. (6-32) and write it in two sweeps

$$\beta(T_{ij}^{n+1/2} - T_{ij}^n) = D\delta_{xx} T_{ij}^{n+1/2} + D\delta_{yy} T_{ij}^n \quad (6\text{-}58)$$

$$\beta(T_{ij}^{n+1} - T_{ij}^{n+1/2}) = D\delta_{xx} T_{ij}^{n+1/2} + D\delta_{yy} T_{ij}^{n+1} \quad (6\text{-}59)$$

It is clear that this is just the same as a time-dependent method, since the iteration parameter β is similar to the inverse of the time step in Eqs. (6-54) and (6-55). If we use a single iteration parameter, though, corresponding to a single time step, the iteration proceeds slowly. It corresponds to integrating to steady state with a small time step, necessitating many time steps. Instead we use a sequence of iteration parameters, which corresponds to a variable time step. The sequence was suggested by Peaceman.[7] We define for an $n \times m$ grid

$$m_1 = \frac{2AX}{AX + AY} \frac{\pi^2}{4n^2} \qquad m_3 = \frac{2AY}{AX + AY} \frac{\pi^2}{4m^2}$$

$$m_2 = \frac{2AX}{AX + AY} \left(1 - \frac{\pi^2}{4n^2}\right) \qquad m_4 = \frac{2AY}{AX + AY} \left(1 - \frac{\pi^2}{4m^2}\right) \quad (6\text{-}60)$$

where $AX = 1/\Delta x^2$ and $AY = 1/\Delta y^2$. The iteration parameter must lie between the values

$$m_1 \leqslant \beta_k \leqslant m_2 \qquad \text{and} \qquad m_3 \leqslant \beta_k \leqslant m_4 \quad (6\text{-}61)$$

If the intervals overlap we choose

$$m_5 + \beta_k \leqslant m_6 \qquad \text{and} \qquad m_5 = \min(m_1, m_3) \qquad m_6 = \min(m_2, m_4) \quad (6\text{-}62)$$

For large n the value of m_2 and m_4 is usually two. We then use the parameters for the sth iteration

$$\beta_s = m_5 \left(\frac{m_6}{m_5}\right)^{s-1/p-1} \qquad s = 1, 2, \ldots, p \quad (6\text{-}63)$$

As an example for $s = 5$, $m_5 = 2$, and $m_6 = 0.0246$ ($n = m = 10$) the iteration parameters are

$$\beta_s = 0.0246, 0.0739, 0.221, 0.666, 2.0 \quad (6\text{-}64)$$

Peaceman[7] suggested using these equations to estimate the iteration parameters, and if divergence occurs to raise m_5 and try again. If m_5 is raised too high the convergence is very slow, so some care must be exercised. Even so, for some problems, especially nonlinear ones or ones with appreciable anisotropy, it may not be possible to find a set of iteration parameters that cause convergence.

When convergence does not result the analyst has one more technique to try: the strongly implicit procedure SIP. The details of this method are beyond the scope of this book, and the interested reader is directed to Peaceman's excellent book.[7] Peaceman pointed out that the alternating direction implicit method is fastest when it works, and it works well for simple, ideal problems. However, it works less well, and maybe not at all, for difficult problems involving complex geometry, or high or low ratios of AX to AY. The strongly implicit procedure, on the other hand, is only a little slower but is more robust, and the iteration parameters are more easily chosen. Peaceman suggested that good computer codes have options for direct solution, strongly implicit procedure and line successive overrelaxation. Weinstein, et al.[16] also compared several methods and mentioned that the point or line iterative methods do not work well for nonlinear problems and may converge slowly or not at all.

6-3 ORTHOGONAL COLLOCATION

The orthogonal collocation method can be applied to two-dimensional problems as is illustrated for flow through a rectangular duct. Reactor problems are more difficult but are illustrated for both packed bed reactors and monolith reactors (i.e. where the reaction occurs only on the wall).

The trial function in two dimensions is just the product of trial functions in each of the dimensions. Consider a problem whose solution is a function of x^2 and y^2. We write Eq. (4-194) for the dependence on x^2 and a similar dependence on y^2

$$T(x^2) = a_0 + (1 - x^2) \sum_{i=1}^{NX} a_i P_{i-1}(x^2) \tag{6-65}$$

$$T(y^2) = b_0 + (1 - y^2) \sum_{j=1}^{NY} b_j P_{j-1}(y^2) \tag{6-66}$$

For the two-dimensional case we need $T(x^2, y^2)$. We multiply Eq. (6-65) by Eq. (6-66) and renumber the coefficients to obtain the trial function

$$T(x^2, y^2) = a_0 b_0 + (1 - x^2) \sum_{i=1}^{NX} b_0 a_i P_{i-1}(x^2) + (1 - y^2) \sum_{j=1}^{NY} a_0 b_j P_{j-1}(y^2)$$

$$+ (1 - x^2)(1 - y^2) \sum_{i=1}^{NX} \sum_{j=1}^{NY} a_i b_j P_{i-1}(x^2) P_{j-1}(y^2) \tag{6-67}$$

We then use $c_0 = a_0 b_0$, $c_i = b_0 a_i$, $c_{j+NX} = a_0 b_j$, and $c_{i+j+NX+NY} = a_i b_j$. This interpolation is not actually used in the solution since we solve the problem in terms of the value of T at the collocation points. The collocation points are chosen as (x_i, y_j) where x_i and y_j are the same collocation points as we used in one dimension. One case is illustrated in Fig. 6-7. We define the temperature at the ith

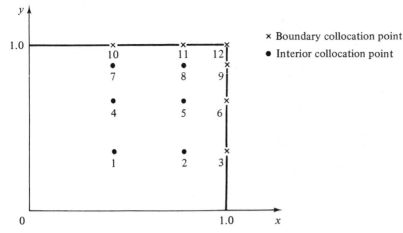

Figure 6-7 Collocation points for $NX = 2$, $NY = 3$, and symmetric polynomials from Table 4-6 and planar geometry. Numbering scheme by Eq. (6-69).

collocation point in x and the jth collocation point in y as

$$T_{IJ} = T(x_i, y_j) \tag{6-68}$$

and must define a local to global numbering scheme to convert the IJ pair to a single index k

$$k = (J - 1)(NX + 1) + I \tag{6-69}$$

We can evaluate x derivatives using the matrices from Table 4-6 for the appropriate geometry, replacing **B** by **BX** or **BY** as required. Both x and y can be planar, or one of them can be cylindrical.

$$\left. \frac{\partial^2 T}{\partial x^2} \right|_{IJ} = \sum_{k=1}^{NX} BX_{IK} T_{KJ} \tag{6-70}$$

$$\left. \frac{\partial^2 T}{\partial y^2} \right|_{IJ} = \sum_{k=1}^{NY} BY_{JK} T_{IK} \tag{6-71}$$

In the collocation method we evaluate the residual at the interior collocation points in Fig. 6-7, and use the boundary conditions at the collocation points on the boundary.

Application of orthogonal collocation in two dimensions is illustrated by solving the problem of flow of a newtonian fluid in a rectangular duct. Figure 6-8 illustrates the geometry. The differential equation is

$$\frac{\partial^2 u}{\partial x'^2} + \frac{\partial^2 u}{\partial y'^2} = -\frac{1}{\mu} \frac{\partial p}{\partial z} \tag{6-72}$$

We define the new coordinates as

$$x = \frac{x'}{L_x} \qquad y = \frac{y'}{L_y} \qquad L = \frac{L_x}{L_y} \tag{6-73}$$

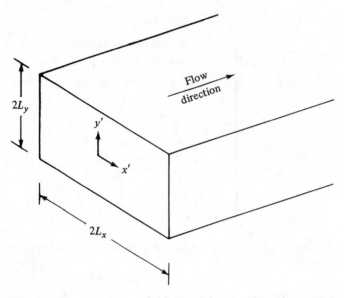

Figure 6-8 Flow in a rectangular duct.

where L is the aspect ratio, and the equation becomes

$$\frac{\partial^2 u}{\partial x^2} + L^2 \frac{\partial^2 u}{\partial y^2} = -\frac{L_x^2}{\mu} \frac{\partial p}{\partial z} \equiv -b \qquad (6\text{-}74)$$

subject to the boundary conditions

$$u(x = \pm 1, y) = u(x, y = \pm 1) = 0 \qquad (6\text{-}75)$$

that imply no slip on the boundary. This problem is symmetric so that poly-nomials in x^2 and y^2 are appropriate.

We apply orthogonal collocation to Eq. (6-74) at the IJ collocation point

$$\sum_{k=1}^{NPX+1} BX_{IK} T_{KJ} + L^2 \sum_{k=1}^{NPY+1} BY_{JK} T_{IK} = -b \qquad (6\text{-}76)$$

The boundary conditions give

$$T_{I,NPY+1} = 0 \qquad I = 1, \dots, NPX+1 \qquad (6\text{-}77)$$
$$T_{NPX+1,J} = 0 \qquad J = 1, \dots, NPY+1 \qquad (6\text{-}78)$$

The average velocity is given by

$$\langle u \rangle = \frac{\displaystyle\sum_{I=1}^{NPX+1} \sum_{J=1}^{NPY+1} WX_I WY_J T_{IJ}}{\displaystyle\sum_{I=1}^{NPX+1} \sum_{J=1}^{NPY+1} WX_I WY_J} \qquad (6\text{-}79)$$

All matrices come from Table 4-6 for planar geometry. We write the equations explicitly for $NPX = NPY = 1$, one interior collocation point, and a square duct $L = 1$ for the nondimensional case $b = 1$

$$BX_{11}u_{11} + BY_{11}u_{11} = -1$$
$$u_{21} = u_{12} = u_{22} = 0 \tag{6-80}$$

Since $BX_{11} = BY_{11} = 2.5$ and $W_1 = \frac{5}{6}$ we get

$$-(2.5 + 2.5)u_{11} = -1$$
$$u_{11} = 0.200 \qquad \langle u \rangle = 0.139 \tag{6-81}$$

This answer is actually very close to the exact solution, with an error of 1.2 percent. The exact solution is found using separation of variables and involves a doubly infinite series. Collocation is much easier to use. Using two collocation points in each direction (i.e. $NPX = NPY = 2$) gives four unknowns and requires solving four equations in four unknowns.

Consider next a chemical reactor with cooling at the wall and axial conduction. The temperature equation can be written as

$$\frac{1}{Pe}\frac{\partial^2 T}{\partial z^2} - \frac{\partial T}{\partial z} + \alpha \frac{1}{r}\frac{\partial}{\partial r}\left(r\frac{\partial T}{\partial r}\right) + \beta R = 0 \tag{6-82}$$

with a similar equation for concentration. The boundary conditions in the radial direction are the usual ones

$$\left.\frac{\partial T}{\partial t}\right|_{r=0} = 0 \qquad \left.-\frac{\partial T}{\partial r}\right|_{r=1} = \text{Bi}_w[T(1,z) - T_w] \tag{6-83}$$

while those in the axial direction are more complicated. Here let us just note that we must specify a boundary condition at $z = 0$ and $z = 1$. Define the solution at the ith collocation point in the r direction and the jth collocation point in the z direction by

$$T_{IJ} = T(r_i, z_j) \tag{6-84}$$

Also we use matrices **BZ**, **AZ**, and **BR** for the collocation matrices in the z and r directions. Since the problem is symmetric in r we use the polynomials in r^2 for symmetric problems. Then the residual evaluated at the ijth collocation point is

$$\sum_{K=1}^{NZ+2}\left(\frac{BZ_{JK}}{Pe} - AZ_{JK}\right)T_{IK} + \alpha'\sum_{K=1}^{NR+1} BR_{IK}T_{KJ} + \beta' R_{IJ} = 0 \tag{6-85}$$

We can solve this system of equations using Newton–Raphson by expanding the reaction rate expression about the current iteration T_{IJ}^s to yield

$$\sum_{K=1}^{NZ+2}\left(\frac{BZ_{JK}}{Pe} - AZ_{JK}\right)T_{IK}^{s+1} + \alpha'\sum_{K=1}^{NR+1} BR_{IK}T_{KJ}^{s+1}$$
$$+\beta'\left[R_{IJ}^s + \left.\frac{dR}{dT}\right|_{IJ}^s (T_{IJ}^{s+1} - T_{IJ}^s) + \left.\frac{dR}{dc}\right|_{IJ}^s (c_{IJ}^{s+1} - c_{IJ}^s)\right] = 0 \tag{6-86}$$

Alternatively the successive substitution method can be used when the reaction rate term is small. In either case we must invert or decompose a matrix of size

$2NZ \times NR$ by $2NZ \times NR$ where there are NZ interior collocation points in the z direction, NR interior collocation points in the r direction, and two variables c and T. (The boundary conditions are eliminated to obtain the number $2NZ \times NR$.) In addition the alternating direction implicit method can be used.

Whether or not one wishes to apply orthogonal collocation depends on the difficulty of the problem. Consider first the cases treated in problems 4-34 and 4-35. Orthogonal collocation is suitable to apply to these cases with axial dispersion. If cooling is provided at the wall the additional radial gradients can be expected to be mild enough to allow use of orthogonal collocation in the radial direction, too. Next consider the case treated in Sec. 5-4 with only radial dispersion. The problem is initial-value in the z direction, and the initial-value techniques use very small steps Δz for an accurate solution. If axial dispersion is introduced into the model then a trial function must be assumed in the axial direction. Orthogonal collocation would not be expected to give a good result since the profile is steep. Either collocation on finite elements or the finite difference method would be appropriate in the z direction. Judging from experience discussed in Secs. 5-8 and 5-9 we would probably choose the finite difference method in the z direction. Even so we can still take advantage of the accurate collocation solutions by using orthogonal collocation in the radial direction.

Young and Finlayson[13] used orthogonal collocation in both directions to solve for the oxidation of sulfur dioxide to sulfur trioxide. One aspect of the study involved determining rate constants from experiments that included significant axial and radial gradients—significant in the sense that they affected the rate of reaction—but not so large that orthogonal collocation could not be used. The rate parameters were determined by performing a numerical nonlinear least squares analysis on the experimental results. This involved solving Eq. (6-82) for a set of assumed rate parameters, for each experimental run, and then resolving the same problem for another set. To determine the best set of rate parameters it was necessary to solve the problem over 300 times, and the efficiency of the orthogonal collocation method was particularly welcome. Once the rate parameters were determined they were used in another simulation to explain the way axial dispersion influenced the experimental results.

Another type of reactor modeling with orthogonal collocation is the monolith. In this problem steep fronts are expected in the axial direction but not in the other two. Thus global polynomials are appropriate in the transverse direction, whereas finite elements or finite differences are necessary in the axial direction.[14]

The last example used finite elements only in one direction. What if the problem is expected to have a steep profile in two or more directions? Then finite elements are indicated in two or more directions. The collocation finite element method can be applied in such cases. Applications indicate that for engineering problems the Galerkin method can handle irregular domains more easily. In addition the first approximation in the collocation method is a cubic trial function, and if the profile is steep a high-order method may not be appropriate. Furthermore, discontinuities often appear in two-dimensional problems, making the high-order accuracy difficult to achieve. For all of these reasons the promise of the orthogonal collocation method shown in one dimension has not yet been realized in two dimensions. The reader is referred to the Bibliography for details of applications of orthogonal collocation in two dimensions.

6-4 GALERKIN FINITE ELEMENT METHOD

The Galerkin finite element method is widely used to solve elliptic boundary-value problems in two and three dimensions. The reason for this widespread use is related to the ease with which the finite element method accommodates complicated and irregular geometries, which are important in engineering applications. We first introduce the ideas for application of the Galerkin finite element method to a two-dimensional heat conduction problem and then give a complete treatment with detailed applications.

The prototype problem is taken as steady-state heat conduction problem with three types of boundaries

$$\mathbf{V} \cdot k\mathbf{V}T = Q \qquad \text{in } A \tag{6-87}$$

$$T = T_1 \qquad \text{on } C_1 \tag{6-88}$$

$$-k\frac{\partial T}{\partial n} = q_2 \qquad \text{on } C_2$$

$$-k\frac{\partial T}{\partial n} = h_3(T - T_3) \qquad \text{on } C_3 \tag{6-89}$$

On C_1 the temperature T_1 is given, on C_2 the heat flux q_2 is given, and on C_3 we have a boundary condition of the third kind involving a heat transfer coefficient, and h_3 and T_3 are given; $Q < 0$ for generation. The trial function is substituted into Eq. (6-87) to form the residual, and the weighting function is taken as δT. The weighted residual gives a system of equations

$$\int_A \delta T \mathbf{V} \cdot k\mathbf{V}T \, dA = \int_A \delta T Q \, dA \tag{6-90}$$

The first term is integrated by parts. Thus

$$\int_A \delta T \mathbf{V} \cdot k\mathbf{V}T \, dA = \int_A \mathbf{V} \cdot (\delta T k)\mathbf{V}T \, dA - \int_A k\mathbf{V}T \cdot \mathbf{V}\delta T \, dA \tag{6-91}$$

The divergence theorem gives

$$\int_A \mathbf{V} \cdot (\delta T k\mathbf{V}T) dA = \int_{C_i} \delta T k \mathbf{n} \cdot \mathbf{V}T \, dC \tag{6-92}$$

where $\mathbf{n}$ is the outward pointing normal. Combination of Eqs. (6-91) and (6-92) in Eq. (6-90) gives

$$-\int_A k\mathbf{V}T \cdot \mathbf{V}\delta T \, dA + \int_C \delta T k \mathbf{n} \cdot \mathbf{V}T \, dC = \int_A \delta T Q \, dA \tag{6-93}$$

The boundary conditions are applied as weighted residuals

$$\int_{C_1} \delta T k \mathbf{n} \cdot \mathbf{\nabla} T \, dC = - \int_{C_2} \delta T q_2 \, dC$$

$$\int_{C_3} \delta T k \mathbf{n} \cdot \mathbf{\nabla} T \, dC = - \int_{C_3} \delta T h_3 (T - T_3) \, dC \quad (6\text{-}94)$$

These are substituted into Eq. (6-93) where $\delta T = 0$ on C_1 and $S = C_1 + C_2 + C_3$

$$- \int_A k \mathbf{\nabla} T \cdot \mathbf{\nabla} \delta T \, dA - \int_{C_2} \delta T q_2 \, dC - \int_{C_3} \delta T h_3 (T - T_3) \, dC = \int_A \delta T Q \, dA \quad (6\text{-}95)$$

Equation (6-95) is the Galerkin statement of the problem. If we apply the same steps in reverse we get

$$\int_A \delta T (\mathbf{\nabla} \cdot k \mathbf{\nabla} T - Q) \, dA - \int_{C_2} (k \mathbf{n} \cdot \mathbf{\nabla} T + q_2) \delta T \, dC$$

$$- \int_{C_3} [k \mathbf{n} \cdot \mathbf{\nabla} T + h_3 (T - T_3)] \delta T \, dC = 0 \quad (6\text{-}96)$$

If this is true for arbitrary δT then the terms in parentheses must be zero. The Euler equation is Eq. (6-87), and the natural boundary conditions are given in Eq. (6-88). The trial function must satisfy the essential boundary conditions of $\delta T = 0$ or $T = T_1$ on C_1, allowing no variation of T on C_1. Thus the value of T is fixed.

The difference between essential and natural boundary conditions is an important one. For an equation that is second-order, any boundary condition involving first derivatives is natural and any boundary condition setting the function value only is essential. The interested reader should read Chapter 7 of Finlayson[4] to see the origin of the terms, which arise for variational principles. What is important for people using a finite element method to realize is that *some* boundary condition will be satisfied on each boundary. The user must do something to specify the temperature on C_1. The user must also do something to specify a given heat flux q_2 on C_2. These conditions are clear enough. However, if the user does nothing then the Galerkin method automatically uses the natural boundary condition, regardless of the user's intent. In the case of Eq. (6-97) if q_2, h_3, and T_3 are not specified, and the value T_1 is not specified on the boundary, then the boundary condition is *automatically*

$$k \mathbf{n} \cdot \mathbf{\nabla} T = 0 \qquad \text{on } C \qquad (6\text{-}97)$$

The user must be careful to know what the natural boundary conditions are for a problem, and if they are not the desired conditions, then the correct ones must be specified.

The finite element part of the method comes in the choice of trial function

$$T(x, y) = \sum_i T_i N_i(x, y) \qquad (6\text{-}98)$$

The basis functions N_i are known, chosen functions of position. The variation of T with respect to T_j is just N_j, so that is the weighting function. Equation (6-95) is then

$$-\sum_i \left(\int_A k\nabla N_i \cdot \nabla N_j \, dA + \int_{C_3} h_3 N_i N_j \, dC \right) T_i$$

$$= \int_{C_2} N_j q_2 \, dC - \int_{C_3} N_j h_3 T_3 \, dC + \int_A N_j Q \, dA \quad (6\text{-}99)$$

We define the element matrices

$$A^e_{IJ} = - \int k\nabla N_I \cdot \nabla N_J \, dA \quad (6\text{-}100)$$

$$F^e_J = \int N_J Q \, dA \quad (6\text{-}101)$$

and elements on the boundary have an additional contribution

$$A^e_{IJ} = - \int_{C_3} h_3 N_I N_J \, dC \quad (6\text{-}102)$$

$$F^e_J = \int_{C_2} N_J q_2 \, dC - \int_{C_3} N_J h_3 T_3 \, dC \quad (6\text{-}103)$$

Then the equations are

$$\sum_e A^e_{IJ} T^e_J = \sum_e F^e_J \quad (6\text{-}104)$$

If an element has boundary conditions on C_1 then the Galerkin equations for that node are replaced by the boundary condition

$$T_i = T_1 \qquad i \text{ on } C_1 \quad (6\text{-}105)$$

In addition to being able to handle irregular geometries, another advantage of the Galerkin method is to be able to refine the mesh. We can use large elements in regions in which the solution has small gradients and small elements in regions in which the solution has large gradients. The mesh is most easily refined if the

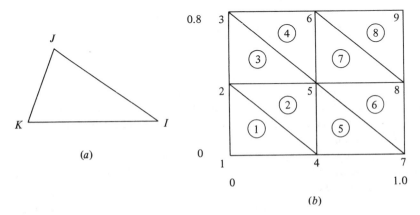

Figure 6-9 Triangular finite elements. (*a*) General element. (*b*) Regular array.

domain is divided into triangles rather than rectangles. Here we present the shape functions for linear functions on the triangle, although quadratic functions can also be used. For the triangle shown in Fig. 6-9a the trial function is

$$T = N_I(x,y)T_I + N_J(x,y)T_J + N_K(x,y)T_K \tag{6-106}$$

and the shape functions are

$$N_I = \frac{a_I + b_I x + c_I y}{2\Delta} \tag{6-107}$$

$$
\begin{aligned}
a_I &= x_J y_K - x_K y_J \\
b_I &= y_J - y_K \qquad \text{plus permutation on } I, K, J \\
c_I &= x_K - x_J
\end{aligned}
\tag{6-108}
$$

$$2 = \det \begin{vmatrix} 1 & x_I & y_I \\ 1 & x_J & y_J \\ 1 & x_K & y_K \end{vmatrix} = 2 \text{ (area of triangle)} \tag{6-109}$$

These parameters obey the restrictions

$$
\begin{aligned}
a_J + a_J + a_K &= 1 \\
b_I + b_J + b_K &= 0 \\
c_I + c_J + c_K &= 0
\end{aligned}
\tag{6-110}
$$

The Galerkin element equations are then (with k and Q constant)

$$A_{IJ}^e = -\frac{k}{4\Delta}(b_I b_J + c_I c_J) \tag{6-111}$$

$$F_I^e = \frac{Q}{2}(a_I + b_I \bar{x} + c_I \bar{y}) = \frac{Q\Delta}{3} \tag{6-112}$$

The centroids of the triangle are given by

$$\bar{x} = \frac{x_I + x_J + x_K}{3}$$

$$\bar{y} = \frac{y_I + y_J + y_K}{3} \tag{6-113}$$

$$a_I + b_I \bar{x} + c_I \bar{y} = \tfrac{2}{3}\Delta$$

As a simple example let us solve Eq. (6-87) on the domain shown in Fig. 6-9b. Take $k = -Q = 1$ and let the boundary condition be $T = T_1 = 0$ on C_1, or nodes 1, 2, 3, 4, 6, 7, 8, and 9. This situation represents uniform heat generation in the rectangle with zero boundary conditions around.

We first note that only T_5 is nonzero, since the other nodes are on the boundary and the temperature is zero there. We compute the terms A_{IJ}^e and F_I^e element by element. For element 2 we have

$$b_5 = y_2 - y_4 = 0.4 \qquad c_5 = x_4 - x_2 = 0.5 \tag{6-114}$$

using Eqs. (6-108) to find b_5 and c_5. When applying Eqs. (6-108) in each element, I is always five and J and K are the other two nodes of the element. For the b the nodes go in counter-clockwise order while those of c go in a clockwise order. The element matrix is then

$$A_{55}^2 = -\frac{0.4^2 + 0.5^2}{4\Delta} \tag{6-115}$$

We repeat this for the other elements

$$
\begin{array}{lll}
\text{Element 3: } b_5 = y_3 - y_2 = 0.4 & c_5 = x_2 - x_3 = 0 & \\
\text{Element 4: } b_5 = y_6 - y_3 = 0 & c_5 = x_3 - x_6 = -0.5 & \\
\text{Element 5: } b_5 = y_4 - y_7 = 0 & c_5 = x_7 - x_4 = 0.5 & (6\text{-}116) \\
\text{Element 6: } b_5 = y_7 - y_8 = -0.4 & c_5 = x_8 - x_7 = 0 & \\
\text{Element 7: } b_5 = y_8 - y_6 = -0.4 & c_5 = x_6 - x_8 = -0.5 &
\end{array}
$$

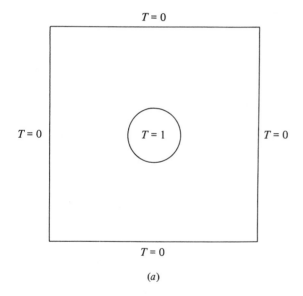

(a)

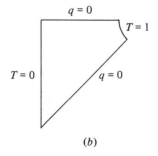

(b)

Figure 6-10 Heat transfer domain. (a) Complete domain. (b) Partial domain including symmetry.

The total matrix $\mathbf{A}$ is the sum of all the element matrices

$$A_{55} = A_{55}^2 + A_{55}^3 + A_{55}^4 + A_{55}^5 + A_{55}^6 + A_{55}^7$$
$$= -\frac{4(0.4)^2 + 4(0.5)^2}{4\Delta} = -\frac{0.41}{\Delta} \tag{6-117}$$

The right-hand-side vector is

$$F_5^3 = \frac{Q\Delta}{3} = -\frac{\Delta}{3} \tag{6-118}$$

and this is the same in each element. The total right-hand side is the sum of these for the elements with node 5

$$F_5 = F_5^2 + F_5^3 + F_5^4 + F_5^5 + F_5^6 + F_5^7 = -2\Delta \tag{6-119}$$

The area of the triangle is 0.1 so the final equation is

$$A_{55}T_5 = F_5$$
$$T_5 = 0.0488 \tag{6-120}$$

It turns out that the same answer is given by orthogonal collocation using quadratic trial functions.

The above example is very simple and is more easily solved using other methods, such as orthogonal collocation. If the shape of the domain is more complicated, however, only the finite element method is easy to use. For the next example we consider a heated cylinder embedded in a square domain. The cylinder is maintained at a dimensionless temperature of one while the outside boundary of the square domain is kept at dimensionless temperature of zero (see Fig. 6-10a). We only need to solve the problem in one-eighth of the domain (see Fig. 6-10b), since the solution in the rest of the domain can be obtained by symmetry. We need to divide the calculation domain into triangular elements. We do this for different meshes, each one more refined than the last, as shown in Figs. 6-11a, 6-11b, and 6-11c. The circular boundary at the corner is approximated by straight-line segments, and as the mesh is refined the segments become smaller and smaller, and represent the circle more accurately. Typical solutions are shown in Fig. 6-12. As the mesh is refined the temperature profile along the diagonal becomes more smooth, as we expected. Temperature contours can be constructed as illustrated in Fig. 6-13. These solutions were obtained using the finite element program[6] and are easily solved. Most other techniques require considerable manipulation to handle the irregular geometry, if they are applicable at all. This particular problem has been solved using the boundary collocation method by Shih.[9] In this method the solution is expanded in trial functions that satisfy the differential equation, and collocation is applied on the boundary. The method is probably more accurate than the finite element method used here, but it must be set up anew for each new problem. Furthermore, the boundary collocation method can only be applied when the differential equation is linear and can be solved analytically. The finite element, in contrast, is applicable when the problem is nonlinear and can be applied equally easily to any arbitrary, irregular domain.

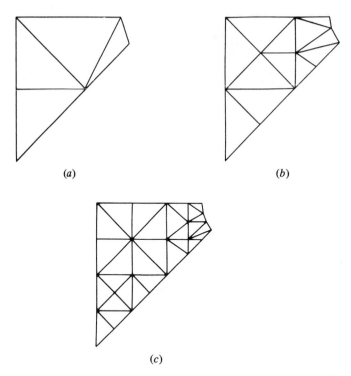

Figure 6-11 Meshes for heat transfer problem in Fig. 6-10. (*a*) Mesh 1. (*b*) Mesh 2. (*c*) Mesh 3.

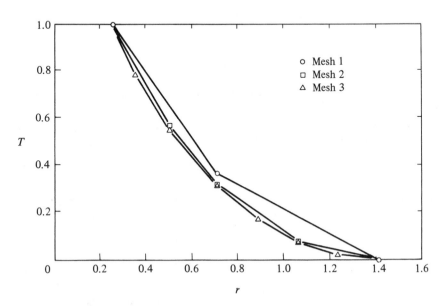

Figure 6-12 Temperature along diagonal.

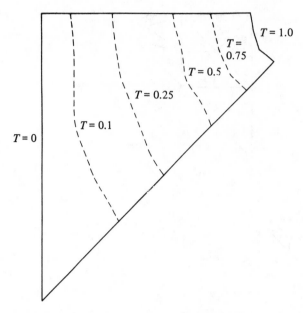

$T = 1.0$

$T = 0.75$

$T = 0.5$

$T = 0.25$

$T = 0.1$

$T = 0$

Figure 6-13 Temperature contours for problem in Fig. 6-12.

Next we consider the nonlinear problem in which the thermal conductivity k, rate of heat generation Q, or heat transfer coefficient h_3, are functions of temperature. We define the element matrices the same way, but with the nonlinear function of T included. (For this illustration the heat transfer coefficient is taken as constant.)

$$A^e_{IJ}(T^e) = - \int_A k(T^e)\nabla N_I \cdot \nabla N_J \, dA \qquad (6\text{-}121)$$

$$F^e_J(T^e) = \int_A N_J Q(T^e) \, dA \qquad (6\text{-}122)$$

Equation (6-104) is now a nonlinear equation since the element matrices depend on the element temperature. The equation can be solved using a successive substitution method. We evaluate the thermal conductivity and heat-generation terms using the old iterate value T^s and we use Eq. (6-104) to solve for T^{s+1}. Thus

$$\sum_e A^e_{IJ}(T^{es})T^{e,s+1}_J = \sum_e F^e_I(T^{es}) \qquad (6\text{-}123)$$

Alternatively Newton–Raphson can be applied to obtain faster convergence

$$\sum_e A^e_{IJ}(T^{es})T^{e,s+1}_J + \sum_{e,K} \frac{dA^e_{IJ}}{dT_K}(T^{es})(T^{e,s+1}_K - T^{es}_K)$$

$$= \sum_e \left(F^e_I(T^{es}) + \sum_J \frac{dF^e_I}{dT_J}(T^{e,s+1}_J - T^{es}_J) \right) \qquad (6\text{-}124)$$

In either case the integrals must be calculated numerically, as shown below.

It is possible to use elements that are rectangles, and to use trial functions that are not linear, but quadratic or cubic. First we divide the domain into rectangular elements and consider linear functions. We define the coordinates on the element to be

$$u = \frac{x - x_k}{\Delta x_k} \qquad v = \frac{y - y_l}{\Delta y_l} \tag{6-125}$$

and u and v go from zero to one on the element as x and y go from x_k to x_{k+1} and y_l to y_{l+1}, respectively. The trial function in u and v is taken as a bilinear function

$$N'_I(u, v) = a + bu + cv + duv \tag{6-126}$$

This shape function is called bilinear in x and y, since it includes the uv or xy term. One trial function is illustrated in Fig. 6-14 and the equations are

$$\begin{aligned} N'_1 &= (1-u)(1-v) & N'_2 &= u(1-v) \\ N'_3 &= uv & N'_4 &= (1-u)v \end{aligned} \tag{6-127}$$

The trial function for N'_4 is obtained by taking the product of two polynomials, each of which is zero along one boundary. The function $1-u$ is zero along $u = 1$, and hence makes $N'_4 = 0$ at nodes 2 and 3, while the function v is zero along $v = 0$, and hence makes N'_4 zero along nodes 1 and 2. The product $v(1-u)$ is zero at nodes 1, 2, and 3, and takes the value one at node 4.

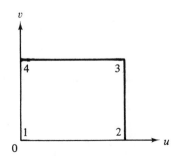

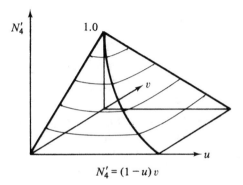

$$N'_4 = (1-u)v$$

Figure 6-14 Bilinear shape function on a rectangular element.

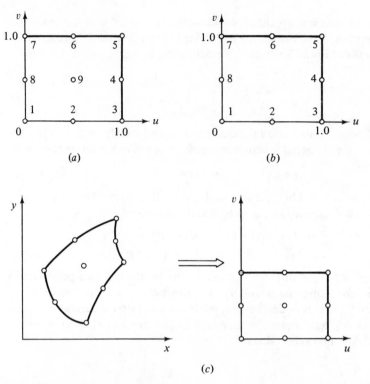

Figure 6-15 Quadratic shape functions on rectangles. (*a*) Lagrangian polynomials. (*b*) Serendipity polynomial. (*c*) Transformed element.

Quadratic functions can also be assumed on the rectangle. Let us consider the case illustrated in Fig. 6-15*a*, and concentrate on node 1. The function $v - \frac{1}{2}$ is zero for nodes 8, 9, and 4, while the function $v - 1$ is zero for nodes 7, 6, and 5. The function $u - \frac{1}{2}$ is zero for nodes 2, 9, and 6, while the function $u - 1$ is zero for nodes 3, 4, and 5. The product of these is zero at all nodes except node 1. We choose a constant multiple of the product as the basis function so that the value is one at node 1

$$N_1 = 4(u - \tfrac{1}{2})(u - 1)(v - \tfrac{1}{2})(v - 1) \tag{6-128}$$

For node 2 similar reasoning leads to the function

$$N_2 = -8u(u - 1)(v - \tfrac{1}{2})(v - 1) \tag{6-129}$$

These functions are referred to as lagrangian quadratic functions since they use the same interpolation as is used in lagrangian interpolation.

An alternative is to use the serendipity elements shown in Fig. 6-15*b*. For node 1 now we need functions which are zero at nodes 2 through 8 but not at the center node. The function $v - 1$ is zero at nodes 7, 6, and 5, while the function $u - 1$ is zero at nodes 3, 4, and 5. The function $u + v - 1$ is zero at nodes 8 and 2, and the product of these functions is zero at the nodes 2 through 8 but not at the center of the element. Typical serendipity functions are

$$N_1 = -(u-1)(v-1)(u+v-1)$$
$$N_2 = -8u(u-1)(v-1)$$

(6-130)

These functions do just about as well as the lagrangian quadratic functions, except in certain fluid mechanics situations. It should be noted that along the line $u = \frac{1}{2}$ the serendipity function is not a complete quadratic function of v, since it only has two unknowns rather than three. Certain problems may have an exact solution, which is a quadratic function of position. The finite element method with serendipity functions will not give exact results. Despite this qualification the economical savings from the serendipity element are substantial (perhaps 20 percent), and the elements are widely used. Once the trial functions are chosen the same element integrals, Eqs. (6-100) to (6-103), must be calculated and assembled into Eq. (6-104).

If the domain is irregular it may be desirable to use elements with an irregular shape (see Fig. 6-15c). In that case we need to transform the terms in the integral from the x–y coordinate system to the u–v coordinate system, which we again take as rectangular. The integrand is transformed using

$$\frac{\partial N_I}{\partial x}\frac{\partial N_J}{\partial x} = \left(\frac{\partial N_I}{\partial u}\frac{\partial u}{\partial x} + \frac{\partial N_I}{\partial v}\frac{\partial v}{\partial x}\right)\left(\frac{\partial N_J}{\partial u}\frac{\partial u}{\partial x} + \frac{\partial N_J}{\partial v}\frac{\partial v}{\partial x}\right)$$

(6-131)

and

$$dxdy = Jdudv$$

(6-132)

$$J = \frac{\partial x}{\partial u}\frac{\partial y}{\partial v} - \frac{\partial x}{\partial v}\frac{\partial y}{\partial u}$$

The part of the integral, Eq. (6-104), is then

$$\int_{A_{xy}}\frac{\partial N_I}{\partial x}\frac{\partial N_J}{\partial x}dxdy = \int_{A_{uv}}\left(\frac{\partial N_I}{\partial u}\frac{\partial u}{\partial x} + \frac{\partial N_I}{\partial v}\frac{\partial v}{\partial x}\right)\left(\frac{\partial N_J}{\partial u}\frac{\partial u}{\partial x} + \frac{\partial N_J}{\partial v}\frac{\partial v}{\partial x}\right)J\,dudv$$

(6-133)

Such an integral must be evaluated using quadrature techniques. Gaussian quadrature is usually used on rectangles, and this is just a combination of the quadrature developed for the orthogonal collocation method. A two-dimensional integral is evaluated as 2 one-dimensional integrals

$$\int_0^1\int_0^1 f(u,v)dudv = \sum_{i=1}^{N+2} W_i \int_0^1 f(u_i,v)dv$$

$$= \sum_{i=1}^{N+2}\sum_{j=1}^{N+2} W_i W_j f(u_i,v_j)$$

(6-134)

Notice that for $N \geqslant 2$ the quadrature weights on the boundaries are zero, so that only interior points are involved in evaluation of the integrals. When $k = k(T)$ the added complication in evaluating Eq. (6-133) is trivial. We just need to know $k(T(u_i,v_j))$ at the quadrature points.

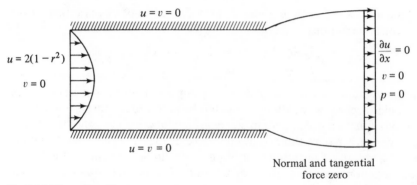

Fig. 6-16 Die swell problem.

The Galerkin finite element method has been used to solve a variety of engineering problems in diverse fields: flow in estuaries, flow in porous media, heat transfer, etc. The power of the method is that irregular geometries are easily handled and the mesh can be easily refined where needed, without refining the mesh over the whole domain. The last feature is a great aid in reducing the computational cost.

To illustrate some features of such applications we consider the flow of a newtonian fluid down a long pipe. The pipe stops and the fluid is emitted into the atmosphere in the form of a jet. For this illustration we neglect gravity and solve the flow problem. One aspect of the problem is the unknown position of the jet. The finite element method is easily applied since irregular geometries can be

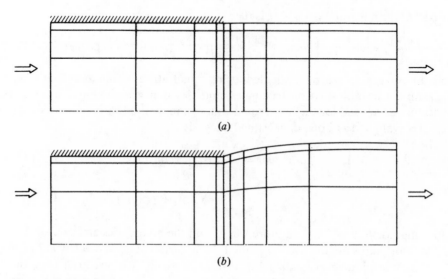

Figure 6-17 Mesh for die swell problem. (*a*) Initial mesh. (*b*) Final mesh.

handled. The equations are those of Navier–Stokes, written here in dimensionless form,

$$\text{Re}\left(u\frac{\partial u}{\partial x} + v\frac{\partial u}{\partial r}\right) = -\frac{\partial p}{\partial x} + 2\frac{\partial^2 u}{\partial x^2} + \frac{1}{r}\frac{\partial}{\partial r}\left(r\frac{\partial u}{\partial r} + r\frac{\partial v}{\partial x}\right)$$

$$\text{Re}\left(u\frac{\partial v}{\partial x} + v\frac{\partial v}{\partial r}\right) = -\frac{\partial p}{\partial r} + 2\frac{1}{r}\frac{\partial}{\partial r}\left(r\frac{\partial v}{\partial r}\right) + \frac{\partial}{\partial z}\left(\frac{\partial u}{\partial r} + \frac{\partial v}{\partial x}\right)$$

(6-135)

In addition we have the continuity equation

$$\frac{\partial u}{\partial x} + \frac{1}{r}\frac{\partial(vr)}{\partial r} = 0$$

(6-136)

The boundary conditions chosen are illustrated in Fig. 6-16. The free surface of the jet is located by requiring a mass balance. The average velocity at every axial location should be the same, and this provides a criterion for either increasing or decreasing the jet radius. We assume a shape, calculate the flow, check the mass balance, and change the shape if need be. Usually only three or four iterations on shape are necessary.

The finite element program FLUID uses lagrangian quadratic functions for the two velocities

$$u^e = \sum_I u_I N_I(x, r)$$

$$v^e = \sum_I v_I N_I(x, r)$$

(6-137)

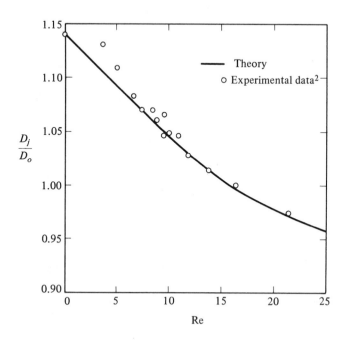

Figure 6-18 Die swell dependence on Reynolds number.

The pressure is expanded in terms of bilinear functions

$$p^e = \sum p_I N'_I(x,r) \tag{6-138}$$

The element shape is rectangular, and these functions are transformed into rectangles to calculate the integrals. The mesh locations are illustrated in Fig. 6-17a before the calculation begins and in Fig. 6-17b after solution. The jet increases in diameter for small Reynolds numbers and decreases in diameter for large Reynolds numbers. The dependence of the final jet diameter on the Reynolds number is compared to experimental data in Fig. 6-18. The agreement is very good. This problem was first solved for Re = 0 in 1974 by the finite element

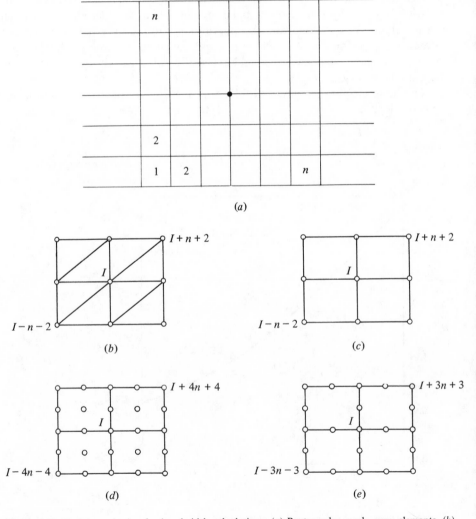

Figure 6-19 Nodal numbering for bandwidth calculations. (a) Rectangular mesh, $n \times n$ elements. (b) Linear triangles. (c) Bilinear rectangles. (d) Lagrangian quadratics: (e) Serendipity quadratics.

method, but has not yet been solved using finite difference methods. It has also been solved using orthogonal collocation on finite elements.[2]

The Galerkin equation, Eq. (6-104), can also be written with a global numbering system

$$\sum_j A_{ij} T_j = F_i \qquad (6-139)$$

The matrix $\mathbf{A}$ is sparse since many of its terms are zero. In a rectangular arrangement of elements, the unknowns can be numbered along one row, then another, etc.* The Galerkin equation with weighting function N_i involves contributions from the trial functions defined in the elements surrounding the ith node, but no others. For an $n \times n$ array, as illustrated in Fig. 6-19a, the contributing nodes are shown for the different basis functions (see Figs. 6-19b to 6-19e). The number of unknowns per row, total number of unknowns, and half-bandwidth of each of the trial functions are listed in Table 6-3. The computation time to perform an LU decomposition on the matrix $\mathbf{A}$ is $BW^2 N$, Eq. (4-262a), where BW is the half-bandwidth and N is the total number of equations. The work estimates for each basis function are given in Table 6-3. This work estimate is only for the solution of the matrix equations, not the calculation of the matrix, which can be from 10 to 50 percent of the total computation time and is typically 20 percent. The work estimate also assumes that pivoting is not necessary, which is valid in most cases.

Changing from linear functions to quadratic functions causes a significant increase in computation time, by a factor of 27 for serendipity functions to 64 for lagrangian quadratics. Of course each quadratic element has more nodes, so that we can perhaps take linear basis functions and $n \times n$ elements, and compare with quadratic basis functions with $n/2 \times n/2$ elements. On this basis the serendipity quadratics are only $27/16$ as expensive as the linear elements. The lagrangian quadratics are four times as expensive as the linear elements. If the error is smaller for the quadratic elements then even fewer elements can be used. The error bounds are the same as listed in Table 4-11 provided that the solution and its derivatives are continuous. Then the errors for linear elements go as

$$\text{Error} = K_1 \Delta x^2 \qquad \Delta x = \Delta y$$

and for quadratic elements $\qquad\qquad\qquad\qquad\qquad\qquad$ (6-140)

$$\text{Error} = K_2 \Delta x^3 \qquad \Delta x = \Delta y$$

For smooth solutions and highly accurate solutions the quadratics are usually cost effective, but for solutions with steep gradients the linear functions are usually cheaper.

The number of entries needed to store the matrix $\mathbf{A}$ is listed in Table 6-3. The storage is roughly twice the half-bandwidth multiplied by the total number of equations, assuming that pivoting is not necessary. For large numbers of elements (even for $n^2 = 10 \times 10$) the storage requirements are large and small computers

---

* A more efficient numbering system is to number along the diagonals rather than rows and columns.

Table 6-3 Comparison of Galerkin finite element methods on $n \times n$ mesh

Basis function	Shape of element	Unknowns per row	Half-bandwidth BW	Total number of unknowns N	Operation count LU	Storage requirements
Linear	Triangular	$n+1$	$n+2$	$(n+1)^2$	n^4	$2n^3$
Bilinear	Rectangular	$n+1$	$n+2$	$(n+1)^2$	n^4	$2n^3$
Serendipity quadratics	Rectangular	$2n+1$ or $n+1$	$3n+3$	$(2n+1)^2 - n^2$	$27n^4$	$18n^3$
Lagrangian quadratics	Rectangular	$2n+1$	$4n+4$	$(2n+1)^2$	$64n^4$	$32n^3$

cannot store the whole matrix. Fortunately, without pivoting the entire matrix does not need to be in the fast storage of the computer at one time, and only the small part that is being processed need be kept there. The remainder can be kept on slower disc storage and called as needed.

In summary, the Galerkin finite element method is a very useful tool for solving engineering problems. Irregular geometries are easily handled, and small mesh sizes can be used in local regions where the solution changes dramatically. For domains with regular boundaries other methods may be possible, but the finite element method is possible for these as well as those with irregular domains.

6-5 COMPARISON

Finite difference, collocation, and Galerkin finite element methods can all be applied to two-dimensional problems, but detailed, comparative studies are rare because of the expense of solving such problems. Some general guidelines have emerged, however. To illustrate them we look at work estimates for the different methods under two conditions. In the first case we require that the number of nodes be the same in all methods. This case is typical of situations with steep gradients where a great many nodes are needed to resolve the front without oscillation. The error in the solution is governed by the solution more than by the method of analysis. In the second case we make the number of elements such that all methods have equivalent error. This case is typical of situations with smooth solutions for which the continuity of the method governs the accuracy, rather than the continuity of the exact solution.

Work estimates are given in Table 6-4 for all methods having the same number of nodes. Specific numbers are given for a 30×30 grid. We see that the alternating direction implicit methods are the best choice, where we have assumed that 20 iterations are needed. The direct methods are less suitable, with the Hermite polynomial collocation on finite element method taking 30 times as long as the finite difference, alternating direction implicit method. This is a case that corresponds to a steep gradient; low-order methods prove superior. There is one qualification: in the alternating direction implicit method, we assume only 20 iterations are necessary. For linear problems, and some nonlinear problems, this may be valid. In other cases, including some linear ones that are very asymmetric, more iterations are necessary and the method may not converge. In those cases direct methods are more suitable, but low-order methods are still preferred.

Next take the case of "equivalent" accuracy. We define equivalent accuracy as when the principal term in the error expression is the same for all methods. The error bounds for two-dimensional problems are similar to Eq. (4-384). Here we assume that $\Delta x = \Delta y$ and make the term Δx^m the same for all methods. We thus use

$$n^2 = NE_1^2 = NE_2^3 = NE_3^4 \tag{6-141}$$

where NE_i is the number of elements in one direction for a method with an ith-degree polynomial for interpolation. For linear, quadratic, or cubic trial functions $i - 1$, 2, or 3 respectively. Now the constant in Eq. (4-384) may not be the same for

Table 6-4 Work estimates for two-dimensional problems*

Method	Estimate	Estimate	n	Operation count $\times 10^{-6}$	Matrix storage locations
	Work for equivalent number of nodes, $n = NE_1 = 2NE_2 = 2NE_3$				
FD, ADI	$6sNE_1^2$	$6sn^2$	30	0.11	$3NE_1 = 90$
OCFE–L, ADI	$64sNE_2^2$	$16sn^2$	15	0.28	$16NE_3 = 240$
GFEM–1 or FD, direct	NE_1^4	n^4	30	0.81	$2NE_1^4 = 1.6 \times 10^6$
GFEM–2S, direct	$27NE_1^4$	$1.7n^4$	15	1.4	$18NE_3^2 = 0.061 \times 10^6$
GFEM–2L, direct	$64NE_2^4$	$4n^4$	15	3.2	$32NE_3^2 = 0.11 \times 10^6$
OCFE–H, direct	$64NE_3^4$	$4n^4$	15	3.2	$128NE_3^3 = 0.43 \times 10^6$
	Work estimate for equivalent accuracy, $n = NE_1$, $NE_1^2 = NE_2^3 = NE_3^4$				
OCFE–H, direct	$64NE_3^4$	$64n^2$	3.16	0.0064	$128NE_3^3 = 4{,}000$
GFEM–1, or FD, direct	NE_1^4	n^4	10	0.010	$2NE_1^4 = 20{,}000$
FD, ADI	$6sNE_1^2$	$6sn^2$	10	0.012	$3NE_1 = 30$
GFEM–2S, direct	$27NE_2^4$	$27n^{8/3}$	4.64	0.013	$18NE_2^2 = 1{,}800$
OCFE–L, ADI	$64sNE_2^2$	$64sn$	3.16	0.013	$16NE_3 = 51$
GFEM–2L, direct	$64NE_2^4$	$64n^{8/3}$	4.64	0.030	$32NE_2^2 = 3{,}200$

* For direct methods the work is the number of multiplications for the LU decomposition using BW^2N, where BW is the half-bandwidth and N is the total number of nodes. For ADI methods the work estimate is for the fore-and-aft sweep; s is the number of iterations and $s = 20$ when needed.

the different methods, since it depends on the solution, perhaps on the second or fourth derivative of the solution. However, using Eq. (6-141) gives an estimate of the number of nodes needed for equivalent accuracy in a case with a smooth solution. The work estimates are given in Table 6-4. Specific values are given for $n = 10$. The smaller value is used because fewer nodes are probably needed when the solution does not have steep gradients. Now the collocation finite element method is the fastest, and the linear and quadratic polynomials require about equivalent work efforts. This example demonstrates that the method of choice may depend on the problem being solved, since the best method for a smooth solution is the worst one for a solution with steep gradients.

Certain qualifications must be imposed for the comparisons just made. The calculations assume a $n \times n$ grid that is uniform, and the finite element methods, in particular the Galerkin finite element methods, usually use a graded mesh, thus saving on the number of elements and, consequently, on the computation time. Furthermore, the comparison assumes a regular domain, and some of the methods, in particular the Galerkin finite element method, are very easily applied to irregular domains. Thus the eventual choice of method must involve a compromise between many important factors.

Another way to compare methods is to solve the problem using several methods and examine the accuracy and computation time, as well as the programming effort. Houstis, et al.[5] have done that for linear elliptic boundary-value problems of the type

$$\alpha \frac{\partial^2 u}{\partial x^2} + 2\beta \frac{\partial^2 u}{\partial x \partial y} + \gamma \frac{\partial^2 u}{\partial y^2} + \delta \frac{\partial u}{\partial x} + \varepsilon \frac{\partial u}{\partial y} + \zeta u = f \qquad (6\text{-}142)$$

The boundaries were often irregular. All finite elements methods used Hermite cubic polynomials; collocation, Galerkin, and least squares methods were tried. Finite difference methods were also applied. The collocation finite element method proved superior to Galerkin and least squares finite element methods and was usually superior to finite difference methods. For good accuracy, the collocation method always was more efficient than the finite difference method. This is the only careful, controlled comparison of methods for two-dimensional problems, and is limited to linear problems.

The collocation and Galerkin finite element methods were compared in application to Eqs. (6-135) to (6-136) illustrated in Fig. 6-16.[1,2] The Galerkin method used lagrangian quadratics for the velocity and bilinear functions for pressure. The collocation method used Hermite cubic functions for velocities and pressure. The Galerkin method used a frontal solution method, whereas the collocation method used a block diagonal LU decomposition. Some of the problems solved had discontinuities in the solution, whereas some of the problems had continuous functions but discontinuous or infinite derivatives of the exact solution. All of the problems had a singularity of some type.

At the outset the Galerkin method looked better: an element had only 22 unknowns whereas the collocation element had 48. Thus the collocation method must use many fewer elements to be competitive. However, because of the singularities, many elements might still be needed. In fact it was found for a newtonian fluid that the collocation method did about as well as the Galerkin method. The collocation method with fewer elements (2×4) gave better integral mass and force balances than did the Galerkin method with more elements (3×9). The Galerkin method, which could use mesh refinement because it had more elements, gave more accurate local properties of the solution. When the fluids were non-newtonian, and the viscosity depended on shear rate, the collocation methods were not competitive. For viscoelastic fluids the collocation method proved to be very much more expensive than the Galerkin method. In addition, the Galerkin method had the capability of refining the mesh in arbitrary ways since it used isoparametric elements. The collocation method treated irregular domains by transforming the problem to a regular domain. Thus the range of possibilities was much less. Based on all these considerations only the Galerkin finite element method was used in the subsequent work because of its greater versatility in treating irregular domains, its ability to use refined meshes, and a desire to solve flow problems that contained singularities so that the high accuracy of high-order method could not be achieved.

Finally we consider the application of different methods to the flow through porous media. The equations governing the pressure, and hence the velocity, are similar to Eq. (5-308) for a two-phase situation. The equations simplify for miscible flooding when only one phase is present. Let us consider solving either the miscible or immiscible flooding case along with the convective diffusion equation, Eq. (5-273), in a two-dimensional domain shaped as a square by injecting fluid at one corner and producing fluid at the opposite corner.

Settari, et al.[8] solved problems like this for miscible displacement using finite difference methods with either upstream weighting or central differences for the convective term. The Galerkin method was applied using either lagrangian quadratic functions or Hermite cubic functions. Enough calculations were made that the accuracy could be assessed and compared to the computation time. Table 6-5 lists sample results. We see that the Galerkin methods are able to use many fewer elements, giving a smaller computation time per time step. The computation time per time step per node is smallest for the finite difference method, but many more nodes are necessary for equivalent accuracy. For this comparison, at least, the most efficient method is the Galerkin finite element method with quadratic trial functions.

Spivak, et al.[10] solved a similar problem for immiscible flooding. They found that a complete simulation using a 5×5 grid with Hermite cubic functions took 2.4 sec per time step on a CDC 6600 computer. A finite difference method with the same number of nodes would need a 12×12 grid, and a typical computation time is 0.72 sec per time step. Thus for an equivalent number of nodes the finite difference method is fastest. These methods were not compared on a basis of equivalent accuracy on this problem because the exact solution was not known.

Young[12] compared a cubic Galerkin method using line successive over-relaxation with a Hermite cubic Galerkin method using a direct method of solution. Typical computation times for a miscible displacement are shown in Table 6-5. For the case with a velocity that must be determined (i.e. we solve both the concentration equation and the pressure equation) the line successive over-

Table 6-5 Comparison of methods for flow through porous media

Method	Grid	CPU W_T ——— Time step	Grid points	W_T ——— Grid point	W_T ——— Element
		Miscible flooding, sharp front[8]*			
FD, upstream	25×25	3.9	625	0.0062	0.0062
FD, central	30×30	7.4	900	0.0083	0.0083
GFEM–2L, direct	5×5	0.8	121	0.0066	0.032
GFEM–3H, direct	5×5	2.2	144	0.015	0.088
		Miscible flooding, smooth front[8]*			
FD, upstream	20×20	1.9	400	0.0048	0.0048
FD, central	11×11	0.35	121	0.0029	0.0029
GFEM–2L, direct	3×3	0.30	49	0.0061	0.033
GFEM–3H, direct	3×3	0.76	64	0.012	0.084
		Miscible flooding, Pe $= 290$[12]†			
GFEM–3L, LSOR					
Known velocity	10×10	0.07	931	0.00075	0.0007
Unknown velocity	10×10	0.93	931	0.0010	0.0093
GFEM–3H, direct					
Unknown velocity	10×10	20	484	0.041	0.20

* CDC 6600 computer.
† IBM 370/168 computer.

relaxation is about 20 times faster than the Hermite cubic direct solution method. The cubic polynomials on a 10×10 grid are more accurate than quadratic polynomials on a 20×20 grid or linear functions on a 50×50 grid. For this case, with Pe $= 290$, the higher-order methods are preferred.

In conclusion, we have seen that there is a great difference between one- and two-dimensional problems, and this difference has important implications on the method of choice. The best method may depend on the problem being solved, its discontinuities or singularities, and whether mesh refinement is useful and/or whether irregular geometries are needed. Even once a method of discretization is chosen the way the algebraic equations are solved may influence the method of choice. While no one method emerges as superior in all cases, we generally expect low-order methods, such as finite difference or Galerkin finite element with linear trial functions, to be best for problems with singularities, and high-order methods, such as quadratic Galerkin or Hermite cubic collocation, to be best for problems with smooth solutions. Even these guidelines may be overturned when an iterative method has difficulty converging and then direct methods are used. Despite the lack of a clearcut decision the reader should be able to assess the methods he or she is using and determine whether they are likely to be better or worse in another application.

PROBLEMS

Finite difference

6-1 Formulate an explicit finite difference algorithm to solve

$$\frac{\partial c}{\partial t} = \frac{\partial}{\partial x}\left[D(c)\frac{\partial c}{\partial x}\right] + \frac{\partial}{\partial y}\left[D(c)\frac{\partial c}{\partial y}\right]$$

Determine the truncation error in Δx, Δy, and Δt for your method. Give a rough guide to a first choice of Δt for a stable solution when $x = 0 \rightarrow 1$, $c = 0 \rightarrow 1$, $D = 1 + \lambda c$, and $\lambda = 2$.

6-2 Determine the truncation error of Eq. (6-57) by expanding the function in a Taylor series.

Orthogonal collocation

6-3 Find the effectiveness factor for a cylindrical pellet of radius R and length $2R$ by solving the problem

$$\frac{1}{r}\frac{\partial}{\partial r}\left(r\frac{\partial y}{\partial r}\right) + \frac{\partial^2 y}{\partial z^2} = \phi^2 y^2$$

The concentration is one on the boundary. Apply orthogonal collocation using different trial functions in the r and z directions. Find η for $\phi = 1$. What is the trial function? How does η compare to the value for an infinite cylinder?

6-4 Solve

$$\frac{\partial^2 u}{\partial x^2} + \frac{\partial^2 u}{\partial y^2} = e^u \qquad 0 \leqslant x, y \leqslant 1$$

$$u = 0 \qquad \text{on boundary}$$

Galerkin finite element method

6-5 Derive Eq. (6-96) from Eq. (6-95).

6-6 Solve the heat conduction problem in the shaded region of Fig. 6-20 when the temperature along the inner curved surface is 100 and on the outer surface is 0. Sketch the geometry, mesh layout, give nodal temperature values, and sketch the isotherms for $T = 10$ and 50.

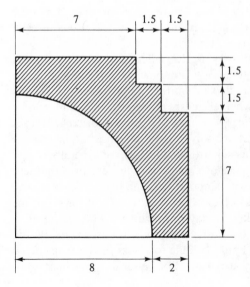

Figure 6-20 Heat transfer region, problem 6-6.

6-7 Solve the heat transfer problem illustrated in Fig. 6-21 using a Galerkin finite element code.

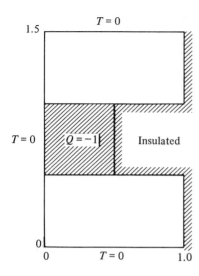

Figure 6-21 Heat transfer with heat generation, problem 6-7.

6-8 Consider the rectangle with two triangles shown in Fig. 6-22. The local array is

$$2A_{IJ} = \begin{pmatrix} -2 & 1 & 1 \\ 1 & -1 & 0 \\ 1 & 0 & -1 \end{pmatrix}$$

Construct the global array A_{ij} with $i, j = 1, 4$.

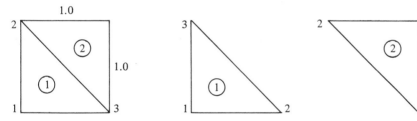

Figure 6-22 Local and global triangles, problem 6-8.

BIBLIOGRAPHY

Comparisons of alternating direction implicit, line successive overrelaxation, and strongly implicit procedure methods as applied in petroleum engineering are provided by Weinstein, *et al.*[16] and by

Briggs, J. E., and T. N. Dixon: "Some Practical Considerations in Numerical Solution of Two-Dimensional Reservoir Problems," *Soc. Pet. Eng. J.*, vol. 8, pp. 185–194, 1968.

Bjordammen, J., and K. H. Coats: "Comparison of Alternating-Direction (ADI) and Successive Overrelaxation (SOR) Techniques in Simulation of Reservoir Fluid Flow," *Soc. Pet. Eng. J.*, vol. 9, pp. 47–58, 1969.

Weinstein, H. G., H. L. Stone, and T. V. Kwan: "Simultaneous Solution of Multiphase Reservoir Flow Equations," *Soc. Pet. Eng. J.*, vol. 10, pp. 99–110, 1970.

Orthogonal collocation on finite elements was first applied to two-dimensional problems by

>Prenter, P. M., and R. D. Russell: "Orthogonal Collocation for Elliptic Partial Differential Equations," *SIAM J. Num. Anal.*, vol. 13, pp. 923–939, 1976.

The alternating direction method of solution was advanced by

>Chang, P. W., and B. A. Finlayson: "Orthogonal Collocation on Finite Elements," *Math. Comp. Sim.*, vol. 20, pp. 83–92, 1978.

One method of treating irregular domains is by truncating the domain in an element and adding collocation points on the boundary, as described by

>Houstis, E. N., and J. R. Rice: "Software for Linear Elliptic Problems on General Two-Dimensional Domains," in R. Vichnevetsky (ed.) *Advances in Computer Methods for Partial Differential Equations—II*, Int. Assoc. for Maths. and Computers in Simulation 1977, pp. 7–12.

Another method of treating irregular domains in collocation is to use deformable elements. Isoparametric elements are used by

>Frind, E. O., and G. F. Pinder: "A Collocation Finite Element Method for Potential Problems in Irregular Domains," *Int. J. Num. Methods Eng.*, vol. 14, pp. 681–701, 1979.

The Galerkin finite element method for two-dimensional problems is widely used in engineering analysis and there are many books describing the details and applications. Two good ones oriented towards fluid mechanics or heat transfer are by Huebner[6] and

>Chung, T. J.: *Finite Element Analysis in Fluid Dynamics*, McGraw-Hill, New York, 1978.

Applications to hydrology are treated by

>Pinder, G. F., and W. G. Gray: *Finite Element Simulation in Surface and Subsurface Hydrology*, Academic Press, New York, 1977.

The classic book, with an emphasis on structural problems in civil engineering, is

>Zienkiewicz, O. C.: *The Finite Element Method in Engineering Science*, McGraw-Hill, New York, 1971.

One of the valuable features of the finite element method is that the computer codes are organized so that information is processed in an element-by-element basis. This allows application to arbitrary geometries that are easily specified by the analyst. The matrices can be processed using an LU decomposition for a banded matrix, but economy is realized using a frontal or profile solver. This also makes possible the storage of only part of the matrix in the computer core at any one time, and permits the solution of larger problems. The frontal routine is given by

>Hood, P.: "Frontal Solution Program for Unsymmetric Matrices," *Int. J. Num. Methods Eng.*, vol. 10, pp. 379–399, 1976; vol. 11, p. 1202, 1977.

REFERENCES

1. Chang, P. W., T. W. Patten, and B. A. Finlayson: "Collocation and Galerkin Finite Element Methods for Viscoelastic Fluid Flow. Part I. Description of Method and Problems with Fixed Geometry," *Comp. Fluids*, vol. 7, pp. 267–283, 1979.
2. Chang, P. W., T. W. Patten, and B. A. Finlayson: "Collocation and Galerkin Finite Element Methods for Viscoelastic Fluid Flow. Part II. Die Swell Problems with a Free Surface," *Comp. Fluids*, vol. 7, pp. 285–293, 1979.
3. Crichlow, H. B.: *Modern Reservoir Engineering—A Simulation Approach*, Prentice-Hall, Englewood Cliffs, 1977.
4. Finlayson, B. A.: *The Methods of Weighted Residuals and Variational Principles*, Academic Press, New York, 1972.
5. Houstis, E. N., R. E. Lynch, T. S. Papatheodorou, and J. R. Rice: "Evaluation of Numerical Methods for Elliptic Partial Differential Equations," *J. Comp. Phys.*, vol. 27, pp. 323–350, 1978.
6. Huebner, K. H.: *The Finite Element Method for Engineers*, John Wiley & Sons, Inc., New York, 1975.
7. Peaceman, D. W.: *Fundamentals of Numerical Reservoir Simulation*, Elsevier, New York, 1977.

8. Settari, A., H. S. Price, and T. Dupont: "Development and Application of Variational Methods for Simulation of Miscible Displacement in Porous Media," in *Fourth Soc. Pet. Eng. Symposium on Numerical Simulation of Reservoir Performance*, Los Angeles, Calif., Feb. 19–20, 1976, pp. 42–67.

9. Shih, F. S.: "On the Temperature Field of a Square Column Embedding a Heating Cylinder," *A.I.Ch.E. J.*, vol. 16, pp. 134–138; vol. 16, p. 1109, 1970.

10. Spivak, A., H. S. Price, and A. Settari: "Solution of the Equations for Multi-Dimensional Two-Phase Immiscible Flow by Variational Methods," in *Fourth Soc. Pet. Eng. Symposium on Numerical Simulation of Reservoir Performance*, Los Angeles, Calif., Feb. 19–20, 1976, pp. 69–91.

11. Young, D.: "Iterative Methods for Solving Partial Difference Equations of Elliptic Type," *Trans. Am. Math. Soc.*, vol. 76, pp. 92–111, 1954.

12. Young, L. C.: "An Efficient Finite Element Method for Reservoir Simulation," in *53rd Annual Fall Meeting, Soc. Pet. Eng.*, Houston, Texas, Oct. 1–3, 1978, paper SPE 7413.

13. Young, L. C., and B. A. Finlayson: "Axial Dispersion in Nonisothermal Packed Bed Chemical Reactors," *Ind. Eng. Chem. Fund.*, vol. 12, pp. 412–422, 1973.

14. Young, L. C., and B. A. Finlayson: "Mathematical Models of the Monolith Catalytic Converter. Part II. Application to Automobile Exhaust," *A.I.Ch.E. J.*, vol. 22, pp. 343–353, 1976.

15. Weinberger, H. F.: *A First Course in Partial Differential Equations with Complex Variables and Transformation Methods*, Blaisdell, New York, 1965.

16. Weinstein, H. G., H. L. Stone, and T. V. Kwan: "Iterative Procedure for Solution of Systems of Parabolic and Elliptic Equations in Three Dimensions," *Ind. Eng. Chem. Fund.*, vol. 8, pp. 281–287, 1969.

APPENDIX

COMPUTER PROGRAMS

```
CCC      SUBROUTINE INVR(A,N,NI)                                          A   1
CCC      THIS SUBROUTINE CALLS DECOMP ONCE AND SOLVE SEVERAL TIMES        A   2
CCC      TO CONSTRUCT THE INVERSE.  IT SOLVES AX=I WHERE EACH MATRIX      A   3
CCC      IS N BY N.                                                       A   4
CCC      INPUT                                                            A   5
CCC         A(N,N) - AN N BY N ARRAY, STORED IN MATRIX WITH DIMENSIONS    A   6
CCC                  NI BY NI                                             A   7
CCC         N - THE SIZE OF THE MATRIX TO BE INVERTED, <=20              A   8
CCC         NI - THE SIZE OF THE DIMENSION OF A                          A   9
CCC      OUTPUT                                                           A  10
CCC         A(N,N) - ON OUTPUT THIS IS THE INVERSE OF THE ORIGINAL A      A  11
CCC                  THE ORIGINAL A IS DESTROYED                          A  12
CCC      DIMENSION A(NI,NI), B(20), C(400)                                A  13
CCC      PACK A DENSELY INTO C                                            A  14
         DO 5 J=1,N                                                       A  15
            IND = N*(J-1)                                                 A  16
            DO 5 I=1,N                                                    A  17
               C(IND+I) = A(I,J)                                          A  18
    5    CONTINUE                                                         A  19
CCC      PERFORM AN LU DECOMPOSITION ON C                                 A  20
         CALL INVERT (N,1,C,B,1)                                          A  21
CCC      SOLVE AX=I                                                       A  22
         DO 20 J=1,N                                                      A  23
            DO 10 I=1,N                                                   A  24
               B(I) = 0.                                                  A  25
   10          B(J) = 1.                                                  A  26
            CALL INVSW (N,1,C,B,1)                                        A  27
CCC      PUT B INTO INVERSE A                                             A  28
            DO 15 I=1,N                                                   A  29
               A(I,J) = B(I)                                              A  30
   15       CONTINUE                                                      A  31
```

```
   20 CONTINUE                                                                      A 32
      RETURN                                                                        A 33
      END                                                                           A 34-
      SUBROUTINE INVERT(N,NE,A,B,ITYPE)                                             B  1
      DIMENSION A(1000,3), B(N), A1(1000), B1(1000), C1(1000)                       B  2
CCC   THIS SUBROUTINE CALLS DECOMP, LUDECOM, OR INVTRI TO PREFORM LOWER             B  3
CCC   DECOMPOSITICN DEPENDING ON WHETHER THE MATRIX A IS                            B  4
CCC        ITYPE        A                                                           B  5
CCC          1        DENSE                                                         B  6
CCC          2    BLOCK TRIDIAGONAL                                                 B  7
CCC          3     TRIDIAGONAL                                                      B  8
CCC   IF A IS DENSE THE MATRIX IS STORED AS A( N,N )                                B  9
CCC   IF A IS BLOCK DIAGONAL THE MATRIX IS STORED AS A ( N,N,NE )                   B 10
CCC   IF A IS TRIDIAGONAL THE ELEMENTS A,B,C ARE STORED AS A(N,1) = A(N),           B 11
CCC        A(N,2) = B(N), A(N,3) = C(N)                                             B 12
CCC   N IS LIMITED TO 2C1 WITHOUT CHANGING DIMENSION STATEMENT FOR A(N),            B 13
CCC        B(N), C(N)                                                               B 14
CCC   THE MATRICES ARE DIMENSIONED AS A(N,N), A(N,N,NE) OR A(N,3)                   B 15
      GO TO (5,10,15), ITYPE                                                        B 16
    5 CALL DECOMP (N,A)                                                             B 17
      RETURN                                                                        B 18
   10 NP = (N-1)/NE+1                                                               B 19
      CALL LUDECOM (NP,NE,A)                                                        B 20
      RETURN                                                                        B 21
   15 DO 20 K=1,N                                                                   B 22
      A1(K) = A(K,1)                                                                B 23
      B1(K) = A(K,2)                                                                B 24
      C1(K) = A(K,3)                                                                B 25
   20 CONTINUE                                                                      B 26
      CALL INVTRI (N,A1,B1,C1)                                                      B 27
      DO 25 K=1,N                                                                   B 28
```

```
          A(K,1) = A1(K)                                              B 29
          A(K,2) = B1(K)                                              B 30
          A(K,3) = C1(K)                                              B 31
   25 CONTINUE                                                        B 32
      RETURN                                                          B 33
      ENTRY INVSW                                                     B 34
CCC THIS PORTION OF SUBROUTINE DOES THE FORWARD AND BACKWARD SWEEP TC B 35
CCC SOLVE AX = B, WITH THE X STORED IN B AND INVR MUST BE CALLED FIRST B 36
      GO TO (30,35,40), ITYPE                                        B 37
   30 CALL SOLVE (N,A,B)                                             B 38
      RETURN                                                          B 39
   35 CALL FAS (NP,NE,N,A,B)                                         B 40
      RETURN                                                          B 41
   40 CALL SWEEP (N,A1,B1,C1,B)                                      B 42
      RETURN                                                          B 43
      END                                                            B 44-
      SUBROUTINE SING(I)                                             C  1
      GO TO (5,15), I                                                C  2
    5 WRITE (6,10)                                                   C  3
   10 FORMAT (////,* MATRIX WITH ZERO ROW IN DECOMPOSE *////)        C  4
      RETURN                                                          C  5
   15 WRITE (6,20)                                                   C  6
   20 FORMAT (////,* SINGULAR MATRIX IN DECOMPOSE. ZERO DIVIDE IN SOLVE C  7
     .  *////)                                                       C  8
      $                                                              C  9
      RETURN                                                         C 10-
      END                                                            D  1
      SUBROUTINE DECOMP(N,A)                                         D  2
      DIMENSION A(N,N)                                               D  3
      COMMON /DENSE/ IPS(201),SC(201)                                D  4
CCC PAGE 68, FORSYTH AND MOLER                                       D  5
CCC INITIALIZE IPS, A AND SCALES
```

319

```
      IF (N.EQ.1) RETURN                                          D  6
      DO 25 I=1,N                                                 D  7
      IPS(I) = I                                                  D  8
      ROWNRM = 0.0                                                D  9
      DO 10 J=1,N                                                 D 10
      IF (ROWNRM-ABS(A(I,J))) 5,10,10                             D 11
      ROWNRM = ABS(A(I,J))                                        D 12
   10 CONTINUE                                                    D 13
      IF (ROWNRM) 15,20,15                                        D 14
   15 SC(I) = 1./ROWNRM                                           D 15
      GO TO 25                                                    D 16
   20 CALL SING (1)                                               D 17
      SC(I) = 0.                                                  D 18
   25 CONTINUE                                                    D 19
CCC GAUSIAN ELIMINATION WITH PARTIAL PIVOTING                     D 20
      NM1 = N-1                                                   D 21
      DO 65 K=1,NM1                                               D 22
      BIG = 0.                                                    D 23
      DO 35 I=K,N                                                 D 24
      IP = IPS(I)                                                 D 25
      SIZE = ABS(A(IP,K)*SC(IP))                                  D 26
      IF (SIZE-BIG) 35,35,30                                      D 27
   30 BIG = SIZE                                                  D 28
      IDXPIV = I                                                  D 29
   35 CONTINUE                                                    D 30
      IF (BIG) 45,40,45                                           D 31
   40 CALL SING (2)                                               D 32
      GO TO 65                                                    D 33
   45 IF (IDXPIV-K) 50,55,50                                      D 34
                                                                  D 35
                                                                  D 36
```

```
 50      J = IPS(K)                                          D 37
         IPS(K) = IPS(IDXPIV)                                D 38
         IPS(IDXPIV) = J                                     D 39
 55      KP = IPS(K)                                         D 40
         PIVOT = A(KP,K)                                     D 41
         KP1 = K+1                                           D 42
         DO 60 I=KP1,N                                       D 43
         IP = IPS(I)                                         D 44
         EM = -A(IP,K)/PIVOT                                 D 45
         A(IP,K) = -EM                                       D 46
         DO 60 J=KP1,N                                       D 47
         A(IP,J) = A(IP,J)+EM*A(KP,J)                        D 48
CCC INNER LOOP                                               D 49
 60      CONTINUE                                            D 50
 65      CONTINUE                                            D 51
         KP = IPS(N)                                         D 52
         IF (A(KP,N)) 75,70,75                               D 53
 70      CALL SING (2)                                       D 54
 75      CONTINUE                                            D 55
         RETURN                                              D 56
         END                                                 D 57
         SUBROUTINE SOLVE(N,A,B)                             E 1
         DIMENSION A(N,N), B(N)                              E 2
         COMMON /DENSE/ IPS(201),SC(201)                     E 3
CCC PAGE 69, FORSYTH AND MOLER                               E 4
CCC FORWARD SWEEP DENSE MATRIX                               E 5
         IF (N.GT.1) GO TO 5                                 E 6
         B(1) = B(1)/A(1,1)                                  E 7
         RETURN                                              E 8
 5       CONTINUE                                            E 9
         NP1 = N+1                                           E 10
```

```
E 11        IP = IPS(1)
E 12        SC(1) = B(IP)
E 13        DO 15 I=2,N
E 14        IP = IPS(I)
E 15        IM1 = I-1
E 16        SUM = 0.
E 17        DO 10 J=1,IM1
E 18        SUM = SUM+A(IP,J)*SC(J)
E 19  10    CONTINUE
E 20        SC(I) = B(IP)-SUM
E 21  15 CONTINUE
E 22 CCC BACK SUBSTITUTION
E 23        IP = IPS(N)
E 24        SC(N) = SC(N)/A(IP,N)
E 25        DO 25 IBACK=2,N
E 26        I = NP1-IBACK
E 27        IP = IPS(I)
E 28        IP1 = I+1
E 29        SUM = 0.
E 30        DO 20 J=IP1,N
E 31        SUM = SUM+A(IP,J)*SC(J)
E 32  20    CONTINUE
E 33        SC(I) = (SC(I)-SUM)/A(IP,I)
E 34  25 CONTINUE
E 35        DO 30 I=1,N
E 36  30    B(I) = SC(I)
E 37     RETURN
E 38-    END
F 1    SUBROUTINE FAS(NP,NE,NT,A,B)
F 2    DIMENSION A(NP,NP,NE), B(NT)
F 3 CCC FORWARD SWEEP BLOCK DIAGONAL MATRIX
```

```
                                                              F 4
      NP1 = NP                                                F 5
      DO 10 L=1,NE                                            F 6
        DO 10 I=2,NP1                                         F 7
          I2 = (L-1)*(NP-1)+I                                 F 8
          S = 0.                                              F 9
          I1 = I-1                                            F 10
          DO 5 J=1,I1                                         F 11
            J2 = I2-I+J                                       F 12
            S = S+A(I,J,L)*B(J2)                              F 13
    5     CONTINUE                                            F 14
          B(I2) = B(I2)-S                                     F 15
   10 CONTINUE                                                F 16
CCC BACK SUBSTITUTION                                         F 17
      DO 25 L1=1,NE                                           F 18
        L = NE-L1+1                                           F 19
        IF (L.NE.NE) GO TO 15                                 F 20
        B(NT) = B(NT)/A(NP,NP,NE)                             F 21
   15   N1 = NP-1                                             F 22
        DO 25 K=1,N1                                          F 23
          I = N1+1-K                                          F 24
          I2 = (L-1)*(NP-1)+I                                 F 25
          M = I+1                                             F 26
          N2 = N1+1                                           F 27
          S = 0.                                              F 28
          DO 20 J=M,N2                                        F 29
   20       J2 = (L-1)*(NP-1)+J                               F 30
            S = S+A(I,J,L)*B(J2)                              F 31
          B(I2) = (B(I2)-S)/A(I,I,L)                          F 32
   25 CONTINUE                                                F 33
      RETURN                                                  F 34-
      END
```

```
      SUBROUTINE LUDECM(NP,NE,A)
      DIMENSION A(NP,NP,NE)
CCC LOWER DECOMPOSITION  BLOCK DIAGONAL MATRIX
      N1 = NP-1
      DO 10 L=1,NE
      DO 5 K=1,N1
      K1 = K+1
      DO 5 I=K1,NP
      S = A(I,K,L)/A(K,K,L)
      A(I,K,L) = S
      DO 5 J=K1,NP
      A(I,J,L) = A(I,J,L)-S*A(K,J,L)
    5 CONTINUE
      IF (L.EQ.NE) RETURN
      A(1,1,L+1) = A(NP,NP,L)
   10 CONTINUE
      END
      SUBROUTINE INVTRI(N,A,B,C)
      DIMENSION A(N), B(N), C(N)
CCC LOWER DECOMPOSITICN  TRIDIAGONAL MATRIX
CCC SOLVES A(I-1)*T(I-1)+B(I)*T(I)+C(I+1)*T(I+1) = D(I)
      DO 5 L=2,N
      S = A(L)/B(L-1)
      B(L) = B(L)-S*C(L-1)
    5 A(L) = S
      RETURN
      END
      SUBROUTINE SWEEP(N,A,B,C,D)
      DIMENSION A(N), B(N), C(N), D(N)
CCC FORWARD SWEEP  TRIDIAGONAL MATRIX
      DO 5 L=2,N
```

G 1
G 2
G 3
G 4
G 5
G 6
G 7
G 8
G 9
G 10
G 11
G 12
G 13
G 14
G 15
G 16
G 17—
H 1
H 2
H 3
H 4
H 5
H 6
H 7
H 8
H 9
H 10—
I 1
I 2
I 3
I 4

```
      5 D(L) = D(L)-A(L)*D(L-1)                                      I  5
CCC   BACK SUBSTITUTION                                             I  6
        D(N) = D(N)/B(N)                                            I  7
        DO 10 L=2,N                                                 I  8
          K = N-L+1                                                 I  9
     10 D(K) = (D(K)-C(K)*D(K+1))/B(K)                              I 10
        RETURN                                                      I 11
        END                                                         I 12-


      SUBROUTINE COLL(A,B,Q,X,W,ND,N,AA)                            A  1
      DIMENSION A(ND,ND), B(ND,ND), Q(ND,ND), X(ND), W(ND)         A  2
      DIMENSION QINV(7,7), Z(7), C(7,7), D(7,7)                    A  3
CCC   THIS SUBROUTINE COMPUTES THE MATRICES FOR ORTHOGONAL         A  4
CCC   COLLOCATION USING SYMMETRIC POLYNOMIALS, TABLE 4-5,4-6       A  5
CCC                                                                A  6
CCC   INPUT VARIABLES                                              A  7
CCC    N = NUMBER OF INTERIOR COLLOCATION POINTS                   A  8
CCC    ND = ARRAY DIMENSION OF MATRICES IN CALLING PROGRAM         A  9
CCC    AA = GEOMETRY FACTOR                                        A 10
CCC       = 1 PLANAR                                               A 11
CCC       = 2 CYLINDRICAL                                          A 12
CCC       = 3 SPHERICAL                                            A 13
CCC                                                                A 14
CCC   OUTPUT VARIABLES                                             A 15
CCC    A = MATRIX FOR FIRST DERIVATIVE, EQ. 4-205                  A 16
CCC    B = MATRIX FOR LAPLACIAN, EQ. 4-205                         A 17
CCC    Q = MATRIX FOR Q INVERSE, EQ. 4-203                         A 18
CCC    X = VECTOR OF COLLOCATION POINTS, TABLE 4-5                 A 19
```

```fortran
CCC    W = VECTOR OF WEIGHTS, EQ. 4-207
CCC

      DIMENSION X1(6,6), X2(6,6), X3(6,6)
      DATA (X1(I),I=1,36)/.4472135955,0.0000000000,0.0000000000,0.
     $0000000000,0.0000000000,0.0000000000,.3399810436,.8611363116,0.
     $6612093865,.9246951142,0.0000000000,0.0000000000,.2386191861,0.
     $1834346425,.5255324099,.7966647774,.9602898565,0.0000000000,0.
     $0000000000,.1488743390,.4333953941,.6794095683,.8650633667,.
     $9739065285,0.0000000C,.1252334085,.3678314990,.5873179543,.
     $7699026742,.9041172564,.9815606342/
      DATA (X2(I),I=1,36)/.5773502692,0.0000000000,0.0000000000,0.
     $0000000000,0.0000000000,0.0000000000,.4597008434,.8880738340,0.
     $0000000000,0.0000000000,0.0000000000,0.0000000000,.3571106870,.
     $7071067812,.9419651451,0.0000000000,0.0000000000,0.0000000000,0.
     $2634992300,.5744645143,.8185294874,.9646596062,0.0000000000,0.
     $0000000000,.2165873427,.4803804169,.7071067812,.8770602346,.
     $9762632447,0.0000000000,.1837532119,.4115766111,.6170011402,.
     $7869622564,.9113751660,.9829724091/
      DATA (X3(I),I=1,36)/.6546536707,0.0000000000,0.0000000000,0.
     $0000000000,0.0000000000,0.0000000000,.5384693101,.9061798459,0.
     $0000000000,0.0000000000,0.0000000000,0.0000000000,.4058451514,.
     $7415311856,.9491079123,0.0000000000,0.0000000000,0.0000000000,0.
     $3242534234,.6133714327,.8360311073,.9681602395,0.0000000000,0.
     $0000000000,.2695431560,.5190961292,.7301520056,.8706625998,.
     $9782286581,0.0000000000,.2304583160,.4484927510,.6423493394,.
     $8015780907,.9175983992,.9841830547/
      WRITE (6,5)
    5 FORMAT (* THE COLLOCATION POINTS ARE*/)
      N1 = N+1
      DO 25 LL=1,N
```

```
      IA = AA+0.001                             A  51
      GO TO (10,15,20), IA                      A  52
   10 X(LL) = X1(LL,N)                          A  53
      GO TO 25                                  A  54
C                                               A  55
   15 X(LL) = X2(LL,N)                          A  56
      GO TO 25                                  A  57
C                                               A  58
   20 X(LL) = X3(LL,N)                          A  59
   25 CONTINUE                                  A  60
      X(N1) = 1.0                               A  61
      DO 30 I=1,N1                              A  62
      AI = I                                    A  63
      Z(I) = 1./(2.*AI+AA-2.)                   A  64
   30 CONTINUE                                  A  65
      DO 35 I=1,N1                              A  66
      Q(I,1) = 1.                               A  67
      QINV(I,1) = 1.                            A  68
      DO 35 J=2,N1                              A  69
      Q(I,J) = X(I)**(2*J-2)                    A  70
   35 QINV(I,J) = Q(I,J)                        A  71
      DO 40 J=1,N1                              A  72
      CA = 2.*J-2.                              A  73
      DA = (2.*J-2.)*(2.*J+AA-4.)               A  74
      DO 40 I=1,N1                              A  75
      C(I,J) = CA*X(I)**(2*J-3)                 A  76
      D(I,J) = DA*X(I)**(2*J-4)                 A  77
   40 CONTINUE                                  A  78
      CALL INVR (QINV,N1,7)                     A  79
      DO 50 I=1,N1                              A  80
      W(I) = 0.0                                A  81
```

```
          DO 50 J=1,N1                                                    A 82
            A(I,J) = B(I,J) = 0.0                                         A 83
            DO 45 K=1,N1                                                  A 84
              A(I,J) = A(I,J)+C(I,K)*QINV(K,J)                            A 85
              B(I,J) = B(I,J)+D(I,K)*QINV(K,J)                            A 86
   45       CONTINUE                                                      A 87
          Q(I,J) = QINV(I,J)                                              A 88
          W(I) = W(I)+Z(J)*QINV(J,I)                                      A 89
   50   CONTINUE                                                          A 90
        WRITE (6,55) (X(I),I=1,N1)                                        A 91
   55   FORMAT (7E15.5//)                                                 A 92
        RETURN                                                            A 93
        END                                                              A 94-
        SUBROUTINE PLANAR(A,B,Q,X,W,N,NX)                                 B 1
        DIMENSION A(N,N), B(N,N), Q(N,N), X(N), W(N), ZU(7)               B 2
        DIMENSION R(30,30), S(30)                                         B 3
        DIMENSION POINT(5,3)                                              B 4
  CCC   THIS SUBROUTINE COMPUTES THE MATRICES FOR COLLOCATION WITHOUT     B 5
  CCC   SYMMETRY, USING COLLOCATION POINTS IN TABLE 4-3.                  B 6
  CCC   INPUT VARIABLES                                                   B 7
  CCC     NX = NUMBER OF COLLOCATION POINTS, INCLUDING TWO END POINTS     B 8
  CCC     N = ARRAY DIMENSIONS OF MATRICES IN CALLING PROGRAM             B 9
  CCC                                                                     B 10
  CCC   OUTPUT VARIABLES                                                  B 11
  CCC     A = MATRIX FOR FIRST DERIVATIVE, EQ. 4-103                      B 12
  CCC     B = MATRIX FOR SECOND DERIVATIVE, EQ. 4-103                     B 13
  CCC     Q = MATRIX FOR Q INVERSE, EQ. 4-101                             B 14
  CCC     X = VECTOR OF COLLOCATIONS POINTS, FROM TABLE 4-3               B 15
  CCC     W = VECTOR OF WEIGHTS, EQ. 4-106                                B 16
  CCC                                                                     B 17
        DATA (POINT(I),I=1,15)/0.,0.57735026918962o,0.77459666924l483,0.  B 18
```

328

```
 8 B 19      $8611363115940053,0.9061798459386644,3*0.,0.3399810435848560.
 8 B 20      $538469310105683,5*0./
 8 B 21      NCOL = NX-2
 8 B 22      IF (NX.GT.7) GO TO 35
 8 B 23      JRT = (NCOL+1)/2
 8 B 24      DO 5 J=1,JRT
 8 B 25    5 ZU(J) = POINT(NCCL,J)
 8 B 26      J = 1
 8 B 27      DO 10 I=1,JRT
 8 B 28      X(J) = (1.0-ZU(I))/2.
 8 B 29      X(J+1) = (1.0+ZU(I))/2.
 8 B 30   10 J = J+2
 8 B 31 C    BUBBLE SORT ON COLLECATION POINTS
 8 B 32      NC1 = NCOL-1
 8 B 33      DO 20 J=1,NC1
 8 B 34      I = J
 8 B 35   15 IF (X(I+1).GT.X(I)) GO TO 20
 8 B 36      STOR = X(I)
 8 B 37      X(I) = X(I+1)
 8 B 38      X(I+1) = STOR
 8 B 39      IF (I.EQ.1) GC TO 20
 8 B 40      I = I-1
 8 B 41      GO TO 15
 8 B 42 C
 8 B 43   20 CONTINUE
 8 B 44      WRITE (6,25) (I,X(I),I=1,NCOL)
 8 B 45   25 FORMAT (3(/),10X,*COLLOCATION PTS*,/,5X,*POINT*,5X,*ORDINATE*,/,
 8 B 46      $(5X,I5,5X,E15.8))
 8 B 47      NC1 = NCOL+1
 8 B 48      DO 30 I=2,NC1
 8 B 49      K = NC1-I+2
```

329

```
30  X(K) = X(K-1)                                      B 50
    X(1) = 0.0                                         B 51
    X(NX) = 1.0                                        B 52
35  DO 50 I=1,NX                                       B 53
    R(I,I) = 0.0                                       B 54
    A(I,I) = 0.0                                       B 55
    S(I) = 1.0                                         B 56
    B(I,I) = 0.0                                       B 57
    DO 40 J=1,NX                                       B 58
    IF (I.EQ.J) GO TO 40                               B 59
    R(I,J) = 1.0/(X(I)-X(J))                           B 60
    S(I) = S(I)*R(I,J)                                 B 61
40      CONTINUE                                       B 62
    DO 45 J=1,NX                                       B 63
    JX = NX-J+1                                        B 64
    IF (JX.LT.J) GO TO 50                              B 65
    IF (JX.EQ.J) A(I,I) = A(I,I)+R(I,J)                B 66
    IF (JX.GT.J) A(I,I) = A(I,I)+R(I,J)+R(I,JX)        B 67
45      CONTINUE                                       B 68
50  CONTINUE                                           B 69
    DO 60 I=1,NX                                       B 70
    DO 55 J=1,NX                                       B 71
    IF (I.EQ.J) GO TO 55                               B 72
    A(I,J) = S(J)*R(I,J)/S(I)                          B 73
    B(I,J) = 2.0*A(I,J)*(A(I,I)-R(I,J))                B 74
    B(I,I) = B(I,I)+R(I,J)*(A(I,I)-R(I,J))             B 75
55      CONTINUE                                       B 76
60  CONTINUE                                           B 77
    DO 85 I=1,NX                                       B 78
    Q(1,I) = S(I)                                      B 79
    K = 1                                              B 80
```

```
      W(I) = 0.0                                                B 81
      DO 75 J=1,NX                                              B 82
      IF (J.EQ.I) GO TO 75                                      B 83
      L = K                                                     B 84
      K = K+1                                                   B 85
      Q(K,I) = Q(L,I)                                           B 86
      IF (L.EQ.1) GO TO 70                                      B 87
      M = L-1                                                   B 88
      Q(L,I) = Q(M,I)-X(J)*Q(L,I)                               B 89
      L = M                                                     B 90
      GO TO 65                                                  B 91
C                                                               B 92
65    Q(1,I) = -X(J)*Q(1,I)                                     B 93
70    CONTINUE                                                  B 94
75    DC 80 J=1,NX                                              B 95
80    W(I) = W(I)+Q(J,I)/FLOAT(J)                               B 96
85    CONTINUE                                                  B 97
      RETURN                                                    B 98
      END                                                       B 99-

      PROGRAM OCRXN (INPUT,OUTPUT,TAPE5=INPUT,TAPE6=OUTPUT)     A 1
      DIMENSION A(7,7), B(7,7), Q(7,7), XC(7), W(7), AA(49), D(7), TH(7)   A 2
      COMMON /RXN/ PAR(8)                                       A 3
CCC   THIS PROGRAM USES ORTHOGONAL COLLOCATION TO SOLVE         A 4
CCC                                                             A 5
CCC   DEL**2 C = PHI*PHI*R(C)                                   A 6
CCC                                                             A 7
CCC   -DC/DR = BIM*(C - 1 ) AT R = 1                            A 8
CCC                                                             A 9
```

331

```
CCC   VARIABLES
CCC
CCC      N = NUMBER OF INTERIOR COLLOCATION POINTS
CCC      AS = GEOMETRY FACTOR
CCC         = 1 PLANAR
CCC         = 2 CYLINDRICAL
CCC         = 3 SPHERICAL
CCC      PHI = THIELE MODULUS
CCC      BIM = BIOT NUMBER FOR MASS TRANSFER
CCC      CGUESS = INITIAL GUESS FOR C(R), A CONSTANT
CCC   INTERACTIVE VERSION
      CALL CONNEC (5)
      CALL CONNEC (6)
    5 WRITE (6,10)
   10 FORMAT (* ENTER N, A, PHI,BIM,CGUESS *)
      READ *,N,AS,PHI,BIM,CGUESS
      WRITE (6,15)
   15 FORMAT (* ENTER FOUR REACTION RATE PARAMETERS *)
      READ *,PAR(1),PAR(2),PAR(3),PAR(4)
      N1 = N+1
      N2 = N*N
      DELTA = PHI*PHI
CCC   SET INITIAL CONDITION
      DO 20 I=1,N1
   20 TH(I) = CGUESS
CCC   CALCULATE MATRICES
      CALL COLL (A,B,Q,XC,W,7,N,AS)
CCC   BEGIN ITERATION
      DO 50 ITER=1,20
CCC   SET THE MATRICES
      DO 25 I=1,N2
```

A 10
A 11
A 12
A 13
A 14
A 15
A 16
A 17
A 18
A 19
A 20
A 21
A 22
A 23
A 24
A 25
A 26
A 27
A 28
A 29
A 30
A 31
A 32
A 33
A 34
A 35
A 36
A 37
A 38
A 39
A 40

```
25          AA(I) = 0.
            DO 35 J=1,N                                                          A 41
            CALL RXN (TH(J),RATE,DR)                                             A 42
            D(J) = DELTA*(RATE-DR*TH(J))                                         A 43
            D(J) = D(J)-BIM*B(J,N1)/(A(N1,N1)+BIM)                               A 44
            DO 30 I=1,N                                                          A 45
            KN = N*(I-1)+J                                                       A 46
            AA(KN) = B(J,I)-B(J,N1)*A(N1,I)/(A(N1,N1)+BIM)                       A 47
            IF (I.EQ.J) AA(KN) = AA(KN)-DELTA*DR                                 A 48
30          CONTINUE                                                            A 49
35          CONTINUE                                                            A 50
CCC    DO THE LU DECOMPOSITION                                                  A 51
            CALL INVERT (N,1,AA,D,1)                                            A 52
CCC    SOLVE FOR THE RIGHT HAND SIDE                                            A 53
            CALL INVSW (N,1,AA,D,1)                                             A 54
CCC    FIND MAXIMUM CHANGE IN SOLUTION                                          A 55
            ER = 0.                                                            A 56
            DO 40 I=1,N                                                         A 57
            ERR = ABS(TH(I)-D(I))                                              A 58
            TH(I) = D(I)                                                        A 59
            IF (ERR.GT.ER) ER = ERR                                            A 60
40          CONTINUE                                                            A 61
            WRITE (6,45) ITER,ER                                                A 62
45          FORMAT (* ITERATION *,I3,*    ERROR IS *,E15.4)                     A 63
            IF (ER.LT.1.E-6) GO TO 55                                           A 64
50          CONTINUE                                                            A 65
CCC    CALCULATE THE EFFECTIVENESS FACTOR                                       A 66
CCC    ETA1 USES EQ. 4-228                                                      A 67
55          SUM = 0.                                                           A 68
            DO 60 I=1,N                                                         A 69
60          SUM = SUM+A(N1,I)*D(I)                                             A 70
                                                                               A 71
```

```
      D(N1) = (BIM-SUM)/(A(N1,N1)+BIM)                         A  72
      SUM = 0.                                                 A  73
      SUM1 = 0.                                                A  74
      DO 65 I=1,N1                                             A  75
      CALL RXN (D(I),RATE,DR)                                  A  76
      SUM1 = SUM1+W(I)                                         A  77
   65 SUM = SUM+W(I)*RATE                                      A  78
      ETA1 = SUM/SUM1                                          A  79
      WRITE (6,70) ETA1                                        A  80
   70 FORMAT (* EFF. FACTOR *,F20.15)                          A  81
      WRITE (6,75)                                             A  82
   75 FORMAT (* DO YOU WANT TO SEE THE SOLUTION */,* IF SO ENTER 1, OTH  A  83
     $ERWISE 0*)                                               A  84
      READ *,KON                                               A  85
      IF (KON.EQ.0) GO TO 5                                    A  86
      WRITE (6,80) (D(I),I=1,N1)                               A  87
   80 FORMAT (25(4F10.6,/))                                    A  88
      GO TO 5                                                  A  89
CCC   EXIT BY ENTERING %A                                      A  90
      STOP                                                     A  91
      END                                                      A  92
      SUBROUTINE RXN(C,R,DR)                                   B   1
      COMMON /RXN/ PAR(8)                                      B   2
CCC   PAR(1) = BETA    PAR(2) = GAMMA                          B   3
      T = 1.+PAR(1)*(1.-C)                                     B   4
      E = EXP(PAR(2)*(1.-1./T))                                B   5
CCC   THIS SUBROUTINE COMPUTES R AND DR/DC, GIVEN C            B   6
      R = C*E                                                  B   7
      DR = E*(1.-PAR(1)*C*PAR(2)/T**2)                         B   8
      RETURN                                                   B   9
      END                                                      B  10
```

334

```
      PROGRAM FDRXN (INPUT,OUTPUT,TAPE5=INPUT,TAPE6=OUTPUT)          A  1
      DIMENSION A(1000), B(1000), C(1000), TH(1000), D(1000), AA(1000,3)  A  2
      DIMENSION R(1000)                                              A  3
      EQUIVALENCE (A(1),AA(1,1)), (B(1),AA(1,2)), (C(1),AA(1,3))      A  4
      COMMON /RXN/ PAR(8)                                            A  5
CCC   INTERACTIVE VERSION                                            A  6
      CALL CONNEC (5)                                                A  7
      CALL CONNEC (6)                                                A  8
      WRITE (6,10)                                                   A  9
   10 FORMAT (* ENTER N, A, PHI,BIM,CGUESS *)                        A 10
      READ *,N,AS,PHI,BIM,CGUESS                                     A 11
      WRITE (6,15)                                                   A 12
   15 FORMAT (* ENTER REACTION RATE PARAMETERS *)                    A 13
      READ *,PAR(1),PAR(2),PAR(3),PAR(4)                             A 14
      N1 = N+1                                                       A 15
      DELX = 1./N                                                    A 16
      DELTA = PHI*PHI                                                A 17
      BB = DELTA*DELX**2                                             A 18
CCC   SET INITIAL CONDITION                                          A 19
      DO 20 I=1,N1                                                   A 20
      R(I) = DELX*(I-1)                                              A 21
   20 TH(I) = CGUESS                                                 A 22
CCC   BEGIN ITERATION                                                A 23
      DO 40 ITER=1,20                                                A 24
CCC   SET THE MATRICES                                               A 25
      C(1) = 2.*AS                                                   A 26
      A(N1) = 2.                                                     A 27
      D(N1) = -BIM*DELX*(2.+DELX*(AS-1.))                            A 28
      B(N1) = -2.+D(N1)                                              A 29
      DO 25 I=1,N                                                    A 30
      CALL RXN (TH(I),RATE,DR)                                       A 31
```

335

```
      B(I)  = -2.-BB*DR                                              A 32
      D(I) = BB*(RATE-DR*TH(I))                                      A 33
      IF (I.EQ.1) B(1) = B(1)+2.-2.*AS                               A 34
      IF (I.EQ.1) GO TO 25                                           A 35
      C(I) = (AS-1.)*DELX/(2.*R(I))                                  A 36
      A(I) = 1.-C(I)                                                 A 37
      C(I) = 1.+C(I)                                                 A 38
CCC 25 CONTINUE                                                      A 39
CCC    DO THE LU DECOMPOSITION                                       A 40
       CALL INVERT (N1,1,AA,D,3)                                     A 41
CCC    SOLVE FOR THE RIGHT HAND SIDE                                 A 42
       CALL INVSW (N1,1,AA,D,3)                                      A 43
CCC    FIND MAXIMUM CHANGE IN SOLUTION                               A 44
       ER = 0.                                                       A 45
       DO 30 I=1,N1                                                  A 46
       ERR = ABS(TH(I)-D(I))                                         A 47
       TH(I) = D(I)                                                  A 48
       IF (ERR.GT.ER) ER = ERR                                       A 49
   30  CONTINUE                                                      A 50
       WRITE (6,35) ITER,ER                                          A 51
   35  FORMAT (* ITERATION *,I3,*   ERROR IS *,E15.4)                A 52
       IF (ER.LT.1.E-6) GO TO 45                                     A 53
   40  CONTINUE                                                      A 54
CCC    CALCULATE THE EFFECTIVENESS FACTOR                            A 55
CCC    ETA1 USES EQ. 4-144,4-49                                      A 56
CCC    ETA2 USES EQ. 4-144,4-137B                                    A 57
   45  SL1 = (D(N-1)-4.*D(N)+3.*D(N1))/(DELX*2.)                     A 58
       SL2 = BIM*(1.-D(N1))                                         A 59
       ETA1 = AS*SL1/DELTA                                           A 60
       ETA2 = AS*SL2/DELTA                                           A 61
       WRITE (6,50) ETA1,ETA2                                        A 62
```

336

```
50 FORMAT (* EFF. FACTOR *,2F20.15)                                    A 63
   WRITE (6,55)                                                        A 64
55 FORMAT (* DO YOU WANT TO SEE THE SOLUTION *,/,* IF SO ENTER 1, OTH  A 65
  $ERWISE 0*)                                                          A 66
   READ *,KON                                                          A 67
   IF (KON.EQ.0) GO TO 5                                               A 68
   WRITE (6,60) (D(I),I=1,N1)                                          A 69
60 FORMAT (25(4F10.6,/))                                               A 70
   GO TO 5                                                             A 71
CCC   EXIT BY ENTERING %A                                              A 72
   STOP                                                                A 73
   END                                                                 A 74
   SUBROUTINE RXN(C,R,DR)                                              B  1
   COMMON /RXN/ PAR(8)                                                 B  2
   T = 1.+PAR(1)*(1.-C)                                                B  3
   E = EXP(PAR(2)*(1.-1./T))                                           B  4
CCC   THIS SUBROUTINE COMPUTES R AND DR/DC, GIVEN C                    B  5
   R = C*E                                                             B  6
   DR = E*(1.-PAR(1)*C*PAR(2)/T**2)                                    B  7
   RETURN                                                              B  8
   END                                                                 B  9

   PROGRAM OCFERXN (INPUT,OUTPUT,TAPE5=INPUT,TAPE6=OUTPUT)             A  1
   DIMENSION A(7,7), B(7,7), Q(7,7), XC(7), W(7), AA(7,7), D(101), TH  A  2
  $(101)                                                               A  3
   DIMENSION HE(40), X(101), AM(1000), C(7), F(7)                      A  4
   COMMON /RXN/ PAR(8)                                                 A  5
CCC   THIS PROGRAM USES ORTHOGONAL COLLOCATION ON FINITE ELEMENTS      A  6
```

337

```
CCC       TO SOLVE                                                    A  7
CCC                                                                   A  8
CCC       DEL**2 C = PHI*PHI*R(C)                                     A  9
CCC                                                                   A 10
CCC       -DC/DR = BIM*(C - 1 ) AT R = 1                              A 11
CCC                                                                   A 12
CCC   VARIABLES                                                       A 13
CCC                                                                   A 14
CCC   NCOL = NUMBER OF INTERIOR COLLOCATION POINTS IN EACH ELEMENT    A 15
CCC          <=5                                                      A 16
CCC   NE = NUMBER OF ELEMENTS, <=40                                   A 17
CCC   NR = 0,UNIFORM SIZE OF ELEMENTS                                 A 18
CCC        1 NON-UNIFORM SIZE  READ X LOCATIONS                       A 19
CCC          OF ELEMENT BOUNDARIES IN SUBROUTINE ELEMENT              A 20
CCC   NT = TOTAL NUMBER OF COLLOCATION POINTS                         A 21
CCC      = NE*(NCOL+1) + 1 <= 101                                     A 22
CCC   KON = 0 DO NOT PRINT MATRICES                                   A 23
CCC         1 PRINT A,B,Q,W MATRICES                                  A 24
CCC   AS = GEOMETRY FACTOR                                            A 25
CCC      = 1 PLANAR                                                   A 26
CCC      = 2 CYLINDRICAL                                              A 27
CCC      = 3 SPHERICAL                                                A 28
CCC   PHI = THIELE MODULUS                                            A 29
CCC   BIM = BIOT NUMBER FOR MASS TRANSFER                             A 30
CCC   CGUESS = INITIAL GUESS FOR C(R), A CONSTANT                     A 31
CCC                                                                   A 32
CCC   INTERACTIVE VERSION                                             A 33
CCC   CALL CONNEC (5)                                                 A 34
CCC   CALL CONNEC (6)                                                 A 35
    5 WRITE (6,10)                                                    A 36
   10 FORMAT (* ENTER NCOL,NE, A, PHI,BIM,CGUESS *)                   A 37
      READ *,NCOL,NE,AS,PHI,BIM,CGUESS
```

```
      KON = 0                                              A 38
      NR = 0                                               A 39
      WRITE (6,15)                                         A 40
   15 FORMAT (* ENTER FOUR REACTION RATE PARAMETERS *)     A 41
      READ *,PAR(1),PAR(2),PAR(3),PAR(4)                   A 42
      NP = NCOL+2                                           A 43
      NT = NE*(NCOL+1)+1                                    A 44
      N2 = NE*NP*NP                                         A 45
      DELTA = PHI*PHI                                       A 46
CCC   SET INITIAL CONDITION                                A 47
      DO 20 I=1,NT                                          A 48
   20 TH(I) = CGUESS                                        A 49
CCC   CALCULATE MATRICES                                   A 50
      CALL PLANAR (A,B,Q,XC,W,7,NP)                         A 51
      IF (KON.EQ.0) GO TO 65                                A 52
      WRITE (6,25)                                          A 53
   25 FORMAT (//,* A-MATRIX*,/)                             A 54
      DO 30 I=1,NP                                          A 55
   30 WRITE (6,60) (A(I,J),J=1,NP)                          A 56
      WRITE (6,35)                                          A 57
   35 FORMAT (//,* B-MATRIX*,/)                             A 58
      DO 40 I=1,NP                                          A 59
   40 WRITE (6,60) (B(I,J),J=1,NP)                          A 60
      WRITE (6,45)                                          A 61
   45 FORMAT (//,* Q-MATRIX*,/)                             A 62
      DO 50 I=1,NP                                          A 63
   50 WRITE (6,60) (Q(I,J),J=1,NP)                          A 64
      WRITE (6,55)                                          A 65
   55 FORMAT (//,* W-MATRIX*,/)                             A 66
      WRITE (6,60) (W(J),J=1,NP)                            A 67
   60 FORMAT (7E17.8)                                       A 68
```

```
  65  CONTINUE                                                              A 69
CCC   CALL ELEMENT (X,HE,XC,NE,NCOL,3,NR)                                   A 70
CCC   BEGIN ITERATION                                                       A 71
      DO 135 ITER=1,20                                                      A 72
CCC   SET THE MATRICES                                                      A 73
      DO 70 I=1,N2                                                          A 74
  70  AM(I) = 0.                                                            A 75
CCC   CALCULATE MATRIX                                                      A 76
      DO 100 K=1,NE                                                         A 77
      N1 = (K-1)*(NP-1)                                                     A 78
      DO 95 I=1,NP                                                          A 79
CCC   N1 + I IS THE GLOBAL NODE NUMBER                                      A 80
      C(I) = TH(N1+I)                                                       A 81
      CALL RXN (C(I),RATE,DR)                                              A 82
      DO 90 J=1,NP                                                          A 83
      IF (I.EQ.NP) GO TO 80                                                 A 84
      IF (I.GT.1) GO TO 75                                                  A 85
CCC   I = 1, FIRST ROW, EQ. 4-269                                          A 86
      AA(I,J) = -A(I,J)/HE(K)                                               A 87
      F(I) = 0.                                                             A 88
      GO TO 85                                                              A 89
CCC   I = 2 THROUGH NP-1, EQ. 4-267                                        A 90
  75  AA(I,J) = B(I,J)+(AS-1.)*HE(K)*A(I,J)/X(N1+I)                        A 91
      F(I) = DELTA*HE(K)*HE(K)*(RATE-DR*TH(N1+I))                          A 92
      IF (I.NE.J) GO TO 85                                                  A 93
      AA(I,J) = AA(I,J)-DELTA*HE(K)*HE(K)*DR                               A 94
      GO TO 85                                                              A 95
CCC   I = NP,EQ. 4-269                                                     A 96
  80  AA(I,J) = A(NP,J)/HE(K)                                               A 97
      F(I) = 0.                                                             A 98
  85  INDEX = NP*NP*(K-1)+NP*(J-1)+I                                       A 99
```

```
            AM(INDEX) = AA(I,J)
   90    CONTINUE
         D(N1+I) = F(I)
   95    CONTINUE
  100 CONTINUE
      IF (NE.EQ.1) GO TO 110
      NE1 = NE-1
      DO 105 KK=1,NE1
      INDEX1 = NP*NP*KK
      INDEX2 = INDEX1+1
      AM(INDEX1) = AM(INDEX1)+AM(INDEX2)
  105 AM(INDEX2) = AM(INDEX1)
  110 CONTINUE
CCC   SET THE BOUNDARY CONDITION AT R = 0.,EQ. 4-270
      DO 115 I=1,NP
  115 AM(NP*(I-1)+1) = A(1,I)
      D(1) = 0.
CCC   SET THE BOUNDARY CONDITION AT R = 1., EQ. 4-271
      DO 120 I=1,NP
      AM(INDEX) = A(NP,I)
      INDEX = NE*NP*NP
  120 INDEX = (NE-1)*NP*NP+NP*I
      AM(INDEX) = AM(INDEX)+HE(NE)*BIM
      D(NT) = BIM*HE(NE)
CCC   DO THE LU DECOMPOSITION
      CALL INVERT (NT,NE,AM,D,2)
CCC   SOLVE FOR THE RIGHT HAND SIDE
      CALL INVSW (NT,NE,AM,D,2)
CCC   FIND MAXIMUM CHANGE IN SOLUTION
      ER = 0.
      DO 125 I=1,NT
```

A 100
A 101
A 102
A 103
A 104
A 105
A 106
A 107
A 108
A 109
A 110
A 111
A 112
A 113
A 114
A 115
A 116
A 117
A 118
A 119
A 120
A 121
A 122
A 123
A 124
A 125
A 126
A 127
A 128
A 129
A 130

341

```
            ERR = ABS(TH(I)-D(I))                                          A 131
            TH(I) = D(I)                                                   A 132
            IF (ERR.GT.ER) ER = ERR                                       A 133
  125     CONTINUE                                                        A 134
          WRITE (6,130) ITER,ER                                          A 135
  130   FORMAT (* ITERATION *,I3,*   ERROR IS *,E15.4)                   A 136
          IF (ER.LT.1.E-6) GO TO 140                                      A 137
  135   CONTINUE                                                          A 138
CCC     CALCULATE THE EFFECTIVENESS FACTOR                                A 139
CCC     ETA1 USES EQ. 4-142, 4-279                                        A 140
CCC     ETA2 USES EQ. 4-144                                              A 141
  140   SUM = 0.                                                          A 142
        DO 145 I=1,NP                                                     A 143
  145   SUM = SUM+A(NP,I)*D((NE-1)*(NP-1)+I)                             A 144
        ETA2 = SUM*AS/(DELTA*HE(NE))                                      A 145
        SUM = 0.                                                          A 146
        SUM1 = 0.                                                         A 147
        DO 150 K=1,NE                                                     A 148
          N1 = (K-1)*(NP-1)                                               A 149
          DO 150 I=1,NP                                                   A 150
            CALL RXN (D(N1+I),RATE,DR)                                    A 151
            SUM = SUM+W(I)*HE(K)*RATE                                     A 152
            SUM1 = SUM1+W(I)*HE(K)                                        A 153
  150   CONTINUE                                                          A 154
        ETA1 = SUM/SUM1                                                   A 155
        WRITE (6,155) ETA1                                               A 156
  155   FORMAT (* EFF. FACTOR *,F20.15)                                   A 157
        WRITE (6,155) ETA2                                               A 158
        WRITE (6,160)                                                    A 159
  160   FORMAT (* DO YOU WANT TO SEE THE SOLUTION *,/,* IF SO ENTER 1, OTH A 160
       $ERWISE 0*)                                                        A 161
```

342

```
      READ *,KON                                                  A 162
      IF (KON.EQ.0) GO TO 5                                        A 163
      WRITE (6,165) (D(I),I=1,NT)                                  A 164
  165 FORMAT (25(4F10.6,/))                                        A 165
      GO TO 5                                                      A 166
CCC   EXIT BY ENTERING %A                                         A 167
      STOP                                                         A 168
      END                                                         A 169-
      SUBROUTINE ELEMENT(X,H,Z,NELTS,NCOL,METH,NR)                B 1
      DIMENSION X(101), H(101), Z(7)                              B 2
CCC   THIS SUBROUTINE READS IN ELEMENT LOCATION IF NR.NE.0 AND MAKES   B 3
CCC   THEM UNIFORM IF NR=0.                                       B 4
CCC                                                               B 5
CCC   INPUT                                                       B 6
CCC      Z(I) = LOCAL COORDINATES IN ELEMENT                      B 7
CCC      NELTS = NUMBER OF ELEMENTS                               B 8
CCC      NCOL = NP - 2                                            B 9
CCC              WHERE NP = NUMBER OF NODES OR POINTS PER ELEMENT  B 10
CCC      METH = 2 GALERKIN                                        B 11
CCC           = 3 COLLOCATION                                     B 12
CCC           = 4 COLLOCATION, HERMITE INTERPOLATION              B 13
CCC      IF NR.NE.0 THE VALUES OF X AT THE END OF EACH ELEMENT    B 14
CCC      ARE TO BE ENTERED. X = 0. IS NOT INSERTED.               B 15
CCC      THERE ARE NELTS NUMBER OF ENTRIES.                       B 16
CCC                                                               B 17
CCC                                                               B 18
CCC   OUTPUT                                                      B 19
CCC      X(I) = X LOCATION OF I-TH GRID POINT (GLOBAL INDEX)      B 20
CCC      H(I) = SIZE OF I-TH ELEMENT                              B 21
CCC                                                               B 22
CCC      NP = NCOL+2                                              B 23
```

343

```
      NT = NELTS*(NCOL+1)+1                              B  24
      X(1) = 0.0                                        B  25
      X(NT) = 1.0                                       B  26
      DO 30 L=1,NELTS                                   B  27
      K = (L-1)*(NP-1)+1                                B  28
      IF (NR.EQ.0) GO TO 10                             B  29
      READ (5,5) X(K+NP-1)                              B  30
    5 FORMAT (F10.5)                                    B  31
      GO TO 15                                          B  32
C                                                       B  33
   10 H(L) = 1.0/NELTS                                  B  34
      X(K) = H(L)*(L-1)                                 B  35
      GO TO 20                                          B  36
C                                                       B  37
   15 CONTINUE                                          B  38
      H(L) = X(K+NP-1)-X(K)                             B  39
   20 IF (METH.LE.2.AND.NP.EQ.2) GO TO 30              B  40
      DO 25 I=1,NCOL                                    B  41
      KI = K+I                                          B  42
   25 X(KI) = X(K)+H(L)*Z(I+1)                          B  43
   30 CONTINUE                                          B  44
      WRITE (6,35)                                      B  45
   35 FORMAT (1H0,/,* ELEMENT LOCATIONS IN X DIRECTION ARE*,//)   B  46
      A = 0.                                            B  47
      WRITE (6,40) A                                    B  48
   40 FORMAT (15X,F15.8)                                B  49
      DO 45 L=1,NELTS                                   B  50
      A = A+H(L)                                        B  51
      WRITE (6,50) L,A                                  B  52
   45 CONTINUE                                          B  53
   50 FORMAT (I5,10X,F15.8)                             B  54
```

```
                                                          B    55
      RETURN                                              B    56-
      END                                                 C     1
      SUBROUTINE RXN(C,R,DR)                              C     2
      COMMON /RXN/ PAR(8)                                 C     3
CCC   PAR(1) = BETA    PAR(2) = GAMMA                     C     4
      T = 1.+PAR(1)*(1.-C)                                C     5
      E = EXP(PAR(2)*(1.-1./T))                           C     6
CCC   THIS SUBROUTINE COMPUTES R AND DR/DC, GIVEN C       C     7
      R = C*E                                             C     8
      DR = E*(1.-PAR(1)*C*PAR(2)/T**2)                    C     9
      RETURN                                              C    10-
      END

      PROGRAM IVRXN (INPUT,OUTPUT,TAPE5=INPUT,TAPE6=OUTPUT)   A     1
      COMMON /RXN/ PAR(8)                                     A     2
      DIMENSION XI(4), XX(4), STORE(4,11)                     A     3
      EXTERNAL RHS                                            A     4
CCC   READ GEOMETRY AA= 1, PLANAR                             A     5
CCC                  2, CYLINDRICAL                           A     6
CCC                  3, SPHERICAL                             A     7
CCC        NUMBER OF CASES, ANUM                              A     8
CCC        ACCURACY = ACCUR                                   A     9
CCC        REACTION PARAMETERS (8F10.0)                       A    10
CCC        ANUM TIMES: PHI,SO = INITIAL GUESS OF C(X=0)       A    11
      READ (5,5) AA,ANUM,ACCUR                                A    12
    5 FORMAT (8F10.0)                                         A    13
      IF (ACCUR.EQ.0.) ACCUR = 1.E-6                          A    14
      READ (5,5) (PAR(I),I=1,8)                               A    15
```

```
      NUM = ANUM
      DO 35 KK=1,NUM                                                  A 16
      READ (5,5) PHI,SO                                               A 17
      PAR(8) = PHI*PHI                                                A 18
      ITER = 0                                                        A 19
      S = SO                                                          A 20
      PAR(7) = AA                                                     A 21
                                                                      A 22
CCC   SET RUNGE-KUTTA PARAMETERS                                      A 23
CCC   SET ON ENTRY                                                    A 24
CCC   RHS = NAME OF SUBROUTINE CALCULATING RIGHT HAND SIDE            A 25
CCC   NV = NUMBER OF FIRST ORDER ORDINARY DIFFERENTIAL EQUATIONS      A 26
CCC   X = INITIAL *TIME*                                              A 27
CCC   XP = FINAL *TIME*                                               A 28
CCC   XI(1-4) = VECTOR OF SOLUTION AT X                               A 29
CCC   ACCUR = DESIRED LOCAL ACCURACY IN INTEGRATION                   A 30
CCC   NR = DIMENSION IN STORE,>=NV                                    A 31
CCC   IPRINT = 0, NO DIAGNOSTICS                                      A 32
CCC   IN = 0, SET = 0 ON FIRST CALL                                   A 33
CCC   ISET = 0                                                        A 34
CCC   RETURNED                                                        A 35
CCC   XX(1-4) = VECTOR OF SOLUTION AT XP                              A 36
CCC   IRR = 0 - NO ERROR                                              A 37
CCC        1 - ERROR TEST COULD NOT BE SATISFIED                      A 38
CCC        2 - EXCESSIVE COMPUTATION TIME WILL RESULT                 A 39
   10 CONTINUE                                                        A 40
      NV = 4                                                          A 41
      X = 0.                                                          A 42
      NR = 4                                                          A 43
      IPRINT = IN = ISET = 0                                          A 44
      XP = 1.                                                         A 45
      IF (S.GE.0.) GO TO 15                                           A 46
```

```
            SO = SO/10.
            S = SO
15          CONTINUE
            XI(1) = S
            XI(2) = 0.
            XI(3) = 1.
            XI(4) = 0.
CCC         XI(1) = U=C,  XI(2) = V = DC/DX
CCC         XI(3) = DU/DS,  XI(4) = DV/DS
CCC         SEE EQ. 4-365
            ITER = ITER+1
            CALL RKINIT (RHS,NV,X,XI,ACCUR,NR,IPRINT,IN,ISET,STORE,XP,XX,
     $      IRR)
            PSI = XX(1)-1.
            DPSI = XX(3)
            SN = S-PSI/DPSI
            IF (ABS(SN-S).LT.ACCUR) GO TO 25
            WRITE (6,20) S,SN,PSI,DPSI
20          FORMAT (4E15.4)
            S = SN
            IF (ITER.GE.100) GO TO 25
            GO TO 10
25          WRITE (6,30) S,PHI,XX(2)
30          FORMAT (//,28(1H+),/,3E15.4,//,28(1H+),/)
35          CONTINUE
            STOP
            END
            SUBROUTINE RHS(X,XI,XF)
            COMMON /RXN/ PAR(8)
            DIMENSION XI(4), XF(4)
```

A47
A48
A49
A50
A51
A52
A53
A54
A55
A56
A57
A58
A59
A60
A61
A62
A63
A64
A65
A66
A67
A68
A69
A70
A71
A72
A73
A74
B1
B2
B3

347

```
CCC     THIS SUBROUTINE COMPUTES F(XI) IN DXI/DX=F(XI,X)         B  4
CCC          PAR(8) = PHI**2   PAR(7) = AA = GEOMETRY FACTOR     B  5
CCC          SEE EQ. 4-365                                       B  6
        CALL RXN (XI(1),R,DR)                                    B  7
        DIV = 1.                                                 B  8
        IF (X.EQ.0.) DIV = PAR(7)                                B  9
        XF(2) = PAR(8)*R/DIV                                     B 10
        XF(4) = PAR(8)*DR*XI(3)/DIV                              B 11
        XF(1) = XI(2)                                            B 12
        XF(3) = XI(4)                                            B 13
        IF (X.EQ.0.) GO TO 5                                     B 14
        XF(2) = XF(2)-(PAR(7)-1.)*XI(2)/X                        B 15
        XF(4) = XF(4)-(PAR(7)-1.)*XI(4)/X                        B 16
      5 RETURN                                                   B 17
        END                                                      B 18
        SUBROUTINE RXN(C,R,DR)                                   C  1
        COMMON /RXN/ PAR(8)                                      C  2
CCC     THIS SUBROUTINE COMPUTES R(C) AND DR/DC, GIVEN C.        C  3
        T = 1.+PAR(1)*(1.-C)                                     C  4
        E = EXP(PAR(2)*(1.-1./T))                                C  5
        R = C*E                                                  C  6
        DR = E*(1.-PAR(1)*C*PAR(2)/T**2)                         C  7
        RETURN                                                   C  8
        END                                                      C  9
```

348

```
      PROGRAM PDE (INPUT,OUTPUT,TAPE5=INPUT,TAPE6=OUTPUT)                  1
      COMMON /SUBFD/ DELX,XX(102)                                         2
      COMMON /SUBOC/ NP,A(7,7),B(7,7),F(7,7),W(7),XC(7)                   3
      COMMON /GEAR9/ HUSED,NQUSED,NSTEP,NFE,NJE                           4
      COMMON /DE/ PAR(7),METH                                             5
C                                                                        6
CCC                                                                      7
CCC                                                                      8
CCC     THIS PROGRAM SOLVES THE PARTIAL DIFFERENTIAL EQUATION            9
CC                                                                       10
CCC            DC/DT = F(C)                                              11
CCC                                                                      12
CCC     WHERE F(C) IS A DIFFERENTIAL OPERATOR. IT USES FINITE            13
CCC     DIFFERENCE OR ORTHOGONAL COLLOCATION, AS CHOSEN BY METH.         14
CCC                                                                      15
CCC     METH - 1 FINITE DIFFERENCE                                       16
CCC          - 2 ORTHOGONAL COLLOCATION, SYMMETRIC, TABLES 4.5,4.6       17
CCC          - 3 ORTHOGONAL COLLOCATION, UNSYMMETRIC, TABLES 4.3,4.4     18
CCC                                                                      19
CCC     DATA INPUT                                                       20
CCC                                                                      21
CCC       1. N,METH,AG      2I5,F10.0                                    22
CCC          N = NUMBER OF INTERIOR FINITE DIFFERENCE POINTS             23
CCC              OR NUMBER OF INTERIOR COLLOCATION POINTS                24
CCC              (<=6 IF METH = 2 AND <=5 IF METH = 3)                   25
CCC          METH = METHOD INDICATOR, AS LISTED ABOVE                    26
CCC          AG = GEOMETRY FACTOR                                        27
CCC               1-PLANAR, 2-CYLINDRICAL, 3-SPHERICAL                   28
CCC       2. PAR(1-7)       7E10.0                                       29
CCC          PARAMETERS THAT CAN BE USED IN THE DIFFERENTIAL EQUATION    30
CCC       3. MF,EPS                                                      31
CCC          MF = METHOD PARAMETER FOR GEARB, USUALLY 22
```

349

```
CCC         EPS = ERROR CRITERION USED IN GEARB                          A  32
CCC      4. C(1-N)      USER SUPPLIES FORMAT IN INITIAL                   A  33
CCC         INITIAL CONDITIONS                                           A  34
CCC      5. TF          E10.0                                            A  35
CCC         TIME YOU WISH TO SAMPLE THE SOLUTION                         A  36
CCC         REPEAT THIS DATA INPUT FOR AS MANY TIMES (UP TO 100)         A  37
CCC         THAT YOU WANT THE SOLUTION PRINTED OUT.                      A  38
CCC         THE LAST DATA INPUT SHOULD BE ZERO OR A NEGATIVE NUMBER      A  39
CCC         TO STOP.                                                     A  40
CCC                                                                      A  41
CCC                                                                      A  42
CCC      THE USER ALSO SUPPLIES SUBROUTINES AS FOLLOWS                   A  43
CCC                                                                      A  44
CCC         SUBROUTINE BC(C,N,METH,C1,CN)                                A  45
CCC         INPUT - N,METH,C(1-N)                                        A  46
CCC         OUTPUT - C1,CN                                               A  47
CCC            C1=C(X=0), USED IF METH = 1 OR 3                          A  48
CCC            CN=C(X=1), USED FOR ALL METH                              A  49
CCC                                                                      A  50
CCC         SUBROUTINE DIFFUN(N,T,C,CDOT)                                A  51
CCC         INPUT - N,T,C(1-N)                                           A  52
CCC            T - TIME                                                  A  53
CCC            C - SOLUTION AT THAT TIME                                 A  54
CCC         OUTPUT - CDOT(1,N)                                           A  55
CCC            CDOT = RIGHT HAND SIDE OF DIFFERENTIAL EQUATICN,          A  56
CCC                   EVALUATED AT T FOR C(1-N)                          A  57
CCC                                                                      A  58
CCC         SUBROUTINE INITIAL(C,N)                                      A  59
CCC         INPUT - N                                                    A  60
CCC         OUTPUT - C(1,N) = INITIAL CONDITIONS                         A  61
CCC         THIS IS WHERE DATA INPUT 4. IS USED                         A  62
```

```
C     IF NO DATA IS READ IN INITIAL THEN SKIP DATA INPUT 4.                  A  63
CCC                                                                          A  64
CCC                                                                          A  65
CCC   DATA OUTPUT                                                            A  66
CCC                                                                          A  67
CCC   METHOD AND NUMBER OF POINTS                                            A  68
CCC   COLLOCATION MATRICS IF METH = 2 OR 3                                   A  69
CCC   PARAMETERS IN DIFFERENTIAL EQUATIONS                                   A  70
CCC   MF,EPS                                                                 A  71
CCC   TF, HUSED, NQUSED, NSTEP, NFE,NJE                                      A  72
CCC   TF = TIME AT WHICH THE FOLLOWING SOLUTION APPLIES                      A  73
CCC   HUSED = LAST STEP SIZE USED                                           A  74
CCC   NQUSED = LAST ORDER USED                                              A  75
CCC   NSTEP = NUMBER OF STEPS TAKEN (TOTAL)                                 A  76
CCC   NFE = NUMBER OF FUNCTION EVALUATIONS (TOTAL), INCLUDIN                A  77
CCC   THOSE TO EVALUATE THE JACOBIAN                                        A  78
CCC   NJE = NUMBER OF JACOBIANS EVALUATED (TOTAL)                           A  79
CCC   I,C(I), I=1,N, THE SOLUTION                                           A  80
CCC   FD - C(1) = C(X=DELX)                                                 A  81
CCC         C(N) = C(1.-DELX)                                               A  82
CCC   OC, SYM - C(1) = C(FIRST INTERIOR COLLOCATION POINT)                  A  83
CCC             C(N) = C(LAST INTERIOR COLLOCATIN POINT)                    A  84
CCC             C(N+1) = C(X=1.)                                            A  85
CCC   OC, UNSYM - C(1) = C(0.)                                              A  86
CCC               C(2) = C(FIRST INTERIOR COLLOCATION POINT)               A  87
CCC               C(N+1) = C(LAST INTERIOR COLLOCATICN POINT               A  88
CCC               C(N+2) = C(X=1.0)                                         A  89
CCC                                                                          A  90
CCC                                                                          A  91
CCC   DIMENSION C(100)                                                      A  92
CCC   READ (5,5) N,METH,AG                                                  A  93
```

```
  5   FORMAT (2I5,F10.0)                                              A   94
      IF (METH.GT.1) GO TO 20                                        A   95
CCC   FINITE DIFFERENCE                                              A   96
      WRITE (6,10) N                                                 A   97
 10   FORMAT (* FINITE DIFFERENCE METHOD WITH *,I5,* GRID POINTS*)   A   98
      DELX = 1./(N+1.)                                               A   99
      XX(1) = 0.                                                     A  100
      N1 = N+1                                                       A  101
      MU = ML = 1                                                    A  102
      DO 15 I=1,N1                                                   A  103
 15   XX(I+1) = DELX*I                                               A  104
      GO TO 90                                                       A  105
CCC   ORTHOGONAL COLLOCATION                                         A  106
 20   WRITE (6,25) N                                                 A  107
 25   FORMAT (* ORTHOGONAL COLLOCATION WITH *,I3,* INTERIOR*,* COLLOCATI A 108
     $ON POINTS *)                                                   A  109
      MU = ML = MAX0(1,N/2)                                          A  110
      IF (METH.GE.3) GO TO 35                                        A  111
      WRITE (6,30)                                                   A  112
 30   FORMAT (* SYMMETRIC POLYNOMIALS, TABLES 4.5,4.6 *)             A  113
      NP = N+1                                                       A  114
      CALL COLL (A,B,F,XC,W,7,N,AG)                                  A  115
      GO TO 45                                                       A  116
C                                                                    A  117
 35   WRITE (6,40)                                                   A  118
 40   FORMAT (* UNSYMMETRIC POLYNOMIALS, TABLES 4.3,4.4 *)           A  119
      NP = N+2                                                       A  120
      CALL PLANAR (A,B,F,XC,W,7,NP)                                  A  121
 45   WRITE (6,50)                                                   A  122
 50   FORMAT (//,* A-MATRIX*,/)                                      A  123
      DO 55 I=1,NP                                                   A  124
```

```
 55    WRITE (6,85) (A(I,J),J=1,NP)                                              A 125
       WRITE (6,60)                                                             A 126
 60    FORMAT (//,* B-MATRIX*,/)                                                A 127
       DO 65 I=1,NP                                                             A 128
 65    WRITE (6,85) (B(I,J),J=1,NP)                                              A 129
       WRITE (6,70)                                                             A 130
 70    FORMAT (//,* Q-MATRIX*,/)                                                A 131
       DO 75 I=1,NP                                                             A 132
 75    WRITE (6,85) (F(I,J),J=1,NP)                                              A 133
       WRITE (6,80)                                                             A 134
 80    FORMAT (//,* W-MATRIX*,/)                                                A 135
       WRITE (6,85) (W(J),J=1,NP)                                               A 136
 85    FORMAT (7E17.8)                                                          A 137
       WRITE (6,95) AG                                                          A 138
 95    FORMAT (* GEOMETRY FACTOR IS *,F5.0)                                     A 139
       READ (5,100) (PAR(I),I=1,7)                                              A 140
100    FORMAT (7E10.0)                                                          A 141
       WRITE (6,105) (PAR(I),I=1,7)                                             A 142
105    FORMAT (//,* PARAMETERS IN DIFFERENTIAL EQUATIONS ARE *,/,4E15.7,/       A 143
      $,3E15.7,//)                                                              A 144
       T = 0.                                                                   A 145
       READ (5,115) MF,EPS                                                      A 146
       INDEX = 1                                                                A 147
       HO = EPS*EPS                                                             A 148
       HUSED = 0.                                                               A 149
       NQUSED = NSTEP = NFE = NJE = 0                                           A 150
       WRITE (6,110) MF,EPS                                                     A 151
110    FORMAT (* GEARB IS USED WITH MF = *,I3,5X,* AND FPS = *,E10.3)           A 152
115    FORMAT (I5,E10.0)                                                        A 153
       CALL INITIAL (C,N)                                                       A 154
       CALL PRINT (C,N,T)                                                       A 155
```

353

```
      DO 125 KK=1,100
      READ (5,120) TF
  120 FORMAT (E10.0)
      IF (TF.LE.0.) GO TO 130
      CALL DRIVEB (N,T,HO,C,TF,EPS,MF,INDEX,MU,ML)
      CALL PRINT (C,N,TF)
  125 CONTINUE
  130 STOP
      END
      SUBROUTINE PRINT(C,N,TF)
      COMMON /SUBFD/ DELX,XX(102)
      COMMON /SUBOC/ NP,A(7,7),B(7,7),F(7,7),W(7),XC(7)
      COMMON /GEAR9/ HUSED,NQUSED,NSTEP,NFE,NJE
      COMMON /DE/ PAR(7),METH
      DIMENSION C(100), CX(102)
      CALL BC (C,N,METH,C1,CN,TF)
      IF (METH.EQ.2) GO TO 10
CCC   FINITE DIFFERENCE AND OC, UNSYMMTRIC
      NX = N+2
      CX(1) = C1
      DO 5 I=1,N
    5 CX(I+1) = C(I)
      GO TO 20
CCC   OC,SYM
   10 NX = N+1
      DO 15 I=1,N
   15 CX(I) = C(I)
CCC   ALL METHODS
   20 CX(NX) = CN
      IF (METH.EQ.1) GO TO 30
      DO 25 I=1,NX
```

354

```
25  XX(I) = XC(I)                                                        B 23
30  WRITE (6,35) TF                                                      B 24
35  FORMAT (//,* TIME = *,E12.5)                                         B 25
    WRITE (6,40) HUSED,NQUSED,NSTEP,NFE,NJE                              B 26
40  FORMAT (/,* LAST STEP SIZE USED = *,E10.3,* FOR *,I3,* -TH*,         B 27
   $* ORDER METHOD*,/,* NUMBER OF STEPS = *,I10,/,* NUMBER OF FUNCTION   B 28
   $ EVALUATIONS = *,I10,/,* NUMBER OF JACOBIAN EVALUATIONS = *,I10,/)   B 29
    DO 50 I=1,NX,2                                                       B 30
    J = I+1                                                              B 31
    IF (J.GT.NX) GO TO 45                                                B 32
    WRITE (6,55) I,XX(I),CX(I),J,XX(J),CX(J)                            B 33
    GO TO 50                                                             B 34
45      WRITE (6,55) I,XX(I),CX(I)                                       B 35
50  CONTINUE                                                             B 36
55  FORMAT (I5,2X,F10.6,F12.9,10X,I5,2X,F10.6,F12.9)                    B 37
    RETURN                                                               B 38
    END                                                                 B 39
    SUBROUTINE INITIAL(C,N)                                              B 40-
    DIMENSION C(N)                                                       C 1
    DO 5 I=1,N                                                           C 2
5   C(I) = 0.                                                            C 3
    RETURN                                                               C 4
    END                                                                 C 5
C                                                                        C 6-
    SUBROUTINE BC(C,N,METH,C1,CN,T)                                      D 1
    DIMENSION C(N)                                                       D 2
CCC SET C1=C(X=0.)    IF METH = 1 OR 3                                   D 3
CCC SET CN=C(X=1.)    FOR ALL METH                                       D 4
    IF (METH.EQ.2) GO TO 5                                               D 5
    C1 = 1.                                                              D 6
5   CN = 0.                                                              D 7
```

```
      RETURN
      END
      SUBROUTINE DIFFUN(N,T,C,CDOT)
      COMMON /SUBFD/ DELX,XX(102)
      COMMON /SUBOC/ NP,A(7,7),B(7,7),F(7,7),W(7),XC(7)
      COMMON /GEAR9/ HUSED,NQUSED,NSTEP,NFE,NJE
      COMMON /DE/ PAR(7),METH
      DIMENSION C(N), CDOT(N), TH(10)
CCC   SOLVE FOR
CCC
CCC       DC/DT = D((1.+PAR(1)*C)DC/DX)/DX
CCC
CCC   IF PAR(2) DIFFERS FROM ZERO ADD TO THE EQUATION
CCC
CCC        /PAR(2) + 2*X*DC/DX
CCC
CCC   CALL BC (C,N,METH,C1,CN,T)
CCC   GO TO (5,15,30), METH
CCC   FD
    5 DO 10 I=1,N
      DELX = 1./(N+1.)
      X = I*DELX
      CI = C(I)
      CO = C1
      IF (I.NE.1) CO = C(I-1)
      CL = CN
      IF (I.NE.N) CL = C(I+1)
      AA = (1.+0.5*PAR(1)*(CL+CI))*(CL-CI)
      AA = AA-(1.+0.5*PAR(1)*(CI+CO))*(CI-CO)
      AA = AA/DELX**2
```

```
         IF (PAR(2).EQ.0.) GO TO 10                              E 30
         AA = AA/PAR(2)+X*(CL-C0)/DELX                           E 31
   10    CDOT(I) = AA                                            E 32
         GO TO 55                                                E 33
CCC   ORTHOGONAL COLLOCATION, SYMMETRIC                          E 34
CCC   THIS IS NOT AN APPROPRIATE CHOICE FOR DIFFUSION PROBLEM.   E 35
CCC   SET UP IS PROVIDED FOR  DC/DT = DEL**2 C + R               E 36
   15    NX = N+1                                                E 37
         NO = 0                                                  E 38
         N1 = 1                                                  E 39
         R = 1.                                                  E 40
         DO 25 J=1,N                                             E 41
         SUM1 = 0.                                               E 42
         DO 20 I=1,NX                                            E 43
   20    SUM1 = SUM1+B(J,I)*C(I)                                 E 44
   25    CDOT(J) = SUM1+R                                        E 45
         GO TO 55                                                E 46
CCC   ORTHOGONAL COLLOCATION, UNSYMMETRIC                        E 47
   30    NX = N+2                                                E 48
         NO = 1                                                  E 49
         N1 = 2                                                  E 50
         DO 35 I=1,N                                             E 51
   35    TH(I+NO) = C(I)                                         E 52
         IF (METH.EQ.3) TH(1) = C1                               E 53
         TH(N+1+NO) = CN                                         E 54
         DO 50 J=1,N                                             E 55
         SUM2 = 0.                                               E 56
         SUM3 = 0.                                               E 57
         JJ = J+NO                                               E 58
         DO 45 K=1,NX                                            E 59
         SUM2 = SUM2+A(JJ,K)*TH(K)                               E 60
```

```
        SUM1 = 0.
        DO 40 I=1,NX
        SUM1 = SUM1+A(K,I)*TH(I)
40
45      SUM3 = SUM1*(1.+TH(K)*PAR(1))*A(JJ,K)+SUM3
        IF (PAR(2).EQ.0.) GO TO 50
        SUM2 = SUM2*2.*XC(JJ)
        SUM3 = SUM3/PAR(2)+SUM2
50  CDOT(J) = SUM3
55  RETURN
        END
```

E 61
E 62
E 63
E 64
E 65
E 66
E 67
E 68
E 69
E 70-

AUTHOR INDEX

SUBJECT INDEX

Acceleration of gravity, 246
Accuracy (*see* Error)
Activation energy, 81, 208
Adams–Bashforth method, 26, 29, 32, 52
Adams–Moulton method, 27, 29, 32, 53
Adaptive mesh, 154–155, 164–166
  (*see also* Mesh refinement)
ADI (*see* Alternating direction implicit method)
Adsorption, 88
Air, 246
Alternating direction explicit method, 284–285
Alternating direction implicit method, 283–286,
    290, 307
  comparisons, 313
Anisotropy, 280, 282, 286
Aquifer, 245
Aspect ration, 288
Asymptotic expansion, 89–94, 98, 120

Backward Euler method, 27, 32, 45, 49, 54, 58,
    137
  application, 197–199, 228–229, 257, 259–261
  error, 198
  in finite difference method, 218–219
  oscillation limit, 38, 44, 53, 46, 47
  rational approximation to exponential, 40, 44
  stability, 38
  truncation error, 53
Basis function, 292
  comparison, 144–149, 232–234, 305–306
  Hermite cubic, 310–311
  linear, 126–127, 131–134, 231, 238, 239, 242

Basis function—*continued*
  linear on rectangles, 299, 304
  linear on triangles, 294
  quadratic, 133–134, 231, 238–240, 242
  quadratic on rectangles, 300, 303
Beta, 81
Biot number for heat transfer, 81
Biot number for heat transfer, 196
Biot number for mass transfer, 81
Boundary collocation method, 296
Boundary condition, 4
  collocation method, 75–76
  essential, 292
  finite element method, 135
  natural, 135, 292
Boundary condition of first kind (*see* Dirichlet
    boundary condition)
Boundary condition of second kind (*see*
    Neumann boundary condition)
Boundary condition of the third kind, 5, 103
Boundary, false, 107
Boundary layer, 158
Boundary-value problem, two-point:
  definition, 4

Calculator, 189
Capillary pressure, 246
Carbon monoxide, 211–213
Carbon monoxide reaction, 88
Catalyst activity, 80
Catalyst diameter, 192
Catalyst pellet, 80–81, 206, 211